T0338252

BROADBAND COMMUNICATIONS VIA HIGH ALTITUDE PLATFORMS

BROADBAND COMMUNICATIONS VIA HIGH ALTITUDE PLATFORMS

David Grace
University of York, UK

Mihael Mohorčič
Jožef Stefan Institute, Slovenia

A John Wiley and Sons, Ltd, Publication

This edition first published 2011
© 2011 John Wiley & Sons Ltd.

Registered office
John Wiley & Sons Ltd, The Atrium, Southern Gate, Chichester, West Sussex, PO19 8SQ, United Kingdom

For details of our global editorial offices, for customer services and for information about how to apply for permission to reuse the copyright material in this book please see our website at www.wiley.com.

Library of Congress Cataloging-in-Publication Data

Grace, David, 1970-
  Broadband communications via high altitude platforms / David Grace, Mihael Mohorčič.
    p. cm.
  Includes bibliographical references and index.
  ISBN 978-0-470-69445-9 (cloth)
  1. High altitude platform systems (Telecommunication) 2. Broadband communication systems–Equipment and supplies. 3. Aeronautics. 4. Artificial satellites in telecommunication. I. Mohorčič, Mihael. II. Title.
  TK7871.67.H54G73 2010
  621.382–dc22

                                                                                    2010018784

A catalogue record for this book is available from the British Library.

Print ISBN: 9780470694459 (HB)
ePDF ISBN: 9780470971833
oBook ISBN: 9780470971840

Typeset in 10/12pt Times by Thomson Digital, Noida, India
Printed and bound in Singapore by Markono Print Media Pte Ltd

Contents

List of Figures

List of Tables

List of Contributors

Debbie Kedhar, Ben Gurion University, Israel

Guanhua Chen, formerly of University of York, UK

Pairoj Likitthanasate, TOT Public Company Limited, Thailand

Roman Novak, Jožef Stefan Institute, Slovenia

Tomaž Javornik, Jožef Stefan Institute, Slovenia

Preface

Aerial platforms taking the form of manned or unmanned airplanes, balloons or airships and equipped with mission-dependent payload, gained worldwide interest in the last decade. Depending on the type of the payload, the platform can be used for various missions from remote monitoring and surveillance to positioning and navigation and the provision of broadband wireless access to fixed or mobile users. Particularly interesting for the latter scenario are platforms able to keep quasistationary position in the lower stratosphere, typically referred to as high altitude platforms (HAPs). They combine some of the best characteristics of terrestrial wireless and satellite communication systems, while avoiding many of their drawbacks. HAPs are believed to be particularly suitable for short-term large-scale events and establishment of ad-hoc networks for disaster relief, but also for the roll-out phase of new services and applications in urban environment and for the provision of basic access to remote and sparsely populated areas. However, in order to represent an acceptable complementary solution to currently existing and developing wireless technologies, HAP-based communication systems will have to efficiently integrate in the future heterogeneous wireless access infrastructure.

The similarity and commonality with terrestrial wireless and satellite communication systems is one of the main reasons why HAPs until recently have not been addressed in a stand-alone book but were only partially covered in some wireless and satellite communications books. Authors of the book, having been involved in the HAP-related research from its earliest times, found this an important problem when introducing newcomers in the research area, which is where the initial idea for writing this book came from. In fact, the current state of development of HAP based communications and their specifics with respect to terrestrial wireless and satellite communications justify, and even call for, a dedicated book with systematic treatment of HAP related issues, problems and solutions. It is aimed to serve as a reference book for the research community while it should also give the up to date status of the on-going research activities, drawing from long-term personal involvement of authors and further contributors in the HAP research activities, mostly carried out within European projects HeliNet and CAPANINA and COST Action 297.

The book is structured in three parts and provides a thorough overview and state of the art of the HAP enabling technologies, describes recent research activities with most promising results, and outlines the roadmap for future development of HAPs. The book is intentionally written in a somewhat unbalanced manner, with particular attention given to the topics that require adaptation of existing or development of new concepts and procedures compared to terrestrial wireless and satellite communications. Most notably these topics include multiple HAP networks, advanced communication and resource management techniques, free space optics and networking implications of using multiple HAPs. Special emphasis is given also to the applications and detailed business modelling, typically neglected or not properly covered in books targeted to technical research community, but deemed of particular importance in this book to show the readers the actual techno-economic potential of different HAP applications and scenarios.

Most of the material covered in the book is a result of long-term research work of authors, contributors and other researchers in the field, many of them worldwide recognised leading experts. When putting the material together we made every effort to adequately reference other publicly available information sources such as journal and conference papers, technical reports and recommendations from various international bodies, which can always be used for even more detailed treatment of the given subject.

It is our hope that this book eases the first steps in this exciting research area and motivates new researchers and PhD students to contribute to and take forward the state-of-the-art in the less developed areas of HAP-based communication systems and eventually enable their successful integration into future heterogeneous wireless access infrastructure.

Although the book is authored by us, we owe special thanks to further contributors, named in the List of Contributors, but also to unnamed other long-time collaborators in several projects, for their contributions, guidance, valuable advice and long-lasting discussions. We hope they will also find the book helpful in their future work related to HAPs.

Finally we would like to thank Ms Polona Anžur for her skills and effort in (re)producing the pictures and drawings, as well as the John Wiley & Sons editorial team, who showed a lot of patience, enthusiasm and support during the preparation of this book, especially Tiina Ruonamaa and Sarah Tilley.

David Grace and Mihael Mohorčič
York, UK and Ljubljana, Slovenia
July 2010

Part One

Basics, Enabling Technologies and Economics

1

Introduction

1.1 Introduction

With an ever increasing demand for capacity for future generation multimedia applications, service providers are looking for novel ways to deliver wireless communications services. In developed countries we are familiar today with seeing mobile phone masts dotted around the countryside, but these can be expensive to deploy and continually service. This patchwork of coverage delivers cellular communications, an efficient way of delivering high-capacity density services. We use the term cellular here to describe the way in which the radio spectrum is reused in order to deliver the high-capacity densities. This concept is now being adopted with a number of technologies, including the widely known 2G and 3G mobile systems, but also new technologies such as WiMAX, and also WiFi, where in this latter case islands of coverage (hot-spots) are provided through spectrum reuse.

An alternative for more rural or less developed areas is to use satellite communications. Satellites today are increasingly sophisticated, and capable of delivering spot beam coverage, with minimal ground infrastructure. However, they are incapable of matching the high-capacity densities seen with terrestrial infrastructure.

A possible third alternative way of delivering communications and other services is to use high altitude platforms (HAPs). HAPs are either airships or planes, which operate in the stratosphere, 17–22 km above the ground [1, 2]. Such platforms will have a rapid roll-out capability and the ability to serve a large number of users, using considerably less communications infrastructure than required by a terrestrial network [3]. Thus, the nearness of HAPs to the ground, while still maintaining wide area coverage, means that they exhibit the best features of terrestrial and satellite communications. We will explore these benefits in more detail in later sections of the book.

Broadband Communications via High Altitude Platforms   David Grace and Mihael Mohorčič
© 2011 John Wiley & Sons, Ltd

The main goal of HAPs is to provide semi-permanent high data rate, high capacity-density communications provision over a wide coverage area, ideally from a fixed point in the sky. In practice due to aeronautical constraints all HAPs present compromises. It is helpful to specify the following HAP 'vital' statistics, and as we shall see, these may radically affect the communications system design and ultimate capabilities:

- payload power, mass and volume;
- station keeping and attitude control;
- endurance.

HAPs can be divided into four categories (as shown in Figure 1.1):

1. Manned plane, e.g. Grob G520 Egrett [4, 5].
2. Unmanned plane (fuel), e.g. AV Global Observer [6].
3. Unmanned plane (solar), e.g. AV/NASA Pathfinder Plus [7].
4. Unmanned airship (solar), e.g. Lockheed Martin HAA [8, 9].

| Manned Planes
–e.g. Grob G520T Egrett
(a) | Unmanned Hydrogen Powered Planes
– e.g. Global Observer
(b) |
| Unmanned Solar Powered Planes
–e.g. NASA/AV Pathfinder Plus
(c) | Unmanned Solar Powered Airships
–e.g. Lockheed Martin HAA
(d) |

Figure 1.1 Examples of the main types of high altitude platforms: (a) manned plane. Reproduced by permission of © Grob Aircraft AG; (b) unmanned plane (fuel). Reproduced courtesy of AeroVironment Inc. www.avinc.com; (c) unmanned plane (solar); Reproduced from NASA - http://www.dfrc.nasa.gov/gallery/photo/index.html; (d) unmanned airship (solar). Reproduced by permission of @Lockheed Martin

To date only the first type of HAP is available for commercial use, although HAPs in categories 2 and 3 have been tested experimentally. The fourth category, the unmanned solar powered airship, still has to be realised.

The HAP most suited to the general communications requirements is the unmanned solar powered airship, in view of its on-station lifetime performance and payload capabilities. However, this type of platform is also the most ambitious, as new materials and designs need to be developed and airship handling techniques re-learned. In the short and medium terms it will be possible to make use of existing or other HAPs in the development phase, e.g. the manned and unmanned planes. All of these have more limited capabilities, but can still be used for missions with more limited requirements. It is also possible through careful system design to ameliorate the effects of some of these constraints as will be discussed later. It is also worth noting the intersection between HAPs and more widely known unmanned aerial vehicles (UAVs), which tend to fly at lower altitudes (with the possible exception of Global Hawk [10]). More information on UAV technologies can be found in [11].

Given the state of maturity of the different HAP vehicles, a step-by-step development approach is now being pursued by organisations, with the aim of generating confidence, develop the technology, and perhaps more importantly provide revenue streams for manufacturers. We will describe some of the projects underway later in Section 1.5. Thus an investment–confidence cycle can be created. Designing payloads, and describing payload characteristics in a modular way, will make them suitable (ideally) for all platform types, reinforcing a commonality of requirements and specifications, and thereby making the technology more accessible to non-specialists [12]. We expect that one or more platform modules will be incrementally deployed to serve a common coverage area, with each platform serving one or more payload modules (telecom or other). Such platforms will be networked, with the detailed operations transparent to the end user.

To aid the eventual deployment of HAPs the ITU has allocated spectrum around 48 GHz worldwide [13] and 31/28 GHz for selected countries [14], with spectrum in the 3G bands also allocated for use with HAPs [15]. There is now an emerging body of work on communications delivery from HAPs both for eventual 3G deployments, e.g. [16–18], as well as for communications deployed in the mm-wave bands. Spectrum sharing studies have been carried out, e.g. [19], since all of these bands will be used by, or adjacent to, other services.

To deliver the best-of-both-worlds of satellite and terrestrial communications systems, efficient spectrum reuse will be required to ensure that such deployments can deliver high spectral efficiencies. We will explore specific cellular techniques and reuse solutions in detail in later chapters, but fundamental to the delivery of cells on the ground is the use of spot-beam antennas on the HAP. One issue not really in common with the terrestrial and satellite counterparts is the relatively poor station-keeping that these HAPs will exhibit. This requires careful design of both the HAP and potentially user terminal equipment, to ensure that the antennas are able to stay pointing continually in the right direction in order to maintain the communications links. One alternative way of coping with such movements is to handoff users from one cell to another, but unlike terrestrial handoff, here it is the cells that move and not necessarily the users.

One big advantage for HAPs over terrestrial systems is that cells can be regularly spaced over an area, so that coverage is substantially unaffected by geography and terrain, and since they all originate from the same HAP this centralisation can be additionally exploited to improve resource utilisation.

The purpose of this book is to focus on how HAPs can be used to deliver broadband communication services over a wide coverage area, typically 60 km wide, based on terrestrial standards. Much of the work is loosely based on the use of the broadband WiMAX standard and exploits the mm-wave bands, e.g. 31/28 GHz, given the bands already approved for HAPs by regulators. However, many of the techniques and principles can be applied more widely, even to lower frequency broadband communications (2–5 GHz) as well as HAP delivery of more conventional 3G mobile services. We have chosen a target data rate of 120 Mbits/s per cell to carry out much of the analyses, which is a fundamental limit constrained by the link budget (e.g. limited by the antenna sizes and transmit powers chosen). This is currently greater than most terrestrial WiMAX systems will support on a per sector basis, but the aim here is to be more technology neutral, hence providing a degree of future proofing.

1.2   History

Like with the start of many new fundamental technologies it is very difficult to pinpoint the inventor or the first time it appeared in print. HAPs have their origins back to 1783 when the Mongolfier brothers launched the first hot air balloon. However, it is not until the early 1960s that we start to find direct references or use of airborne craft capable of providing a semi-permanent presence to deliver communications. One example was Echo which was a balloon that was used to bounce radio signals from the Bell Laboratories facility at Crawford Hill to long distance telephone call users [1]. At a similar time the Communications Research Laboratory of Japan published a study on the use of airships to deliver communications. To our knowledge these were not taken much further, and there are other anecdotal references to projects over the years since then. The next public reference that we have come across appears in an editorial in 1992 [20], again proposing a similar concept.

It was 1997/8 when HAPs really started generating interest. This was catalysed by SkyStation International who put forward the concept of a 200 m long solar powered airship HAP, capable of flying at 20 km altitude for a period of years. Their aim was to provide 3G and broadband communications, both in their infancy at that time. Coverage was planned to be upwards of 300 km diameter, as shown in Figure 1.2, delivered from 700 cells produced from a phased array antenna system. They had a number of credible backers including Alexander Haig former US Secretary of State, and Y.C. Lee as its Chief Technology Officer. This project was taken seriously and much of the initial work within the International Telecommunications Union – Radiocommunication Sector (ITU-R) was undertaken on behalf of SkyStation, with ITU-R Recommendation F.1500 based on their design. They successfully managed to get 47/48 GHz band for HAPs use at the World Radiocommunication Conference (WRC) in 1997, with further frequencies at subsequent WRC gatherings.

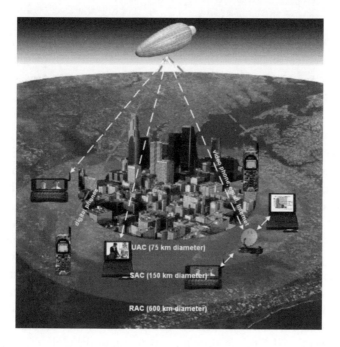

Figure 1.2 The original SkyStation HAP broadband and 3G communications concept. Reproduced from © SkyStation

At a similar period of time, as a result of this spurt of activity, other projects commenced. Another major project was put forward by Angel Technologies [21] based on a manned stratospheric plane, the Proteus, and developed by Scaled

Composites. A photograph of their plane is shown in Figure 1.3, complete with antenna pod beneath the main fuselage. They planned to deliver high capacity communications services over areas of high population, again similar to SkyStation.

Figure 1.3   The Angel Technologies – HALO plane with antenna pod below. Reproduced from © Angel Technologies

Both of these projects failed. The SkyStation concept was ahead of the technology, especially the airship technology, of the time. Angel Technologies aircraft technology was more conventional, but the communications technology and capability claims were over ambitious, resulting in a failure of the business model. Both the technology and business model design are discussed in later chapters.

In Asia at the same time, both Japan and Korea decided to start up their own projects.

Japan put significant funding into their activities through a millennium project initiative that commenced in 1998, with the primary splits being between aeronautics, telecommunications, and earth observation. The Japan Aerospace Exploration Agency (JAXA) coordinated the aeronautics aspects, with the telecommunications activities under the coordination of the National Institute of Information and Communications Technology (NICT). The project and its accomplishments are described in Section 1.5.6, but the aim was to again develop a 200 m long solar powered airship, capable of delivering telecommunications and remote sensing.

Similarly, a smaller project was started by the Korean Government, split across aeronautics and telecommunications, run by the Korean Aerospace Research Institute (KARI) and Electronics and Telecommunications Research Institute (ETRI).

Activities in Europe took longer. The European Space Agency undertook an initial study on high altitude long endurance (HALE) [22], but did not put significant investment in a full scale project. They later commissioned a second study a number of years later [23], but still remain cool to the concept. In 1999 Professor Bernd Kröplin of the University of Stuttgart and Per Lindstrand of Lindstrand Technologies [24] shared the Körber Prize [25] for innovation for putting forward the outline aeronautical designs for an unmanned solar powered airship. The prize provided only modest funding, and insufficient external backing meant a full scale airship was never achieved.

In 1999/2000 a Consortium, coordinated by Politecnico di Torino, Italy, was awarded the 3-year long HeliNet project [26, 27], funded under the 5th Framework Programme of the European Community, and which kick-started the authors' activities in the field. This project was to develop a scale-sized prototype solar powered plane and three pilot applications: broadband communications; remote sensing; and traffic localisation. The was followed 3 years later by the CAPANINA project [28, 29], funded under the 6th Framework Programme, and coordinated by the University of York, UK. This aimed to capitalise on HeliNet, but now in the more focused area of HAP delivery of broadband communications for fixed and high-speed users (and also the main focus of this book). Again more details are discussed in Section 1.5.3.

In the USA over this same period there were a number of other activities, led by NASA and latterly AV Inc [30]. NASA's ERAST programme had already successfully developed the Pathfinder, Pathfinder Plus and Helios unmanned stratospheric planes, each capable of flying modest payloads. NICT of Japan saw this potential, and given that JAXA's airship programme was running more slowly than the NICT communications programme, they decided to undertake pioneering telecoms trials with Pathfinder Plus in 2001. In Hawaii, they successfully demonstrated 3G and HDTV. Helios, NASA's most futuristic stratospheric craft, suffered a mishap in 2003 [31], and crashed in the Pacific Ocean while testing a regenerative fuel cell design, which prevented NASA and NICT carrying out a further round of stratospheric tests.

Following, the Helios mishap, AV Inc (a NASA spin-off) started developing the Global Observer, with a 1/5 scaled prototype flying in 2005. This was successfully used at the end of 2006 for a NASA/NICT/CAPANINA joint test in California.

Lockheed Martin [8, 9, 32] in the mean time had also received US Defense Department funding to develop an airship HAP for the military. To the authors' knowledge this activity is still underway.

As of 2008 there are still a number of activities ongoing, each building on previous developments. One of the most significant is the StratXX [33] project in Switzerland that is developing a solar-powered airship HAP. Key personnel worked on both the HeliNet and CAPANINA projects. There are activities underway with manned stratospheric aircraft, e.g. ERS srl [34] in Italy, using the Grob family of planes. There is also COST 297–HAPCOS scientific cooperation action [35], which was an international discussion forum on HAPs for Communications and Other Services which brings together radio and optical communications, and aeronautical experts on a bi-annual basis. Experts are based in 20 signatory countries, with meetings typically hosting 50–60 delegates. Regulatory activities at ITU-R still continue with much of the work set on studying the multiple system sharing in bands around 5–7 GHz.

1.3 Wireless Communications in a HAP Environment

With the HAP characteristics in mind that we discussed earlier, this section describes in more detail the general concepts and system level design issues relating to the use of HAP(s) to deliver segments of a broadband communications system. HAPs are ideally placed to deliver 'first/last mile' and 'second' mile segments (referred to later as fronthaul), interfacing to more conventional terrestrial and satellite segments via backhaul link(s) structure.

There are a number of interlinked system design issues, ultimately constrained by the platform characteristics.

1.3.1 Comparison of HAPs Capabilities when Compared with Terrestrial and Satellite Systems

The fundamental point is that HAPs can and should exploit the advantages of both terrestrial and satellite systems. Owing to the similar link lengths, maximum link data rates can be comparable with terrestrial wireless links. Fundamentally HAPs can provide regional coverage – a much wider coverage area than a terrestrial base station – owing to the high look angle reducing the attenuation caused by terrain and buildings, etc. Compared with geostationary satellites there is a fundamental path loss advantage of up to 69 dB, enabling HAPs to offer higher data rates and/or use smaller antennas. Thus, if the next generation of satellites deliver their promised link data rates in the hundreds of Mbps range, this HAP link budget advantage can always be exploited in the future to increase link data rates above that of corresponding satellites.

1.3.1.1 Capacity and Coverage

The total capacity of a single HAP-based system is ultimately limited by the HAP and the size, weight and power that can be reserved for the payload. The size of the service area is constrained by the architectural and HAP payload configurations. Three main architectural configurations have been analysed in depth by researchers (these are also discussed in more detail in later chapters):

- **Cellular ubiquitous coverage over a service area** [see Figure 1.4(a)] [2]. The capacity is determined by the number of cells and data rate per cell, with

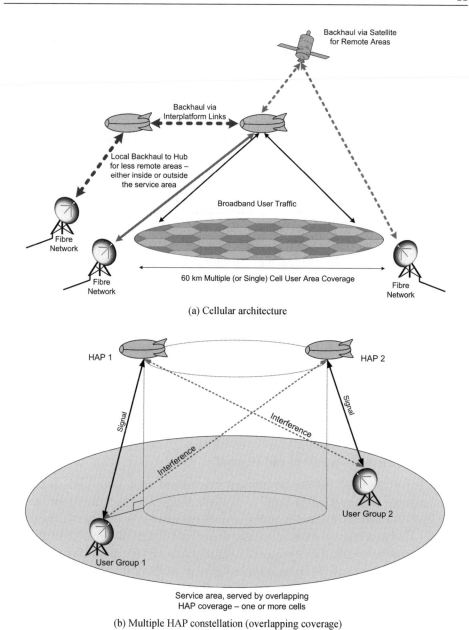

(a) Cellular architecture

(b) Multiple HAP constellation (overlapping coverage)

Figure 1.4   HAP architecture examples

the cell size and number of cells controlling the size of the service area. The cell size is dominated by the beamwidth of the HAP antenna, and this beamwidth is ultimately controlled by elements of the link budget, including, path loss, ground terminal antenna gain and transmit power. Interference between co-channel HAP antenna beams is an important factor affecting the capacity (as does the number of beams on a specific spectrum assignment) [26].

- **Spot beam islands of coverage** [36]. This configuration is probably best suited for specialist broadband connections. HAP capacity is still constrained by the size of payload, and will be similar to the previous case, but now the size of the service area can be decoupled from the capacity constraints, subject to satisfying the link budget for any specific link. The maximum number of spot beams is again affected by the HAP payload capacity, and the interbeam interference caused by the HAP antenna power profiles. It has been shown that 1 Gbps per spotbeam is theoretically achievable in clear air [36] with steerable ground and HAP antennas compensating for HAP attitude movements.
- **Multiple HAP constellations to provide capacity enhancement** [see Figure 1.4(b)] [37]. Operating a constellation of multiple HAPs can increase capacity within a service area. Studies have shown that it is possible to support 16 HAPs on the same spectrum assignment by exploiting the directionality of the ground-based antennas to reduce the interference from other HAPs in the constellation. Such a strategy can be used in conjunction with both architectures above. Multiple HAPs can additionally be used in a coverage enhancement configuration by moving them further apart.

1.3.1.2 HAP Fleet Management and Handoff

In the early days of HAPs, developers considered that long endurance was a fundamental requirement in order to provide long endurance missions, and an enhanced cost-benefit. However, today several companies/researchers have developed detailed fleet management schedules that would enable multiple short duration HAPs to provide a continuous presence over multiple service areas cost effectively [38]. Such concepts have been used for many years with military AWACS systems. However for continuous broadband delivery, fleet management must be accompanied by inter-HAP handoff strategies that are capable of switching over all the communications from one HAP to another without loss of service. This may require specific developments in the case of wireless broadband equipment, which is more accustomed to operation with fixed point-to-point terrestrial operations. In the case of more conventional user protocols, operating at lower data rates, such as WCDMA and WiMAX (IEEE802.16-2004), existing handoff strategies should suffice, even though they have been developed for a moving user and fixed base station configuration (here the HAP base station is moving).

Intra-HAP handoff can be used with the cellular architecture to switch traffic between spot beams to limit the need for payload stabilisation to compensate for HAP attitude movements, especially yaw (rotation). Compensation for pitch and roll (and drift) is still required, to keep the coverage over the service area, or alternatively the size of offered coverage must exceed the size of required service.

While the technical feasibility of all of the above has been assessed positively, practical trials relating to these above issues have yet to be carried out.

1.3.1.3 Radio Propagation Environment

The radio propagation environment is somewhat different from terrestrial (and to a lesser extent satellite) scenarios. The frequencies above 10 GHz will in general require line of sight paths and be increasingly prone to rain attenuation, but due to relatively high slant path angles, such attenuation is restricted to the first few kilometres of the link from the ground. Typical attenuation factors are 20–30 dB for 99.9% availability at 31/28 GHz and approximately 10 dB higher for 48/47 GHz [39, 40]. It is important that link budgets include suitable figures, and this attenuation can normally be compensated for through higher antenna gain. Bands below 10 GHz can be operated increasingly as non-line-of-sight paths as the frequency is decreased, but will be prone to attenuation due to shadowing and multipath. These effects will in general be less than a corresponding terrestrial link at a given frequency.

Propagation models relating to the above frequencies are in development [41–43], but in general lack full practical verification from HAPs. Given the similarity with satellite slant path links, some models, especially relating to rain outage can be accurately extrapolated.

1.3.2 Regulatory Environment and Restrictions

The regulatory environment is highly complex as it relates to HAPs, but falls into two main areas:

- **Radio regulation** – mainly dealing with use of the radio spectrum, and prevention of HAPs and corresponding ground equipment from causing harmful interference to other user types sharing the same frequency band.
- **Aeronautical regulation** – mainly dealing with aspects of safety in controlled civilian airspace.

The plethora of regulations that impact on HAPs can present a major challenge to the operation of HAPs depending on one's viewpoint. One perspective is that HAPs are a completely new technology, which requires a whole new regulatory framework. In fact in the early years of HAPs development this was seen as an advantage, as putting such a regulatory framework in place provided an extra degree of credibility for this emerging technology. SkyStation (USA), NICT (Japan) and ETRI (Korea) spent significant time pursuing new radio regulations (as discussed below) under the auspices of the ITU-R. This did indeed build momentum, but by focusing on regulation also drew attention to the fact that aeronautical regulations pertaining to HAPs were far from clear. For example even today it is not possible to fly an unmanned craft in civilian airspace for safety reasons, although as described below work is underway to develop the necessary regulatory framework. Also anomalies have emerged, e.g. within the ITU-R a high altitude platform station (HAPS) is defined as flying above 20 km altitude, whereas many of the HAPs will fly between 17 km and 20 km, and as such the regulations as developed do not strictly apply. Also, controlled airspace stops above approximately 15 km, so from an aeronautical perspective the airspace is currently unregulated, but it is necessary of course to fly the HAP to this height.

Thus, today there are increasing moves to now focus on the similarity with existing systems and services and apply and or extrapolate existing regulations. Thus, from a radio perspective it may be possible to consider a HAP flying below 20 km as a tall terrestrial mast, making many more potential radio frequency bands available to use. Also, to circumvent the aeronautical regulatory issues of unmanned flight in civilian airspace, manned stratospheric HAPs are attracting considerable attention, as an intermediate evolutionary step.

A third line of thinking, especially by entrepreneurs, is to just go ahead and build HAPs and fly them in a suitable environment (or friendly country) and let the regulations catch up with reality, giving them a head start on any competition. There is some precedent for this; anecdotally the development of the direct broadcast by satellite (DBS) market in the UK by Sky TV in the 1980s shows what is possible by just going ahead and developing a product. A company in Luxembourg was responsible for owning and launching the satellites, and in conjunction with the local regulatory authorities, Sky TV selected an unused frequency that suited their needs from a commercial perspective, but which was not in a band set aside for DBS to the home use. By the time international regulators tried to force Sky TV to pick an alternative frequency, it was no longer really feasible as Sky TV had a customer base in the hundreds of thousands. In parallel, British Satellite Broadcasting (BSB) was following a more conventional regulatory route, and chose to operate in the band set aside for DBS, but which required development of brand new technology, which was subsequently delayed. This delay eventually caused them to be less commercially successful than Sky TV, resulting in their takeover by Sky a couple of years after launch.

So, to sum up, the regulatory environment is in some confusion, which is only to be expected with a brand new technology that has yet to be fully developed. We now discuss in more detail the two main regulatory areas of radio and aeronautical regulation.

1.3.2.1 Radio Regulation

The ITU-R has wide ranging activities concerning spectrum regulation for HAPs. A number of frequency bands have been specified by ITU-R for HAPS (with a narrower definition from that used in this book) [13], and these are included in the successive WRC resolutions:

- 48/47 GHz [13] – 300 MHz bandwidth in both directions – worldwide.
- 31/28 GHz [14] – revised at WRC 07 to 300 MHz in both directions – for use in over 40 countries worldwide (include all countries in North and South America but excluding all of Europe).
- 2 GHz [15] worldwide to support IMT-2000 from HAPS.
- 6 GHz [44] is also under consideration as a WRC 11 Agenda item for Gateway link use for IMT-2000 use.

These activities take place in working parties (e.g. WP9B, WP4-9S, and to a lesser extent WP8F). A list of main current ITU-R recommendations is included in Table 1.1.

Table 1.1 List of the main ITU-R recommendations on HAPS in January 2010 [45]

Number	Title
F.1500 (05/00)	Preferred characteristics of systems in the fixed service using high altitude platforms operating in the bands 47.2–47.5 GHz and 47.9–48.2 GHz. In force
F.1569 (05/02)	Technical and operational characteristics for the fixed service using high altitude platform stations in the bands 27.5–28.35 GHz and 31.0–31.3 GHz
F.1570 (05/02) F.1570-1 (02/03)	Impact of uplink transmission in the fixed service using high altitude platform stations on the Earth exploration-satellite service (passive) in the 31.3–31.8 GHz band
F.1607 (02/03)	Interference mitigation techniques for use by high altitude platform stations in the 27.5–28.35 GHz and 31.0–31.3 GHz bands
F.1608 (02/03)	Frequency sharing between systems in the fixed service using high altitude platform stations and conventional systems in the fixed service in the bands 47.2–47.5 GHz and 47.9–48.2 GHz
F.1609 (02/03) F.1609-1 (04/06)	Interference evaluation from fixed service systems using high altitude platform stations to conventional fixed service systems in the bands 27.5–28.35 GHz and 31.0–31.3 GHz

(*continued*)

Table 1.1 (*Continued*)

Number	Title
F.1612 (02/03)	Interference evaluation of the fixed service using high altitude platform stations to protect the radio astronomy service from uplink transmission in high altitude platform station systems in the 31.3–31.8 GHz band
F.1764 (04/06)	Methodology to evaluate interference from fixed service systems using high altitude platform stations to fixed wireless systems in the bands above 3 GHz
F.1819 (09/07)	Protection of the radio astronomy service in the 48.94–49.04 GHz band from unwanted emissions from HAPS in the 47.2–47.5 GHz and 47.9–48.2 GHz bands
F.1820 (09/07)	Power flux-density at international borders for high altitude platform stations providing fixed wireless access services to protect the fixed service in neighbouring countries in the 47.2–47.5 GHz and 47.9–48.2 GHz bands
M.1456 (05/00)	Minimum performance characteristics and operational conditions for high altitude platform stations providing IMT-2000 in the bands 1885–1980 MHz, 2010–2025 MHz and 2110–2170 MHz in Regions 1 and 3 and 1885–1980 MHz and 2110–2160 MHz in Region 2
M.1641 (06/03)	A methodology for co-channel interference evaluation to determine separation distance from a system using high altitude platform stations to a cellular system to provide IMT-2000 service within the boundary of an administration
P.1409 (10/99)	Propagation data and prediction methods required for the design of systems using high altitude platform stations at about 47 GHz
SF.1601 (02/03), SF.1601-1 (04/05) SF.1601-2 (02/07)	A methodology for interference evaluation from the downlink of the fixed service using high altitude platform stations to the uplink of the fixed-satellite service using the geostationary satellites within the band 27.5–28.35 GHz
SF.1843 (2007)	Methodology for determining the power level for high altitude platform stations ground terminals to facilitate sharing with space station receivers in the bands 47.2–47.5 GHz and 47.9–48.2 GHz
SM.1633 (06/03)	Compatibility analysis between a passive service and an active service allocated in adjacent and nearby bands

1.3.2.2 Aeronautical Regulation

The original vision for HAPs was that they would be unmanned and capable of fully autonomous control, without a continuous link to Air Traffic Control (ATC). However, this is currently far in advance of current Air Traffic Management (ATM) regulations. They require that one or more pilots are on board an aircraft who must be capable of flying the plane at all times, and while airborne the pilot must have voice and data communications capability with ATC. The main reason for this is that regulated civilian airspace has become increasingly crowded, so from a safety perspective, pilots are a tried and trusted means of avoiding collisions [46].

Regulators such as the Joint Aviation Authority (JAA) in Europe are currently grappling with this problem for all types of UAV. One possible half-way solution is to have a remote pilot on the ground, who is capable of providing continuous control via a radio/satellite link. However, this may still provide some autonomous control in the event of link failure. Importantly, the trend is to create regulations for unmanned aircraft that will be similar to existing manned aircraft regulations in order to ensure that an unmanned aircraft does not create a greater safety risk to the population or properties or to other manned or unmanned aircraft sharing common airspace, than existing aircraft.

A second issue already alluded to above is the height of current civilian regulated airspace (currently approximately 15 km) [46], which means that currently these stringent regulations do not apply except for launch and recovery phases. A way round the current regulations in launch and recovery phases is also possible if they could take place outside of conventional civilian airspace, e.g. in special designated zones or even military airspace. However, it remains to be seen whether the current height for regulated airspace will remain unchanged if a large number of aircraft start to use it.

Despite the alternative strategies above it is realistically anticipated that the aeronautical regulations will eventually have to cover the following areas [46]:

- System Safety Objectives.
- Tailored 'JAR' code adjusted to the HAP class and application.
- Communication Data Link (electromagnetic protection, security from external intrusion, failure handling and return home procedures).
- Ground control station (Human Machine Interface minimising human errors, safety warnings and indications, failure handling, redundancy, minimum crew required).
- Flight Management System (flight control and navigation, fail safe principles).
- Flight Termination System (autonomous control, predefined emergency recovery sites, gliding capability).
- Emergency electrical power (capacity for emergency landing).
- Hardware and software qualification.

Operational regulations (besides maintenance aspects) should cover [46]:

- Qualification requirements for HAP operators and crew.
- ATC communication capability within the relevant ATM environment.

- Required safety operational equipment (in particular covering 'Sense/See and Avoid').
- Emergency procedures.

To help with the above the UK Government is currently financing the ASTRAEA project [47] which aims to help define these regulations and also hardware and software for civil operation of UAVs to facilitate routine UAV operations in civilian airspace in the UK. One of the objectives of the ASTRAEA project is to overcome resistance to operation of unmanned aircraft by the regulators, who want to maintain their excellent safety records. Therefore, it will be required that such craft must prove that they are even safer than conventional manned commercial aircraft.

1.4   Candidate Standards for Provision of Services and Applications from HAPs

HAPs are seen as a candidate infrastructure for the provision of various types of services and applications to fixed and mobile, individual and group terminals. Their complementary role, or a role in disaster recovery scenarios, requires the use of existing or developing wireless standards, with minimum adaptations to HAP specifics, if needed. This enables the opportunity to reuse existing (off-the-shelf) equipment on HAPs, while on the other hand making use of widespread user terminals. Clearly not the same standard meets optimally all the needs of different services and applications nor operating scenarios, but the capability of HAPs for being landed for payload maintenance, upgrade and/or replacement allows for modular development of complementary payloads that fit different missions.

Currently there are no definite standards that need to be used on HAPs or that would be developed taking into account specific characteristics of HAPs. Provision has been made by ITU-R regarding the use of frequencies in 2 GHz band for the delivery of 3G (IMT-2000) based communications directly from HAPs, whereas true broadband fixed and/or mobile wireless access is confined to millimetre wave band, more precisely to the frequencies at 31/28 GHz and 48/47 GHz with some restrictions. For these frequency bands there are several candidate standards that could serve the purpose [48], in particular IEEE 802 standards (IEEE 802.11, IEEE 802.16 and IEEE 802.20), DOCSIS standards (MMDS and LMDS) and DVB standards (DVB-S/S2, DVB-RCS).

None of the above standards suits perfectly for the service provision via HAPs, hence the choice is driven not only by HAP characteristics but also by targeted services and applications to be provided, business model, operating frequency band, etc. Different frequency bands also infer different link budgets and the use of techniques to mitigate propagation impairments. Furthermore, with respect to mobile services the respective standard used should support mobility management and operation in the presence of Doppler frequency shifts.

Important criteria in selection of a communication standard are also its flexibility, i.e. the degree of freedom it allows to adapt the system to the specific needs while staying within the predefined specifications, and the openness of the standardisation process, giving an opportunity to influence the decisions of standardisation body or propose potential adaptation of specifications for specific operating environment. In the following some candidate standards are briefly reviewed focusing on broadband wireless access standards and their suitability for delivery of services and applications from HAPs.

1.4.1 Mobile Cellular Standards

The use of cellular standards from HAPs is generally seen as one of the more acceptable and feasible scenarios, particularly in the light of allocated frequencies in the 2 GHz frequency band to support IMT-2000 services from HAPs. This actually spurred a number of studies investigating the performance of 3G standards on HAPs starting from the earlier ones focusing on UMTS and W-CDMA in general [18, 49–53] to more recent investigating the performance of HSDPA from HAPs [54, 55]. These studies started in parallel or superseded studies on the use of 2G technologies, most notably GSM [56–58], mainly conducted in the frame of or related to the European FP5 IST project HeliNet [59, 60]. Obviously, the focus in these studies shifted from cellular radio network planning and efficient management of radio resources in GSM related studies to interference and system capacity investigation in 3G related studies.

In order to become a technology of choice, HAPs will have to make their way into next generation mobile technologies known as IMT Advanced. The best known representatives of the IMT Advanced set of standards are the 3rd Generation Partnership Project (3GPP) Long Term Evolution (LTE) and LTE-Advanced (LTE-A). LTE is focusing on the evolution of the 3GPP radio technologies, concretely UMTS, towards 4G, whereas LTE-A is intended to adapt LTE to the requirements of 4G [61]. For these standards no particular radio frequency provision has been made for the use on HAPs. Moreover, the radio regulation does not appear to be keen in making any further spectrum allocations for HAP systems due to the absence of commercially available aerial platforms. For HAP systems it is thus important that radio regulation adopts a sufficiently flexible spectrum licensing strategy to ensure that the future potential of HAPs, or any other technically sound technology, is not prevented from entering into operation once aerial platforms become commercially viable [62].

In general, a standalone implementation of any cellular standard on HAP should not represent a major obstacle for instance for initial roll-out of a service or in case of major disaster relief, since an aerial platform flying at the altitude around 20 km and equipped with a base station can be seen as a tall mast. However, coexistence of HAP and terrestrial systems renders problems in radio network planning and interference avoidance. The coverage of terrestrial cells is largely influenced by various obstacles such as buildings, trees, hills, etc., whereas comparatively large HAP cells are only shaped by the antenna pattern. Thus a HAP-based base station, although representing a secondary system, can potentially cause much higher interference to terrestrial primary systems than vice versa, which calls for the development and utilisation of interference avoidance techniques. These problems are recently being investigated from the perspective of cognitive radio and dynamic spectrum assignment, which may play an important role in interference avoidance and consequently adoption of HAP systems.

1.4.2   IEEE 802 Wireless Standards

IEEE 802 has developed an alternative series of wireless Internet standards [63]. The main intent is to bring to market low-cost products that serve customer needs. Much of the work involves license-exempt spectrum, which removes the spectrum acquisition costs from the economic picture. Furthermore, it weakens the concept of a monolithic 'operator' with strong control over the provided services. Instead, it opens up the market to enterprise and innovation.

IEEE 802 wireless Internet technologies mostly offer data rates much higher than those provided by even the fixed user case in IMT-2000. However, the basic structures of IEEE standards were not intended to offer the mobility of IMT-2000 in the sense of providing services to moving vehicles. However, extensions to high mobility have been defined for IEEE 802.16 whereas IEEE 802.20 was from the start developed as a standard for mobile broadband wireless access.

IEEE 802 LAN/MAN Standards Committee [64] has developed separate standardisation branches for Wireless Metropolitan Area Network (WMAN), Wireless Local Area Network (WLAN) and Wireless Personal Area Network (WPAN), and more recently with the transition towards digital TV and subsequent release of analogue TV frequency bands also Wireless Regional Area Network (WRAN):

- WMAN includes IEEE 802.16 and IEEE 802.20 standards, which support high-rate broadband-wireless-access services to fixed and mobile users from central base stations.
- WLAN denotes the IEEE 802.11 set of standards, which support users roaming within homes, office buildings, campuses, hotels, airports, restaurants, cafes, etc.

- WPAN, represented by IEEE 802.15 standards, is intended for the support of short-range links among computers, mobile telephones, peripherals, and other consumer electronics devices that are worn or carried.
- WRAN is defined by IEEE 802.22 standard for a cognitive radio-based PHY/MAC/air interface for use by license-exempt devices on a noninterfering basis in spectrum that is allocated to the TV Broadcast Service.

IEEE 802.11 and 802.15 have worked particularly closely since they both address unlicensed bands. IEEE 802.16 has historically dealt with licensed bands and been more independent. The work on IEEE 802.20 originally started almost in parallel with the amendment to IEEE 802.16e to enable connections for mobile devices, but aimed at higher speeds of terminals and higher throughputs. IEEE 802.22 is technically the most advanced standard and the first to include the concept of cognitive radio for better exploitation of shared frequency spectrum.

1.4.2.1 IEEE 802.11

The IEEE 802.11 working group is working on an evolving family of specifications using the Ethernet protocol and Carrier Sense Multiple Access with Collision Avoidance (CSMA/CA) for path sharing. Originally it started with a standard describing MAC (Medium Access Control) sublayer and three PHYs (PHYsical layers), which are now obsolete, but served as a baseline for subsequent amendments that:

- Describe several Wireless LAN PHYs, which differ in transmission techniques and consequently in maximum throughput, all operating in the unlicensed band at 2.4 GHz or at 5–6 GHz.
- Enhance original MAC and MAC-management functionality to provide expanded international operation and roaming, improved support for quality of service, dynamic channel selection, transmit power control, and an architecture and protocol that support both broadcast/multicast and unicast delivery and mesh networking.
- Offer additional security for WLAN applications and define more robust encryption, authentication and key exchange.

The crucial role in widespread adoption and interworking of devices based on the IEEE 802.11 standards played the Wi-Fi Alliance [65]. The term Wi-Fi is often used as a synonym for IEEE 802.11 technology, because of the close relationship with its underlying standard.

IEEE 802.11 standards are mainly intended for use in short-range point-to-point configurations, so they are not particularly suitable for HAPs. On the other hand, Wi-Fi certification guaranteeing compatibility and interoperability of devices based on the 802.11 specification, low equipment price, the availability of off-the-shelf components and easiness of set-up may be appealing for the use of adapted IEEE 802.11-based solutions with extended range. For broadband wireless access provision via HAPs the signal may need to be upconverted to mm-wave frequencies, whereas for rapid establishment of ad hoc networks in emergency scenarios the equipment may as well be used in its nominal band. There are several studies reporting on range extension of WLAN networks, however, data rates achieved in such adapted systems are rather low and are not likely to justify system implementation and HAP deployment. The IEEE 802.11b protocol, in particular, seems to be adaptable to HAP operating environment. This was demonstrated by the CAPANINA project during its second year trial [66] using an 802.11 base station with carrier frequency conversion to the 28 GHz and a high gain tracking reflector antenna at the ground station. The achieved bit rate, however, was low, in the CAPANINA trial most of the time below 1 Mbit/s, and the standard is not suitable for high-speed mobility environments. This low data rate was achieved by adjusting the acknowledgement time-outs, in order to cope with long range links. These longer round-trip times results in a much larger proportion of the transmit–receive duty cycle being wasted, resulting in the lower throughputs than seen with conventionally short-range Wi-Fi systems.

1.4.2.2 IEEE 802.16

IEEE standard 802.16 was designed to evolve as a set of air interfaces based on a common MAC protocol, while PHY layer specifications depend on the used spectrum and associated regulations [67]. Typical applications that can be supported by the IEEE 802.16 access system include applications and services such as digital audio/video multicast, digital telephony, ATM cell relay, IP datagram transfer, backhaul service for cellular or digital wireless telephone networks, virtual point-to-point connections, and frame relay service.

The initial standard, based on a single carrier technology with adaptive modulation and coding schemes only addressed frequencies from 10 to 66 GHz, where extensive spectrum is currently still available worldwide. It assumed line-of-sight (LOS) propagation with no significant concern over multipath propagation. Offered capacities and addressed frequency bands make this standard particularly well suited for use on HAPs. The main concerns about this standard derive from the fact that it was developed for LOS conditions and fixed wireless access only.

Amendment IEEE 802.16a (WirelessMAN) to the initial standard has extended the air interface support to lower frequencies in the 2–11 GHz band so as to allow non-LOS operation and robustness to significant multipath propagation. Compared with the higher frequencies, such spectra offer the opportunity to reach many more customers less expensively, although at generally lower data rates. Operating in the 2–11 GHz band three PHYs have been developed, one again based on a single carrier technology, one using OFDM with a 256-point FFT and one based on OFDMA with a 2040-point FFT. The combined the initial standard for 10–66 GHz and WirelessMAN amendment formed the Air Interface for Fixed Broadband Wireless Access System denoted as IEEE 802.16-2004.

A further amendment, IEEE 802.16e, extended PHY and MAC layers of Wireless-MAN-OFDM for combined fixed and mobile operation in the 2–6 GHz licensed frequency bands. It is aimed at supporting bit rates up to 15 Mbit/s to mobile subscriber station with nomadic mobility up to approximately 100 km/h. Fixed and mobile standards along with the management plane related amendments were eventually brought together in the standard IEEE 802.16-2009 entitled 'Air Interface for Fixed and Mobile Broadband Wireless Access System'.

A particularly important role for the acceptance and implementation of IEEE 802.16 standard equipment is played by the global standard and interoperability forum WiMAX [68]. This forum is also certifying the compliance of equipment with adopted specifications. Thus, WiMAX is to IEEE 802.16 what Wi-Fi is to IEEE 802.11. In fact, the terms WiMAX and mobile WiMAX became synonyms for the fixed and mobile versions of WirelessMAN standard.

Being developed for broadband wireless access the IEEE 802.16 protocol family appears as the most suitable option for broadband service delivery via HAPs. It can be used in a point-to-multipoint topology to link commercial and residential buildings to high-rate core networks, thus resolving the 'last mile' problem of the Internet connection. Furthermore, IEEE 802.16 variants specify different air interfaces for operating in frequency bands between 2 GHz and 66 GHz, thus covering all HAPS allocated bands. However, the upper frequency bands are only covered by the variant for fixed services, whereas extensions for mobility management are only standardised for the frequencies below 6 GHz, thus only available for exploitation from HAPs should frequencies below 6 GHz become available also for HAPs. In fact, there has been a significant body of research work on WiMAX provision from HAPs conducted in the European COST Action 297 [35], many studies focusing on the coexistence of terrestrial and HAP-based WiMAX services.

1.4.2.3 IEEE 802.20

IEEE 802.20 Working Group was concerned with the development of a IEEE 802 based standard for the Mobile Broadband Wireless Access (MBWA) that meets the needs of business and residential end user markets. The 802.20 technology evolved as pure mobile and optimised for IP-based services. The standard was eventually approved by the IEEE as IEEE 802.20-2008 and its scope consists of the physical, medium access control (MAC) and logical link control (LLC) layers. The new air interface was developed for the operation in licensed bands below 3.5 GHz with peak data rates per user in excess of 1 Mbit/s. It supports various vehicular mobility classes up to 250 km/h, and spectral efficiencies, sustained user data rates and numbers of active users that are all significantly higher than achieved by existing mobile systems. This standard thus seems attractive, should frequencies below 3.5 GHz ever become available for the use on HAPs. An important drawback is the fact that the technology is not widely spread, so user terminals are not broadly available to date.

1.4.3   Multipoint Distribution Services for Multimedia Applications – MMDS and LMDS

By restricting the scope of HAPs to delivery of asymmetric broadband services for fixed users only, two further standards appear interesting for the use on HAPs, the Local Multipoint Distribution Services (LMDS) and Multichannel Multipoint Distribution Services (MMDS). LMDS and MMDS have adapted the DOCSIS (Data Over Cable Service Inferface Specification) from the cable modem world. The version of DOCSIS modified for wireless broadband is known as DOCSIS + . LMDS and MMDS wireless modems utilise the DOCSIS + key-management protocol to obtain authorisation and traffic encryption material, and to support periodic reauthorisation and key refresh.

DOCSIS + incorporates the DOCSIS standard in both the MAC and PHY layers to support robust wireless operation. The air interface enables service providers to increase subscriber coverage through near-LOS operation and supports assured quality of service.

The unidirectional MMDS is a wireless network for delivery of broadcast video programs or data information. The bidirectional MMDS can be used to transport two-way data, video, and voice information. MMDS is primarily a LOS microwave-based transport technology, in which the antenna broadcasts the signals towards the end users, typically within a small regional area in the order of up to 50 km. The customer's premise is equipped with an MMDS antenna, a radio frequency transceiver, and a Set Top Box. The frequency spectrum for MMDS system is below 10 GHz, typically in the range of 2–3 GHz. The bandwidth for MMDS is typically in the order of 200 MHz.

The LMDS is a one- or two-way wireless system for data, video and telephony. The unidirectional LMDS is mainly addressing video broadcasting applications, whereas a two-way LMDS system allows for interactive services. LMDS is also a LOS microwave-based transport technology, typically serving a very small regional area in the order of a few kilometres. However, LMDS uses much higher radio frequencies than MMDS, since these systems operate at frequencies above 10 GHz, typically between 26 GHz and 31 GHz; as a consequence LMDS has more bandwidth, typically in the order of 1 or 2 GHz.

Being developed for multipoint distribution services, MMDS and LMDS are particularly appealing for the provision of asymmetric broadcast/multicast services to fixed users via HAPs, LMDS in particular due to the operating frequency band similar to that of HAPs. However, both MMDS and LMDS are suitable for the provision of services and applications to fixed users only, and they appear to have been made obsolete by WiMAX and IEEE 802.16-SC standards.

1.4.4 DVB Standards

DVB specifications cover all aspects related to the audio, video and data signals broadcasting over a variety of transmission physical media. With the use of an interaction channel for return link these standards, originally developed for broadcasting services, also appear potential candidates for delivery of broadband communications via HAPs. Due to some similarities between satellites and HAPs a combination of DVB-S [69] or DVB-S2 [70] and DVB-RCS [71] standards developed for single carrier systems operating at frequencies above 10 GHz proves to be a particularly interesting candidate.

The satellite DVB standards have several appealing characteristics for the use on HAPs, particularly the supported data rates on the forward link (downlink), support for adaptive coding and modulation, native point-to-multipoint operation, and high operating frequency bands including those covering HAP allocated frequencies. The use of satellite DVB standards on HAPs would be particularly suitable in the case of integrated satellite and HAPs networks [72, 73]. The main drawbacks are the relatively low data rate supported on the return link (uplink), thus only asymmetric applications can be supported, requirement for the LOS channel conditions and no explicit support for mobility management. The combination of DVB-S2 and DVB-RCS standards [74] for the use on HAPs has been investigated in several studies with particular attention on the implementation and performance evaluation of adaptive coding and modulation [75–79].

1.4.4.1 DVB-S/S2

The Digital Video Broadcasting-Satellite (DVB-S) determines the standard for satellite broadcasting and for the supply of multimedia services [69]. DVB-S is the oldest and most widespread of the DVB specifications family, making the basic DVB-S techniques proven and mature. DVB-S is being used for point-to-point,

point-to-multipoint and mobile satellite data communications systems. A major advantage of the digital DVB-S platform is its ability to provide broadcast transmission of a large volume of data at a very high rate and with an excellent protection against a variety of transmission errors.

The outdated DVB-S standard has been replaced by DVB-S2 standard [70]. It is characterised by variable coding and modulation functionality, which allows different modulations and error protection levels to be used and changed on a frame-by-frame basis. Combined with the use of a return channel this may be used for closed-loop adaptive coding and modulation (ACM), allowing the transmission parameters to be optimised for each individual user and his propagation path conditions. DVB-S2 is capable of accommodating any input stream format, including continuous bit streams, single or multiple MPEG Transport Streams, native IP as well as ATM packets.

DVB-S2 does not explicitly specify symbol rate range. However, implementations by different vendors can cover maximum transmission rates as high as 300 Mbit/s. The combination of DVB-S2 and DVB-RCS can also be adapted to support full mesh network topology.

1.4.4.2 DVB-RCS

In order to support interactive satellite applications the DVB Project has defined a specification for the return channel via satellite, the DVB-RCS [71]. It supports symbol rates in the range from 128 ksymbols/s to 4 Msymbol/s, resulting in return link information bit rates beyond 5 Mbit/s, for some modes even over 10 Mbit/s.

The DVB-RCS specification was designed to support quality of service (QoS) for a variety of applications over the satellite link and defines a variety of bandwidth allocation mechanisms. Thus, each terminal can have multiple virtual channel assignments with different QoS parameters. Supported bandwidth allocation mechanisms include: continuous rate assignment, rate based dynamic capacity, volume based dynamic capacity, absolute volume based dynamic capacity, and free capacity assignment.

The DVB-RCS specification also defines provisions for security, in particular DVB common scrambling in the forward link, satellite interactive network individual user scrambling in the forward and return link, IP network security (IPsec) and higher layer application security mechanisms.

1.5 Overview of Past and Present HAP Related Projects, Trials and Development Plans

This section provides more detailed information of the major HAPs projects and players, some of which have already been briefly discussed above in Section 1.2. We focus here on their major achievements, including trials and demonstrations, and if they are still ongoing their future plans, where these are publicly available.

1.5.1 StratXX AG – X-Station

This project is developing a solar powered airship based HAP, the X-Station [33]. They are 'striving to commercialise stratospheric communication platforms through the innovative application of advanced technologies', including the rapid provision of communications systems using HAPs in regions currently not well served or lacking infrastructure. They plan to deploy a number of services on their platform, capable of providing, TV and radio, broadcast, mobile telephony, Voice over Internet Protocol (VoIP), remote sensing and local GPS. To date they have developed lower altitude airship-based platforms, including PhoeniXX and X-BUGS, and a photograph of one of their airships in development is shown in Figure 1.5. The test programme involved testing both the craft and the communications and remote sensing payloads.

Figure 1.5 StratXX X-Station Airship in development in 2007 [33]. Reproduced by permission of © StratXX Near Space Technologies AG

StratXX is located in Switzerland, and it uses a number of European strategic partners from industry and academia, including CSEM (Switzerland), DLR (Germany), EPFL (Switzerland), RUAG Aerospace (Switzerland), and University of York (UK), as well as its own in-house expertise.

1.5.2   ERS srl

ERS srl [34] is a small Italian company that is planning to initially use stratospheric manned plane-based technologies, including the Grob G520T and G600, to deliver a range of services from communications and remote sensing payloads. ERS is one of the few companies with actual experience of flying payloads in the stratosphere. Over the last decade they have been responsible for a number of scientific experiments using the M55 Geophysica, a Russian built stratospheric plane.

1.5.3   CAPANINA

The European Commission supported the CAPANINA project (FP6-IST-2003-506745) [28] as part of the 6th Framework Programme, to further develop the state-of-the-art in broadband from aerial platforms. The project ran from November 2003 until January 2007, involving a consortium of 13 partners, representing a mixture of large industry, SMEs, and academia/research organisations.[1]

CAPANINA focused on development of low-cost broadband technology from HAPs aimed at providing efficient coverage to users who may be marginalised by geography, distance from infrastructure, or those travelling inside high-speed public transport vehicles (e.g. trains travelling up to 300 km/h) [28], as shown in Figure 1.6. The aim was to exploit this future wireless technology to deliver burst data rates to users of up to 120 Mbps anywhere within a 60 km coverage area. Both mm-wave band and free space optical communications technologies were considered.

The project adopted a three-strand approach:

- Identification of appropriate applications and services and associated business models. This included establishing the most appropriate integrated network architectures, and included wireless and free space optical link technologies, and multiple platform technologies and spectrum sharing.
- The development of a system testbed that allowed nearer-term tests of broadband services/applications to fixed users, including backhaul for terrestrial WLAN, corporate communications and video-on-demand, along with an evaluation of free space optical technology.
- Longer-term state-of-the-art research and innovation that examined advanced mobile broadband wireless access. Outline system design and critical hardware

---

[1] The CAPANINA Partners were: University of York (UK), Jožef Stefan Institute (Slovenia), Politecnico di Torino (Italy), Universitat Politecnica de Catalunya (Spain), Carlo Gavazzi Space (Italy), Budapest University of Technology & Economics (Hungary), DLR (Germany), BTexact (UK), EuroConcepts Srl (Italy), CSEM (Switzerland), Contraves AG (Switzerland), National Institute of Information & Communications Technology (Japan), and Japan Stratosphere Communications (Japan).

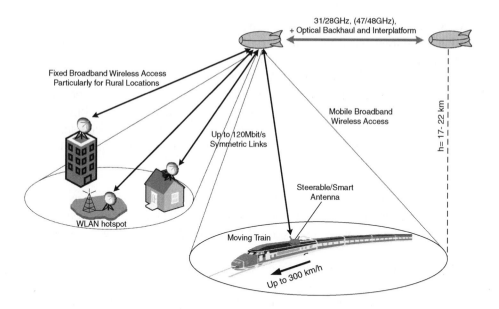

Figure 1.6 CAPANINA project scenario

was developed for a scenario that delivered broadband to trains, integrating with on-board wireless LAN base stations.

The project delivered a number of trials with different types of HAP.

Trial 1 [28, 80] took place between August 2004 and October 2004 in Pershore, UK. A spherical aerostat, capable of operating at an altitude of 300 m was used. The following aspects were successfully demonstrated:

- Broadband Fixed Wireless Access (BFWA) up to a fixed user using 28 GHz band.
- Demonstration of end-to-end network connectivity.
- Demonstration of services such as high speed internet and video-on-demand.
- Optical communications – HAP to ground – simplified overall system without beacons to perform first ground station tracking tests.

These tests were carried out at low wind speeds since the spherical aerostat was less aerodynamically stable than the more conventional teardrop shape.

Trial 2 [80] took place in August 2005 near Kiruna, Sweden using a stratospheric balloon for a short term, one off mission of 9 h. The following aspects were demonstrated:

- A broadband wireless link and applications. Using an uprated mm-wave WLAN payload with on board server, a number of broadband applications were tested, such as video download and web-page download. Channel rates were limited by the IEEE 802.11b technology used at the time. This payload used the 28/29 GHz bands, within band allocations permitted by the Swedish authorities. To overcome the problems of the lack of station-keeping on a free flying balloon, wide area coverage from the balloon was achieved using a lens antenna. A tracking antenna on the ground was used to achieve the link budget.
- Optical Communications
  - The first known 1.25 Gbit/s HAP to ground link was tested. This was achieved using on board optical laser tracking to cope with the balloon's station keeping, and a tracking optical receiver on the ground, based on an astronomical telescope [81].
  - Measurements of the turbulence of the optical channel were tested, and compared with simulation
- Integration of a multi-payload system on a Stratospheric Carrier, subject to low temperature, low air density, higher levels of radiation than on the ground, challenging weight and power constraints.

Trial 3 [82] took place in November 2006 in Paso Robles, California, USA using the AeroVironment (AV) Global Observer 1/5 scaled-size prototype, with direct support from AV and NICT, Japan. This trial was limited in scope owing to the fact that the prototype was limited to low altitude flight. The project used it to test station-keeping and altitude models and also examine the capabilities of low altitude UAV to support localisation of signals from the ground, such as may be required in disaster scenarios.

1.5.4   USEHAAS

USEHAAS [83] was a European FP6 Specific Support Action under Aeronautics and Space Priority that ran from March 2005 to October 2006. It aimed to examine prospective aeronautical research agendas in High Altitude Aircraft and Airships (HAAS), along with associated mission/applications in conjunction with the European aeronautical and space industry.

Sub-objectives included:

- To analyse the world state of the art including European work relating to HAAS aeronautical uses, as well as the programmes and tests underway in USA, Europe, Japan, China and Korea.

- To develop tentative Research Objectives for European/Global HAAS Deployment regarding a variety of end-user services and prepare an outline of a potential aeronautical research programme.
- To disseminate recommendations on the Objectives and the Aeronautical Research Agenda.
- To issue a final version of the Specific Research agenda, including the impact on regulations and recommendations on such a call. It may assemble the activities in the sector, and provide appropriate technological roadmaps based on inputs made during the workshops and working groups with end users and possible industrial partners

1.5.5   COST 297

COST 297 [35], which has recently concluded, provided a forum for research and technological development into HAP communications and other services. Its main objective was: 'To increase knowledge and understanding of the use of High Altitude Platforms (HAPs) for delivery of communications and other services, by exploring, researching and developing new methods, analyses, techniques and strategies for developers, service providers, system integrators and regulators.'

Activity was divided into three working groups:

- WG1 Radio Communications;
- WG2 Optical Communications;
- WG3 Aerial Platforms.

1.5.6   The Japanese National Project

This government funded millennium project commenced in 1998, with the primary splits being between aeronautics, telecommunications, and earth observation. The Japan Aerospace Exploration Agency (JAXA) was coordinating the aeronautics aspects, with the telecommunications activities under the coordination of National Institute of Information and Communications Technology (NICT). The aeronautics and telecommunications activities are described separately below.

1.5.6.1 Japan Aerospace Exploration Agency (JAXA)

JAXA (formerly the National Aerospace Laboratory) has had several significant technical achievements. These have mainly been centred on the launch and landing phases of the stratospheric airship, which are seen as the most difficult aspects, airship materials and construction, station-keeping and power management. More specifically:

- Basic research into the properties of airship materials and construction techniques suitable for long-term stratospheric flights.
- Regenerative fuel cell (RFC) and solar power research and development. They currently have a RFC working prototype capable of delivering 400 Wh/kg and have a design on the drawing board to deliver 700 Wh/kg.
- Airship deflation tests using small explosive charges, to open up a hole in the envelope to ensure controlled decent.

This research and development was used in two major experiments:

- August 2003 – they successfully launched vertically an unmanned 47 m long airship shaped balloon (using ATG's patented launch sequence), at Hitachi Port in Ibaraki, Japan. Under test were the buoyancy control and deflation strategies. The balloon reached the 16.4 km test altitude, where the deflation strategy described previously was used to successfully bring the balloon back to earth in conjunction with a parachute. Additionally, the experiments enabled JAXA to operate a simple payload that was responsible for the collection and analysis of greenhouse gases in the atmosphere.
- November 2004 – they successfully completed tests of an unmanned remotely/ automatically piloted 67 m long airship carried out at 4 km altitude, at the Taiki test range in Hokkaido, Japan. The craft had a payload weight capability of 400 kg, including the mission battery. Tests included:
  ○ heat and buoyancy control;
  ○ station keeping;
  ○ ground handling;
  ○ establishment of a flight operations system;
  ○ observation of vegetation and traffic system from the airship;
  ○ telecommunications tests (described below).

1.5.6.2 National Institute of Information and Communications Technology (NICT)

NICT, formerly Communications Research Laboratory of Japan (CRL) and Telecommunications Advancement Organisation (TAO), has been active in the National Project since its inception in 1998.

NICT has concentrated on communications, broadcasting and radiolocation payloads [84] on HAPs, with the end applications being:

- Disaster relief/event servicing, both 3G and broadband.
- Broadcast HDTV.
- Broadband fixed access to users.
- Broadband mobile access to long distance trains and other vehicles.

These applications were selected primarily because HAPs have the potential to provide users over a wide area with various services at reasonable cost. For example, the project has calculated that 15 000 HDTV terrestrial transmitters are needed to deliver broadcast TV over Japan's mountainous terrain, which can be delivered with around 15 HAPs.

In close coordination and cooperation with JAXA, they have examined tracking and control techniques for stratospheric flight [85], and developed the following hardware for use with experiments:

- Mechanical antenna beamsteering for a multibeam horn antenna at 47/48 GHz.
- A 31/28 GHz digital beamforming (DBF) antenna, capable of delivering 9 Mbps to nine fixed beams and three adaptive beams. They are currently working on a new version that will deliver data rates in excess of 50 Mbps.
- UHF antenna for broadcast HDTV.
- Radiolocation payload.
- Free space optical transmitter and receiver.

These have formed the core of a number of experiments:

- June–July 2002 – World first telecommunications trials using NASA's Pathfinder-Plus solar powered aircraft in Kauai, Hawaii, USA at approximately 20 km altitude, as shown in Figure 1.7. 3 G [86] and HDTV [87] applications were demonstrated and specific tests included:
  o Technical data for the operation of equipment in the stratosphere, including temperature, wind speed, and station-keeping behaviour
  o Interference suppression using on-board array antenna at 2 GHz.
  o HDTV transmission with 1 W transmission power.
  o Video connection to off-the-shelf terrestrial 3G (W-CDMA) videophone.
- November 2002 – Video-on-demand, IP phone, Web access and HDTV video transmission were demonstrated [88]. Tests of both the DBF and multibeam horn antennas on a helicopter at 3 km altitude. Specific tests included:
  o multibeam forming;
  o remote array calibration;
  o beam steering and tracking;
  o beam stabilisation;
  o link performance.

NASA Dryden Flight Research Center Photo Collection
http://www.dfrc.nasa.gov/gallery/photo/index.html
NASA photo: ED02-0161-2 Date: June 24,2002 Photo by: Nick Galante

Pathfinder-Plus flight in Hawaii

Figure 1.7 Pathfinder-Plus carrying onboard equipment. Reproduced from NASA – http://www. dfrc.nasa.gov/gallery/photo/index.html

- September–November 2004 – Digital broadcasting [89], radiolocation [90] and optical communications [91] were demonstrated from a low altitude airship (see Section 1.5.6.1 for more details on the aeronautics aspects). Specific tests included:
  - Evaluation of airship to ground channel at 2 GHz.
  - Evaluation of coverage, using conventional digital HDTV receivers.
  - Radio location from platform to ground.
  - Free space optical communications ground-platform (acquisition and tracking only).

NICT used the AV Global Observer 1/5 scaled-prototype in November 2006 (also in conjunction with CAPANINA), as discussed previously.

They have also continued to test on a smaller scale a number of aspects using a Zeppelin NT in Japan, including:

- New version of DBF antenna for the 31/28 GHz band.
- Radiolocation.

Spectrum regulation is the second area where the Japanese project has had significant impact. NICT is active on several ITU-R Working Parties, e.g. WP9B, and has been responsible for originating six ITU-R Recommendations, and contributing to others originated by, for example, SkyStation and ETRI Korea. It has been instrumental in opening up the 31/28 GHz spectrum to HAPs for secondary status at the WRC in 2000 [92].

1.5.6.3 Discussion

The Japanese National Project has spent in excess of $100M since its inception in 1998, although funding is now at a low level. It is the undisputed world leader in HAP communications hardware development and trials, probably achieving more than all other HAP projects put together. The key personnel involved in the project believe HAPs will only succeed in Japan and Asia if they succeed on a worldwide basis. They see the ITU-R and spectrum regulation as being a significant building block. Their abilities to get spectrum licensed in the 31/28 GHz bands originally in Region 3, but later extending to 40 countries worldwide is an important way of HAPs gaining credibility. They have also used partnerships, e.g. with NASA/ Japan Stratosphere Communications (JSC) and CAPANINA to help fill in gaps in their expertise. Collaboration with JSC enabled stratospheric trials to be carried out where several world firsts have been achieved. This approach has been used because Japan's airship programme was at a different stage of maturity to communications.

1.5.7   The Korean National Project

The Korean national activities have been running for a number of years. They are split into the aeronautics part coordinated by the Korean Aerospace Research Institute (KARI), which has been running since December 2000, research and development aspects of communications coordinated by the Electronics and Telecommunications Research Institute (ETRI) which has been running since February 2002, and analysis of communications services by SK Telecom [93]. We now describe each aspect of the project.

1.5.7.1 Korean Aerospace Research Institute (KARI)

The Korean project has a 10-year plan to build an unmanned stratospheric airship capable of supplying 10 kW of power to a payload, weighing up to 1 t. The activities commenced in December 2000.

The project included the development and test of a 50 m unmanned airship, capable of flying to an altitude of 3 km height and carrying a 100 kg payload, as shown in Figure 1.8. This phase of development also included:

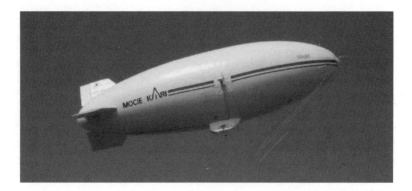

Figure 1.8   The 50 m long unmanned airship developed by KARI for Phase 1. Photograph courtesy of Korean Aerospace Research Institute, Airship Department [94]. Reproduced by permission of © Korean Aerospace Research Institute

- Building of ground facilities including a large hanger in Goheung in the far South of Korea.
- The development and test of electric motors to power the vehicle.
- Launch, landing and station keeping.
- Safety critical systems, including the transfer of control between automatic and manual control.
- Communication system tests.
- Power management of a regenerative fuel cell.
- Basic research into high strength materials, heat management and operational technology for large airships.

1.5.7.2 Electronics and Communications Research Institute (ETRI)

ETRI's activities commenced in February 2002. They included analysis of possible HAPs services and business models [95] for the Korean market, as well as the development of core technologies [96]. The Institute has been very active in the ITU-R working parties 9B, 9D, and 4-9S, which cover HAPs and has originated (and contributed to) several of the ITU-R Recommendations in force on HAPs. Additionally it has played an important role within WRC 2003, where it was instrumental in extending the use of 28/31 GHz for secondary use within Region 2. Additionally it has also produced material to support WRC 2003 Resolution #734 to extend the frequency bands beyond those currently allocated [97].

Core technologies and studies developed include:

- Development of a direct beam forming multibeam receiving system for S and Ka band.
- Development of mechanical beam steering for 47/48 GHz horn antennas.
- Interference analysis and sharing studies for WCDMA.
- Interference analysis and sharing studies for 28/31 GHz and 47/48 GHz.

1.5.7.3 Discussion

The Korean National Project is one of the most significant research led projects around the world. Its work so far has contributed on a technical basis at the highest level, and its activity within the ITU is extremely significant, with ETRI responsible for much of the pressure to get the 28/31 GHz bands licensed in Asian countries. Additionally, its work in trying to extend the allocated frequencies beyond the existing bands, in order to support WRC 2003 Resolution #734 is world leading.

1.5.8   NASA Activity

NASA has been responsible for building a variety of high altitude plane-based craft, including the Pathfinder, the Pathfinder Plus and Helios, which were developed as part of the ERAST programme [98]. Helios currently has the altitude record of some 98 863 ft established in July 2001. As mentioned earlier it subsequently suffered a 'mishap' in which the vehicle was lost. At the time of loss the vehicle was carrying a 250 kg fuel cell beneath the central part of the wing. The resulting investigation has attributed the loss of the craft to a flexing of the wing, caused by the payload being too heavy for the particular design of the craft. This flexing caused lift to be lost and the vehicle to stall, causing the loss of the craft [31]. The investigation has resulted in new recommendations for the design of future vehicles, with the outcome that future designs will be based on the more conventional fuselage, wing and control surfaces (see below), and less reliant on the unique 'flying wing' designs used in the Pathfinder Plus and Helios. More details can be found in [98].

1.5.8.1 Discussion

NASA's track record on the development of specialist aeronautical vehicles is strong, and is one of a few organisations worldwide that have stratospheric vehicles. They are capable of supporting communications applications, with event servicing/ disaster relief applications being possible in near term, with more permanent applications being possible in the longer term.

1.5.9   AV Inc

AV Inc (formerly AeroVironment) [30] started life as a spin-off company of NASA whose objective is to exploit and commercialise NASA technology. It has for a number of years been commercialising NASA's UAV activities. It has specific interests in high altitude planes and is currently developing the energy technology to provide around the clock powered flight, based on hydrogen fuel cell technology. It has built a scaled-size prototype of the latest craft, the Global Observer, and has demonstrated flights of up to 5 days using fuel cell technology. It is also interested in providing round the clock solar powered flight based on a combination of solar powered arrays and regenerative fuel cells. AV Inc has a full size fuel cell and electrolyser storage system for such craft operating in its test facilities.

The craft themselves are aimed at telecommunications, remote sensing and atmospheric measurement. AV Inc has a long term interest, via the spin-off company SkyTower Inc, to deliver a range of applications such as:

- Fixed Broadband
- 3G-Mobile
- Narrowband
- Direct Broadcast
- Event servicing.

It has been pursuing market opportunities on a worldwide basis, and has been specifically targeting developing countries, where there is an absence of infrastructure.

In July 2002, it was jointly responsible for providing the successful tests of various communications for the Japanese National Project (as discussed in Section 1.5.6).

1.5.9.1 Discussion

AV Inc is a significant player in the commercialisation of HAPs. It has detailed business models and excellent technical capabilities. It is also one of the few companies worldwide that has the proven capabilities to launch such a mission and its on-going research programmes into energy storage could prove critical in the development of HAPs.

The Helios mishap has slowed down their marketing of this technology. The fuel cell critical to round the clock deployment was being tested when Helios failed. This delay has also allowed the business models to be revised towards shorter-range missions such as event servicing and disaster relief communications, which are less reliant on solar-powered flight and advanced energy storage, and as such providing a less risky initial exploitation path.

1.5.10   Lockheed Martin, Boeing and Worldwide Aeros

Lockheed Martin, Boeing and Worldwide Aeros (USA) have completed Phase 1 of a large development programme as a result of the US Homeland Defense initiative [8, 9, 32]. Lockheed is now proceeding with Phase 2 of the project [99] and has $40M to develop a solar-powered High Altitude Airship (HAA). The airship is planned to have a mission life of 1 month, operating at 65 000 ft (approximately 20 km), while providing 10 kW of power to a 4000 pound (approximately 2 t) payload. An artist's impression of the craft is shown in Figure 1.9. It is intended that it will become part of the Ballistic Missile Defense System Test bed. Some of the early activities will be to demonstrate station-keeping and autonomous flight. It will be used for military and civilian activities including:

- weather and environmental monitoring;
- short and long range missile warning;
- surveillance;
- target acquisition.

Figure 1.9   Artist's impression of Lockheed Martin High Altitude Airship [99]. Reproduced by permission of © Lockheed Martin

1.5.10.1 Discussion

This programme is one of the more credible developing HAPs. Lockheed Martin has significant experience; the development site at Akron has been developing lighter than air vehicles since 1928. Being funded from the US military budget also provides further security that there will be significant financial resources to ensure that the craft will be built. However, there have been recent rumours that in order to keep to the tight development timescales the specification of the craft has been loosened, including providing a much reduced payload of 250 kg.

1.5.11 Advanced Technologies Group (ATG)

In the early 2000s ATG was active in its Stratsat programme. Stratsat was a solar powered airship-based platform. However, ATG has since gone bankrupt and its technical team is in the process of reforming to pursue its many airship interests. They had some unique technology and deployment methods for their craft. Although it was proposed to power the craft using solar power, in conjunction with regenerative fuel cells/batteries for much of the time, they have also incorporated a diesel engine into some of their designs, which can be operated for limited periods to keep the craft on station in the event of wind gusts. The craft was designed for a 5 year mission and sufficient diesel will be taken on board for the full length of the mission. They have thought about the launch phase of the Stratsat, and ways to overcome the sheer size of such vehicles (up to 200 m long). They intended to launch the craft vertically out of the hangar in such a manner that it is outdoors and close to the ground for only a minimum period of time [100]. They had test facilities and hangars capable of housing such vehicles in Cardington, UK.

They have spent significant effort in pursuing Stratsat Malaysia, a project to be funded by the Malaysian government, which had BT and EADS as partners. They have been looking at a range of applications:

- Mobile Telephony;
- Civil Broadband;
- Civil Communications;
- Maritime Communications;
- Military Communications, Data Transmission and Surveillance;
- EOIR Imagery;
- Radar Imagery;
- Civil Tracking/Road Usage.

1.5.12 European Space Agency (ESA) Activity

ESA has studied broadband delivery from HAPs, as part of its long-term low-level investigations into HAPs. This collaboration was between Airobotics GmbH (Germany), Booz Allen and Hamilton (Netherlands), University of Surrey (UK),

Politechnico di Torino (Italy), Fraunhofer Institute (Germany) and the German Aerospace Center (DLR) [23]. The remit of the study was to examine both the aeronautics and applications. Four categories of platform were compared:

- aerodynamic, solar powered;
- aerodynamic, fuel powered;
- aerostatic, solar powered;
- aerostatic, fuel powered.

Emphasis was placed on the use of solar power, but it was found that the eventual choice of platform depended heavily on the choice of payload, and the conceptual design for a platform has been developed.

The feasibility of a range of communications applications, including broadband, 3G, and DAB/DVB-T broadcasting have been studied. The choice of these payloads was based on technical feasibility, benefits for operators and market demands. The payloads themselves were based on three technical strategies: transparent; radio frequency switching; and regenerative.

This activity carries on from earlier ESA HALE study [22] which was completed a few years ago.

1.5.13   Flemish Institute for Technological Research (VITO)

One of the most interesting activities has been the development of the Zephyr which is a small stratospheric UAV. The Flemish Institute for Technological Research (VITO) has its ongoing Pegasus project [101] that is using a modified version of the Zephyr craft developed by QinetiQ [102] called the Mercator, and plans to deliver real-time remote sensing type applications, via a high speed communications link. The project timeline extends to 2011, by which time they hope to have a fully fledged remote sensing system operating.

1.5.14   QinetiQ Ltd

QinetiQ Ltd based in the UK has a number of disparate interests in HAPs, stemming from the days when it was the UK's Defence Research and Evaluation Agency (DERA). It has the following activities:

- Zephyr 3 unmanned plane and other UAV activity;
- airships;
- communications;
- surveillance/remote sensing.

1.5.15 Space Data Corporation

Space Data Corporation [103] is a US company that delivers wireless data services to rural and remote areas. One such example is the provision of telemetry services to the oil and gas industries. It is currently doing this using free flying balloon-based technology, their SkySite® network. These balloons are launched every 8–12 h from sites close to the coverage area, with the payload recovered on landing after around 24 h. They operate up to a height of 80 000–100 000 ft and provide at that altitude a coverage area of 350 miles (approximately 580 km) in diameter. Coverage is currently restricted to Texas, Oklahoma and Louisiana with limited coverage in neighbouring US states.

They can offer for example:

- SCADA communications;
- RTU/EFM/PLC monitoring;
- pump-off controller monitoring;
- kW meter reading;
- compressor/tank alarming;
- pipeline monitoring.

These operate in the 900 MHz band where Space Data holds a 1.7 MHz allocation of spectrum in the US. The company additionally offers services to the transportation sector, field communications and location services.

1.5.16 HeliNet

One of the early HAP projects was HeliNet (IST-1999-11214) [26, 27], which ran between January 2000 and March 2003 funded by the European Commission's FP5 programme. The project examined aeronautical issues and three prototype applications: broadband telecommunications; environmental monitoring; and vehicle localisation. A design for a scale size prototype stratospheric aircraft, 'Heliplat', was developed and key components built.

The University of York (UK) undertook the majority of the work in the broadband application, with input from Politecnico di Torino (Italy - also overall coordinator), Jožef Stefan Institute (Slovenia), the Budapest University of Technology and Economics (Hungary), and the Technical University of Catalonia (Spain). Barclay Associates (run by Professor Les Barclay, a former Deputy Director of the UK Radio Communications Agency) provided input on regulatory matters. Other members of the consortium were: CASA (Spain); Enigma Technology (UK); Carlo Gavazzi Space (Italy); Fastcom (Switzerland); and Ecole Polytechnique Federale de Lausanne (Switzerland).

1.5.17   Lindstrand Technologies Ltd (UK)/University of Stuttgart

Lindstrand Technologies Ltd [24], a UK company led by Per Lindstrand, has been involved in a project funded by the Koeber Institute and ESA. Lindstrand has established LBL as a 50/50 partnership with DASA (Daimler Chrysler Aerospace). The company has significant expertise in advanced materials required for airship-based HAPs, and production facilities capable of manufacturing airship envelopes.

The University of Stuttgart and Lindstrand won the Körber prize in 1999 [25]. Professor Bernd Kröplin has significant experience of airships and the group has a number of innovative designs for HAP vehicles. Their knowledge of the aerodynamic properties of large airships is probably the best in Europe.

1.5.18   SkyStation

This project was responsible for much of the initial work on the radio regulations within ITU-R. It is no longer active.

1.5.19   Angel Technologies – HALO

Angel Technologies [1, 21] developed a manned stratospheric plane in the late 1990s. It was planning to deliver continuous services to users using a fleet of these craft that would have a mission duration of 8 h, where they would circle above the coverage area. The project is not thought to be currently active, and was almost certainly a casualty of the dot.com boom and bust. One of the main reasons for inactivity could be down to the fact that the business model was not appropriate to the type of HAP vehicle. Such a vehicle may be much better suited to short-range missions, such as event servicing and disaster relief.

References

1. G. M. Djuknic, J. Freidenfelds and Y. Okunev, *Establishing Wireless Communications Services via High-Altitude Aeronautical Platforms: A Concept Whose Time Has Come?* IEEE Commun. Mag., September 1997, pp. 128–135.
2. T. C. Tozer and D. Grace, *High-Altitude Platforms for Wireless Communications*, IEE Electron. Commun. Eng. J., June 2001, pp. 127–137.
3. D. Grace, N. E. Daly, T. C. Tozer, A. G. Burr and D. A. J. Pearce, *Providing Multimedia Communications from High Altitude Platforms*, Int. J. Sat. Commun., November 2001, No. 19, pp. 559–580.
4. Grob, G520T, Egrett Platform, www.globalsecurity.org/military/world/europe/egrett.htm, January 2008.
5. Grob G520T Egrett Platform on Grob Website, www.grob-aerospace.net/index.php?id=137, January 2008.

6. AeroVironment launches Global Observer Program for the US Special Operations Command, www.defense-update.com/newscast/0907/news/260907_hale_av.htm, 26 September 2007.

7. NASA Pathfinder Plus Platform, www.dfrc.nasa.gov/gallery/photo/Pathfinder-Plus/index.html, January 2008.

8. Lockheed Martin High Altitude Airship, www.globalsecurity.org/intell/systems/haa.htm, January 2008.

9. Lockheed Martin HAA Program, www.lockheedmartin.com/products/HighAltitudeAirship/index.html, January 2008.

10. Global Hawk Platform, www.globalsecurity.org/intell/systems/global_hawk.htm, January 2008.

11. UAVNet Thematic Network, www.uavnet.com, January 2008.

12. D. Grace, *et al.*, *Applications and Business Models for HAP Delivery*, FP6 CAPANINA Project, Doc Ref: CAP-D26-W10-BUT-PUB-01 www.capanina.org/documents/CAP-D26-WP11-BT-PUB-01.pdf, January 2007.

13. ITU Recommendation ITU-R F.1500, *Preferred Characteristics of Systems in the Fixed Service Using High Altitude Platforms Operating in the Bands 47.2-47.5 GHz and 47.9-48.2 GHz*, International Telecommunications Union, Geneva, Switzerland, 2000.

14. ITU Recommendation ITU-R F.1569, *Technical and Operational Characteristics for the Fixed Service Using High Altitude Platform Stations in the Bands 27.5-28.35 GHz and 31-31.3 GHz*, International Telecommunications Union, Geneva, Switzerland, 2002.

15. ITU Recommendation M.1456, *Minimum Performance Characteristics and Operational Conditions for HAPS Providing IMT-2000 in the Bands 1885-1980 MHz, 2010-2025 MHz and 2110-2170 MHz in Regions 1 and 3 and 1885-1980 MHz and 2110-2160 MHz in Region 2*, International Telecommunications Union, Geneva, Switzerland, 2000.

16. F. Dovis, R. Fantini, M. Mondin and P. Savi, *Small-Scale Fading for High-Altitude Platform (HAP) Propagation Channels*, IEEE J. Selected Areas Commun., April 2002, Vol. 20, No. 3, pp. 641–647.

17. E. Falletti, M. Mondin, F. Dovis and D. Grace, *Integration of a HAP within a Terrestrial UMTS Network: Interference Analysis and Cell Dimensioning*, Int. J. Wireless Personal Communications - Special Issue on Broadband Mobile Terrestrial-Satellite Integrated Systems, February 2003, Vol. 24, No. 2.

18. S. Masumura and M. Nakagawa, *Joint System of Terrestrial and High Altitude Platform Stations (HAPS) Cellular for W-CDMA Mobile Communications*, IEICE Trans. Commun., October 2002, Vol. E85-B, No. 10, pp. 2051–2058.

19. M. Oodo, R. Miura, T. Hori, T. Morisaki, K. Kashiki and M. Suzuki, *Sharing and Compatibility Study Between Fixed Service Using High Altitude Platform Stations (HAPs) and Other Services in 31/28 GHz Bands*, Wireless Personal Communications, 2002, Vol. 23, pp. 3–14.

20. R. Steele, *Guest editorial - An Update on Personal Communications*, IEEE Commun. Mag., December 1992, pp. 30–31.

21. N. J. Collela, J. N. Martin and I.F Akyildiz, *The HALO Network*, IEEE Commun. Mag., June 2000, pp. 142–148.

22. ESA HALE Study, www.lindstrand.co.uk/hale.html.

23. Stratospheric Platforms - a definition study for an ESA system, European Space Agency, April 2004, telecom. esa.int/telecom/www/object/index.cfm?fobjectid=8188.

24. Lindstrand Technologies Website, www.hiflyer.com.

25. *Körber Prize Winner 1999 - High-Altitude Platforms to Supplement Satellites*, www.koerber-stiftung.de/english/publications/press_archive/pm_1999/pm_12_07_99.html.

26. J. Thornton, D. Grace, M. H. Capstick and T. C. Tozer, *Optimising an Array of Antennas for Cellular Coverage from a High Altitude Platform*, IEEE Trans. Wireless Commun., May 2003, Vol. 2, No. 3, pp. 484–492.

27. D. Grace, J. Thornton, T. Konefal, C. Spillard and T. C. Tozer, *Broadband Communications from High Altitude Platforms - The HeliNet Solution*, Invited Paper for Wireless Personal Mobile Conference, Aalborg, Denmark, September 2001, Vol. 1, pp. 75–80.

28. D. Grace, M. Mohorčič, M. H. Capstick, M. Bobbio Pallavicini and M. Fitch, *Integrating Users into the Wider Broadband Network via High Altitude Platforms*, IEEE Wireless Commun., October 2005, Vol. 12, No. 5, pp. 98–105.

29. CAPANINA Website, www.capanina.org.
30. AV Inc Website, www.av-inc.com.
31. T. E. Noll, J. M. Brown, M. E. Perez-Davis, S. D. Ishmael, G. C. Tiffany and M. Gaier, *Investigation of the Helios Prototype Aircraft Mishap*, Vol. I Mishap Report, NASA, January 2004, www.nasa.gov/pdf/64317main_ helios.pdf.
32. S. Konyavko, *HAA and Lighter than Air Development in the US*, International HAPS Workshop, Seoul, Korea, Korean Institute of Industrial Engineers, and Electronics and Telecommunications Research Institute, November 2004.
33. StratXX AG Website, www.stratxx.com.
34. ERS srl Website, www.ers-srl.com.
35. COST297 Website, www.hapcos.org.
36. Z. Peng and D. Grace, *Coexistence Performance of High Altitude Platform and Terrestrial Systems Using Gigabit Communication Links to Serve Specialist Users*, EURASIP J. Wireless Commun. Network – Special Issue: Advanced Communication Techniques and Applications for High Altitude Platforms, October 2008, Vol. 2008, article ID 892512, 11 pages, 2008. doi: 10.1155/2008/892512.
37. D. Grace, J. Thornton, G. Chen, G. P. White and T. C. Tozer, *Improving the System Capacity of Broadband Services Using Multiple High Altitude Platforms*, IEEE Trans. Wireless Commun., March 2005, Vol. 4, No. 2, pp. 700–709.
38. ERS srl Website, www.ers-srl.com, January 2008.
39. D. Grace, N. E. Daly, T. C. Tozer, A. G. Burr and D. A. J. Pearce, *Providing Multimedia Communications Services from High Altitude Platforms*, Int. J. Satellite Commun., November 2001, Vol. 19, No. 19, pp. 559–580.
40. T. Konefal, C. Spillard and D. Grace, *Site Diversity for High Altitude Platforms: A Method for the Prediction of Joint Site Attenuation Statistics*, IEE Proc. Ant. Prop., April 2002, Vol. 149, No. 2, pp. 124–128.
41. E. Falletti, F. Sellone, C. L. Spillard and D. Grace, *A Transmit and Receive Multi-Antenna Channel Model and Simulator for Communications from High Altitude Platforms*, Int. J. Wireless Information Networks, January 2006, Vol. 13, No. 1, pp. 59–75.
42. Iskandar and S. Shimamoto, *Channel Characterization and Performance Evaluation of Mobile Communication Employing Stratospheric Platforms*, IEICE Trans. Commun., March 2006, Vol. E89-B, No. 3.
43. T. Celcer, G. Kandus, T. Javornik, M. H. Capstick and M. Mohorčič, *Analysis of HAP Propagation Channel Measurement Data*, Proc. 15th IST Mobile and Wireless Communications Summit 2006, Myconos, Greece, 4–8 June 2006.
44. Resolution 734 (Rev. WRC-07), *Studies for Spectrum Identification for Gateway Links for High Altitude Platform Stations in the Range from 5850 to 7250 MHz*, World Radiocommunication Conference, Geneva, 2007.
45. International Telecommunications Union Website, www.itu.int.
46. *USEHAAS Strategic Research Agenda – Developing a EU Research Strategy in the High Altitude Aircraft (HAAS) Sector – Volume 1*, Technical Report, FP6 USEHAAS Specific Support Action, ASA4-CT-2005-516081, www.usehaas.org.
47. ASTRAEA Project Website, www.projectastraea.co.uk.
48. M. Mohorčič, T. Javornik, A. Lavric, I. Jelovcan, E. Falletti, M. Mondin, A. Boch, L. Feletti, L. Lo Presti, D. Merino, D. Borio, J. A. Delgado Penin, F. Arino and E. Bertran, *Selection of Broadband Communication Standard for High-Speed Mobile Scenario*, FP6 CAPANINA Project, Doc Ref: CAP-D09-WP21-JSI-PUB-01, www.capanina.org/documents/CAP-D09-WP21-JSI-PUB-01.pdf, January 2005.
49. Y. C. Foo, W. L. Lim and R. Tafazolli, *Performance of High Altitude Platform Station (HAPS) in Delivery of IMT-2000 W-CDMA*, Proc. Stratospheric Platform Systems Workshop (SPSW 2000), Tokyo, Japan, September 2000.
50. B. El-Jabu and R. Steele, *Cellular Communications Using Aerial Platforms*, IEEE Trans. Vehicular Technol., May 2001, Vol. 50, No. 3, pp. 686–700.
51. E. Falletti, M. Mondin, F. Dovis and D. Grace, *Integration of a HAP within a Terrestrial UMTS Network: Interference Analysis and Cell Dimensioning*, Wireless Personal Communications, February 2003, Vol. 24, No. 2, pp. 291–325.

52. D. I. Axiotis, M. E. Theologou and E. D. Sykas, *The Effect of Platform Instability on the System Level Performance of HAPS UMTS*, IEEE Commun. Lett., February 2004, Vol. 8, No. 2, pp. 111–113.
53. T. Hult, D. Grace and A. Mohammed, *WCDMA Uplink Interference Assessment from Multiple High Altitude Platform Configurations*, EURASIP J. Wireless Commun. Networking, 2008, Vol. 2008, Article ID 182042, 10.1155/2008/182042.
54. T. V. Do, N. H. Do and R. Chakka, *Performance Evaluation of the High Speed Downlink Packet Access in Communications Networks Based on High Altitude Platforms*, Lecture Notes in Computer Science 5055, Springer, 2008, pp. 310–322.
55. B. Taha-Ahmed and M. Calvo-Ramón, *UMTS-HSDPA in High Altitude Platforms (HAPs) Communications with Finite Transmitted Power and Unequal Cell's Load*, Computer Commun., March 2009, Vol. 32, Issue 5, pp. 828–833.
56. M. Pent, L. Lo Presti, M. Mondin and S. Orsi, *HELIPLAT as a GSM base station: a feasibility study*, Proc. Data System in Aerospace (DASIA 99), Lisbon, Portugal, 17–21 May 1999.
57. F. Dovis and F. Sellone, *Smart Antenna System Design for Airborne GSM Base-Stations*, Proc. 2000 IEEE Sensor Array and Multichannel Signal Processing Workshop, Cambridge, UK, 16–17 March 2000, pp. 429–433.
58. M. Mondin, F. Dovis and P. Mulassano, *On the Use of HALE Platforms as GSM Base Stations*, IEEE Personal Commun., April 2001, Vol. 8, No. 2, pp. 37–44.
59. M. Bobbio Pallavicini, F. Dovis, E. Magli and P. Mulassano, *HeliNet Project: the Current Status and the Road Ahead*, Proc. Data Systems in Aerospace (DASIA 2001), Nice, France, May 2001.
60. J. Thornton, D. Grace, C. Spillard, T. Konefal and T. C. Tozer, *Broadband Communications from a High-Altitude Platform: the European HeliNet Programme*, Electron Commun. Eng. J., June 2001, pp. 138–144.
61. D. Martín-Sacristán, J. F. Monserrat, J. Cabrejas-Peñuelas, D. Calabuig, S. Garrigas and N. Cardona, *On the Way Towards Fourth-Generation Mobile: 3GPP LTE and LTE-Advanced*, EURASIP J. Wireless Commun. Networking, 2009, Vol. 2009, Article ID 354089, 10.1155/2009/354089.
62. R. Prasad and A. Mihovska (Eds), *New Horizons in Mobile and Wireless Communications, Volume 2: Networks, Services, and Applications,* Artech House, 2009.
63. R. B. Marks, I. C. Gifford and B. O'Hara, *Standards in IEEE 802 Unleash the Wireless Internet*, Microwave Mag., June 2001, pp. 46–56.
64. IEEE 802 LAN/MAN Standards Committee Website, http://grouper.ieee.org/groups/802/.
65. The Wi-Fi Aliance Website, http://www.wi-fi.org/.
66. J. Thornton, A. D. White and T. C. Tozer, *A WiMAX Payload for High Altitude Platform Experimental Trials*, EURASIP J. Wireless Commun. Networking, 2008, Vol. 2008, Article ID 498517, 10.1155/2008/498517.
67. C. Eklund, R. B. Marks, K. L. Stanwood and S. Wang, *IEEE Standard 802.16: A Technical Overview of the Wireless MAN Air Interface for Broadband Wireless Access*, IEEE Commun. Mag., June 2002, pp. 98–107.
68. The WiMAX Forum Website, http://www.wimaxforum.org/.
69. ETSI EN 300 421 V1.1.2 (1997-08), *European Standard (Telecommunications Series), Digital Video Broadcasting (DVB); Framing Structure, Channel Coding and Modulation for 11/12 GHz Satellite Services*
70. ETSI EN 302 307 V1.1.1 (2004-06), *European Standard (Telecommunications series), Digital Video Broadcasting (DVB); Second Generation Framing Structure, Channel Coding and Modulation Systems for Broadcasting, Interactive Services, News Gathering and Other Broadband Satellite Applications*
71. ETSI EN 301 790 V1.3.1 (2003-03), *European Standard (Telecommunications Series), Digital Video Broadcasting (DVB); Interaction Channel For Satellite Distribution Systems*
72. E. Cianca, M. De Sanctis, A. De Luise, M. Antonini, D. Teotino, M. Ruggieri and R. Prasad, *Integrated Satellite-HAP Systems*, IEEE Radio Commun., December 2005, pp. S33–S39.
73. F. De Rango, M. Tropea and S. Marano, *Integrated Services on High Altitude Platform: Receiver Driven Smart Selection of HAP-Geo Satellite Wireless Access Segment and Performance Evaluation*, Int. J. Wireless Inform. Networks, January 2006, Vol. 13, No. 1, pp. 77–94.

74. B. Bennett, K. Quock, E. Summers, M. Difrancisco and B. A. Hamilton, *DVB-S2 Technology Development for DoD IP SATCOM*, Proc. Military Communications Conference (MILCOM 2006), Washington, DC, USA, 23–25 October 2006.

75. M. Smolnikar, M. Mohorčič and T. Javornik, *Utilisation of LDPC Decoder Parameters in DVB-S2 ACM Procedures*, Proc. 2007 International Workshop on Satellite and Space Communication (IWSSC'07), Salzburg, Austria, 12–14 September 2007, pp. 194–198.

76. M. Smolnikar, A. Aroumont, M. Mohorčič, T. Javornik and L. Castanet, *On Transmission Modes Subset Selection in DVB-S2/RCS Satellite Systems*, Proc. 2008 International Workshop on Satellite and Space Communications (IWSSC'08), Toulouse, France, 1–3 October 2008, pp. 263–267.

77. M. Smolnikar, T. Javornik and M. Mohorčič, *Target BER Driven Adaptive Coding and Modulation in HAP Based DVB-S2 System*, Proc. 4th Advanced Satellite Mobile Systems (ASMS'08), Bologna, Italy, 25–28 August 2008.

78. M. Mohorčič, M. Smolnikar and T. Javornik, *Performance Comparison of Adaptive Coding and Modulation in HAP Based IEEE 802.16 and DVB-S2 Systems*, Proc. 26th AIAA International Communications Satellite Systems (ICSSC'08), San Diego, CA, USA, 10–12 June 2008.

79. M. Smolnikar, T. Javornik, M. Mohorčič and M. Berioli, *DVB-S2 Adaptive Coding and Modulation for HAP Communication System*, Proc. 2008 IEEE 67th Vehicular Technology Conference (VTC2008-spring), Marina Bay, Singapore, 11–14 May 2008, pp. 2947–2951.

80. M. Bobbio Pallavcini, B. Cahill, J. Horwath and M. Capstick *Test Results Summary Report - Part 1*, FP6 CAPANINA Project, Doc Ref: CAP-D22a-W40-CGS-PUB-01, www.capanina.org/documents/CAP-D22a-WP44-CGS-PUB-01.pdf, September 2006.

81. J. Horwath, *Final Optical Terminal Design Report*, CAPANINA Project, Doc Ref: CAP-D21-WP34-DLR-PUB-01, www.capanina.org/documents/CAP-D21-WP34-DLR-PUB-01.pdf, August 2006.

82. M. Bobbio Pallavcini, *Test Results Summary Report - Part 2*, FP6 CAPANINA Project, Doc Ref: CAP-D22b-W40-CGS-PUB-01, www.capanina.org/documents/CAP-D22b-WP44-CGS-PUB-01.pdf, January 2007.

83. USEHAAS Website, www.usehaas.org.

84. R. Miura and M. Suzuki, *R&D Achievements and Future Perspective on Telecomunications and Broadcasting System Using Stratospheric Platform*, 5th Stratospheric Platforms Systems Workshop, Tokyo, Japan, February 2005 (in Japanese/English).

85. K. Ohasi, *Promising Features of the Integrated SPF Tracking and Control System*, 5th Stratospheric Platforms Systems Workshop, Tokyo, Japan, February 2005 (in Japanese/English).

86. M. Oodo, H. Tsuji, R. Miura, M. Maruyama and M. Suzuki, *Experiment of IMT-2000 Using Stratospheric-Flying Solar-Powered Airplane*, GLOBECOM 2003 - IEEE Global Telecommunications Conference, December 2003, Vol. 22, No. 1, pp. 1152–1156.

87. M. Nagatsuka, Y. Morishita, M. Suzuki and R. Miura, *Testing of Digital Television Broadcasting from the Stratosphere*, IEEE Trans. Broadcasting, December 2004, Vol. 50, No. 4.

88. Y. Arakaki, R. Miura, M. Oodo, Y. Morishita and M. Suzuki, *Preliminary Flight Tests for Stratospheric Platform Systems - mm-Wave Multi-Beam Hone (MBH) Antenna Systems Using Helicopter*, AIAA-2003-2270 21st International Communications Satellite Systems Conference and Exhibit, Yokohama, Japan, 15–19 April 2003

89. M. Nagatsuka et al., *Experiment for Wide-Area Digital Broadcasting*, 5th Stratospheric Platforms Systems Workshop, Tokyo, Japan, February 2005 (in Japanese/English).

90. H. Tsuji et al., *Fundamental Experiment of Real-Time Estimation of Mobile-Terminal Localization*, 5th Stratospheric Platforms Systems Workshop, Tokyo, Japan, February 2005 (in Japanese/English).

91. Y. Arimoto et al., *Laser Tracking Trial between the Airship and Ground Station*, 5th Stratospheric Platforms Systems Workshop, Tokyo, Japan, February 2005 (in Japanese/English).

92. Article 5, Radio Regulation, ITU, 2004 Edition.

93. D.-S. Ahn, B.-J. Ku, T. C. Hong, J. M. Park, B.-S. Kang and Y.-S. Kim, *Core Technology Development and Standardization for HAPS in Korea*, 5th Stratospheric Platforms Systems Workshop, Tokyo, Japan, February 2005 (in Japanese/English).

94. www.kari.go.kr.

95. D.-S. Ahn, B.-J. Ku, T. C. Hong, J. M. Park, B.-S. Kang and Y.-S. Kim, *Core Technology Development and Standardization for HAPS in Korea*, 5th Stratospheric Platforms Systems Workshop, Tokyo, Japan, February 2005 (in Japanese/English).

96. J.-M. Park, B.-J. Ku, Y. S. Kim and D.-S. Ahn, *Technology Development for Wireless Communications System Using Stratospheric Platform in Korea*, PIMRC 2002, Lisbon, Portugal, September 2002, pp. 162–170.

97. Preliminary Draft New Recommendation ITU-R F.[HAPS-RRS], *Methodology to Evaluate Interference from Fixed Service System Using High Altitude Platform Stations (HAPS) To Radio-Relay System in the Bands above 3 GHz*, October 2004.

98. J. Del Frate, *Developing Technologies for High Altitude Platforms and Long Endurance*, 5th Stratospheric Platforms Systems Workshop, Tokyo, Japan, February 2005 (in Japanese/English).

99. Lockheed Martin High Altitude Airship, www.lockheedmartin.com/data/assets/7966.pdf.

100. J. Miller, *STRATSAT ATG Progress Update 2005*, 5th Stratospheric Platforms Systems Workshop, Tokyo, Japan, February 2005. (in Japanese/English).

101. Pegasus for Europe Website, www.pegasus4europe.com.

102. QinetiQ Website, www.qinetiq.com.

103. Space Data Corporation Website, www.spacedata.net.

2

Aeronautics and Energetics

2.1 Operating Environment and Related Challenges

The choice of platform type, materials used, design and operation are largely driven by the operating environment. And while the idea of exploiting the stratosphere is not entirely new, previously being of particular interest for long-haul aerial transport, its potential use for the provision of telecommunication services raised a number of research issues also in the area of aeronautics. For long-haul sub- and hyper-sonic aerial transport stratospheric altitudes are interesting because they are above most weather related phenomena, and are characterised by quite predictable and steady horizontal winds and temperature for a given location and season. Airborne platforms for provision of communication services, on the other hand, should preferably be able to keep a quasi-stationary position above the served area on the ground. During the operation they need to withstand stratospheric wind at velocities averaging 30–40 m/s and sudden wind gusts. Stratospheric wind velocity, however, changes with the season of the year, geographic location and the altitude above the earth, and is a result of intense interactions among radiative, dynamical, and chemical processes. Also, little is known about the wind velocity on shorter timescales on the order of minutes or hours, which gives rise to concerns about the ability of airborne platforms' station keeping and thus directly impacts the design of the platform.

2.1.1 The Layers of the Atmosphere

The atmosphere surrounding the Earth's surface consists of several layers that are determined by physical properties such as temperature, atmospheric pressure, wind

Broadband Communications via High Altitude Platforms   David Grace and Mihael Mohorčič
© 2011 John Wiley & Sons, Ltd

speed and the density of the air. These layers are depicted in Figure 2.1 and include the troposphere, stratosphere, mesosphere, thermosphere and exosphere.

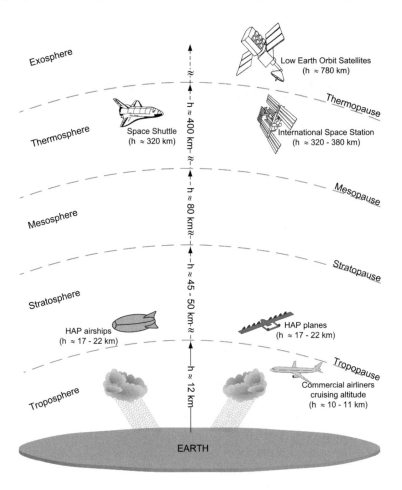

Figure 2.1   The layers of the atmosphere (altitudes are not on a linear scale)

The pressure and the density of the air decrease monotonously with increasing altitude because the air is compressible. The pressure of the air decreases according to Equation (2.1) [1]:

$$p(\mathrm{z}) = p_0 \cdot e^{-\frac{g}{R^* \cdot T_m} \cdot z} \tag{2.1}$$

where z stands for altitude, p_0 and $p(z)$ for pressure at the ground and at altitude z, R^* for the gas constant for dry air, g for the gravity acceleration constant and T_m for mid temperature between the temperature at the ground T_0 and the temperature $T(z)$ at altitude z, as shown in Equation (2.2).

$$T_m = \frac{T_0 + T(z)}{2} \tag{2.2}$$

Clearly, Equation (2.1) is only valid for dry air, while for humid air the molecular weight and consequently the gas constant for air changes. As a simple rule of thumb, the pressure halves every 5.5 km of altitude and decreases from approximately 1000 hPa at the Earth's surface to about 225 hPa at the tropopause, that is the border between the lowest layer troposphere and the second layer stratosphere (see Figure 2.2).

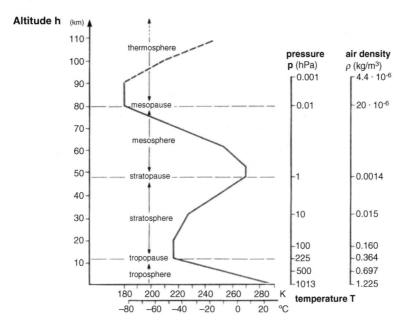

Figure 2.2 The temperature profile, pressure and density of the air with increasing altitude [1]. Reproduced by permission of © 1989 Bibliographisches Institut AG

As shown in Figure 2.2, the ambient temperature does not change monotonously. In fact, it changes due to absorption of sunlight and UV light in different layers of the atmosphere. The sunlight absorbed by the surface of the Earth heats the low layers of the air, making air parcels move upwards and cool with the increasing altitude at approximately 6.5 K/km [2]. However, molecules of the air also absorb part of the solar radiation and heat up the air. This phenomenon is most pronounced in the upper stratosphere, where the ozone layer absorbs UV light and causes a local maximum in temperature of about 270 K. At approximately 50 km above the Earth's surface this local maximum defines the border between the stratosphere and the mesosphere, which is called the stratopause. Above and below this border the temperature decreases, in the stratosphere towards the tropopause, i.e. the border with the troposphere, and in the mesosphere towards the mesopause, which defines the border with the thermosphere. The temperature in the upper stratosphere decreases with decreasing altitude at approximately 2.8 K/km until it becomes nearly constant in the lower stratosphere, whereas in the mesosphere it decreases with increasing altitude at 2–4 K/km [2]. In the thermosphere there is again pronounced absorption of short UV light on nitrogen and oxygen which become partially ionised, hence this layer is also called the ionosphere. In spite of this absorption, however, the ambient temperature does not increase in this layer due to extremely low air density.

The most interesting part of the stratosphere for HAPs is the lower stratosphere where the temperature inversion begins. Due to this temperature inversion the ascending air in the troposphere cannot go higher, which is important for the dynamics and the chemistry in the troposphere, for the cloud formation and the weather. At mid-latitudes this phenomenon typically starts at an altitude of approximately 13 km, while at the Poles it could be as low as 7 km, and at the Equator as high as 18 km as a consequence of different ground temperature.

In the stratosphere there are practically no weather phenomena due to the low water content. Consequently, the atmosphere is dynamically stable with no or little turbulence and the mildest winds. In fact, with increasing altitude above the ground the wind speed increases until reaching a maximum near the tropopause. In the lower stratosphere, however, the wind speed decreases significantly and, after reaching a local minimum at altitudes between 20 km and 25 km, again slowly increases with altitude through the upper stratosphere into the mesosphere. A typical wind profile for different altitudes is depicted in Figure 2.3.

Generally, wind speeds at a given altitude increase with latitude from the Equator to the Poles. Considering the seasons of the year, wind speed is at the calmest during summer and considerably higher during winter. Twice a year the wind changes direction by 180°. In winter the typical circulation pattern is from West to East and in summer from East to West. The slow transition to the opposite wind direction happens in spring and autumn which are thus characterised by the lightest winds [3].

Another important phenomenon for the design and subsequent operation of airborne vehicles is the air turbulence, which again changes with location and the season. Based on true gust velocity measurements up to the lower stratosphere the air turbulence appears to have its maximum near the tropopause altitudes and

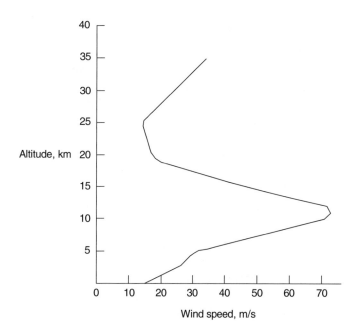

Figure 2.3 Typical wind profile in the stratosphere based on rawinsonde observation data. Reproduced from NASA - http://www.dfrc.nasa.gov/gallery/photo/index.html [2]

then decreases to a minimum near altitudes between 20 km and 25 km. In the upper stratosphere, however, wind speeds again increase, indicating that the amount and intensity of turbulence may be expected to increase at altitudes of 30–50 km [3].

As discussed at the beginning of this chapter, for provision of communication services aerial platforms should be kept relatively stationary with respect to the service area on the ground. This requirement makes the altitudes around 20 km most suitable for the operation of HAPs, requiring the minimum effort to compensate for unpredictable movements due to wind and wind gusts, exhibiting as horizontal and vertical displacement of the platform and change of inclination. The vertical winds causing vertical displacement are almost insignificant, whereas the change of inclination can be compensated by the beamforming antenna and a gimbaling system at the bottom of the platform. The most important component is thus the horizontal displacement, which needs to be compensated with a propulsion mechanism [4], to provide reliable services without temporal outages. Of course, in case of a plane type of aerial platform this horizontal displacement is added to the flight trajectory of the plane, which is nominally of small circular shape so as to keep a relatively stationary position above the ground.

2.2 Types of Airborne Vehicles Used for HAPs

The most fundamental classification of aerial platforms is based on the physical law that lifts and keeps the platform in the air. In this respect we distinguish aerostatic platforms or aerostats making use of buoyancy to float in the air, and aerodynamic platforms or aerodynes that use dynamic forces created by the movement through the air. For aerostatic platforms we can thus say that they are lighter than air, whereas aerodynamic platforms are heavier than air.

2.2.1 Aerostatic Aerial Platforms

A platform capable of providing a lifting force without movement through the surrounding air mass is referred to as an aerostat. Aerostats include balloons and airships and make use of a lifting gas in an envelope to provide buoyancy, making them lighter than air platforms. The buoyancy or lifting force equals the weight of displaced air, which is, assuming uniform density of the surrounding air (ρ_{air}), directly proportional to the volume (V) of the platform or the displaced air. The buoyant force (F_B) is given by Equation (2.3), where g is the gravitational acceleration:

$$F_B = \rho_{\text{air}} \cdot V \cdot g \tag{2.3}$$

The resulting lifting force on an aerostatic aerial platform is thus a sum of buoyant force and the overall weight of the platform. In the case of aerostats (e.g. balloon or airship) the lifting gas in an envelope makes the entire platform, on average, lighter than air. Typically used lifting gases with lower density that the surrounding air include hydrogen, helium, ammonia and methane with the density at sea level pressure and $0\,^{\circ}\text{C}$ (273 K) as summarised in Table 2.1.

Table 2.1 Density of the air and typically used lifting gases at sea level pressure and $0\,^{\circ}\text{C}$ (273 K)

Gas	Density (kg/m^3)
Air	1.292
Hydrogen, H_2	0.08988
Helium, He	0.1786
Methane, CH_4	0.717
Ammonia, NH_3	0.73

The most commonly used lifting gases for aerostats, besides hot air, are diatomic hydrogen H_2 and helium He. Helium is twice as heavy as diatomic hydrogen, but hydrogen has only about 7% more buoyancy than helium because they are both much lighter with respect to the weight of the displaced air and buoyancy depends on the difference of the densities of the air and gas and not upon their ratio. At room temperature and sea level pressure they both provide the lifting force of about 9.8 N per cubic metre of gas. The main advantage of hydrogen is its low cost as it can be obtained inexpensively from the electrolysis of water; however, it is highly combustible and poses high safety risk. Helium is thus preferred as it is nonflammable and not combustible, but shares hydrogen's disadvantage of leaking through the envelope due to having small molecules. Still, its main disadvantage is the cost of production. Although it is the second most abundant element in the observable universe, next to hydrogen, it is relatively rare on Earth, present only in low concentrations in the atmosphere and in some natural gas fields as a result of radioactive decay.

When hydrogen and helium are not available one can use methane as a lifting gas. Compared with hydrogen and helium it has larger molecules and thus the advantage of not leaking through the envelope as rapidly. Another alternative for a lifting gas is ammonia, which has sometimes been used as a lifting gas for weather balloons. Its main advantage in comparison with helium and hydrogen is its relatively high boiling point at $-33.34\,°C$ (239.81 K), so it could potentially be liquefied aboard an airship to reduce lift and add ballast, or returned to a gas to add lift and reduce ballast.

According to Equation (2.3) the buoyant force is directly proportional to the volume and the density of the displaced air, which changes with the temperature. The aerial platform stops rising when buoyant force and the weight of the platform reach equilibrium. If the volume of the platform in equilibrium expands, the buoyancy increases, and if it is compressed, the buoyancy reduces. Because the atmospheric pressure decreases with altitude (cf. Figure 2.2), the altitude of an aerostat in equilibrium tends to be stable. If an aerostat rises it tends to increase in volume due to reduced atmospheric pressure, but the construction and the payload of an aerostat do not expand. Thus the average density of the platform decreases less than that of the surrounding air. Consequently, the buoyancy decreases because the weight of the displaced air is reduced and the platform tends to stop rising. Similarly, a sinking aerostat tends to stop sinking due to higher atmospheric pressure at lower altitudes. Taking this into account, an aerostatic HAP needs to be designed so as to reach equilibrium in lower stratosphere, where the air density decreases to less than a tenth of its sea level value. Thus, according to Equation (2.3), the stratospheric aerostats need to be designed with a huge volume to compensate for thinner air. Such huge platforms develop a great deal of dynamic drag, when in motion, and represent a challenge for ground operations and maintenance, as well as for take-off and landing phases of the flight.

As to the shape, aerostats can take the form of balloons or airships. Nowadays, balloons are unpowered aerostats, so it is difficult to control their flight and for security reasons they need to be manned or tethered. As such they are not suitable candidates for HAPs except for some short term missions and technology demonstration purposes. Their main advantage is their availability, especially in the form of hot air balloons, and loose power and weight limitations for the payload.

Airships, on the other hand, are typically huge powered aerostats with lengths up to 200 m, manned or unmanned, capable of staying in the air for long periods of time. Typically they come with a rigid or semi-rigid outer framework in an aerodynamic shape and with fins for stabilisation purposes, thus in addition to buoyancy providing also some aerodynamic lift as explained in the following. Unmanned powered airships are of particular interest to the HAP community due to their station keeping capability using electric motors and propellers. Due to the size of an airship the limitations regarding the weight, size and power consumption of the payload are rather loose, and the large surface of the airship can be used for power generation using solar cells. The main drawbacks appear to be requirements for expensive ground infrastructure, ground operations centre and high-strain envelope material. Nevertheless, there have been several projects worldwide from the late 1990s developing and prototyping airships for HAP systems, and the main representatives are listed in Section 1.5.

2.2.2 Aerodynamic Aerial Platforms

For flying in the air the aerodynes exploit the aerodynamic lift created by two primary principles, Bernoulli's Principle and Newton's Third Law of Physics. According to Bernoulli's Principle the total energy in a steadily flowing fluid system is constant along the flow path. Thus, simultaneously with the increasing speed of the flow the pressure decreases and vice versa. This principle is exploited by the cross-section shape of the plane's wing depicted in Figure 2.4, typically referred to as airfoil. The top surface of the airfoil is more curved and hence longer than the bottom surface. This causes the air to flow faster over the upper surface, which creates on the top surface of the wing lower pressure than on the bottom surface. The difference in pressures provides the lifting force that keeps the plane in the air.

The other contribution to the aerodynamic lift is making use of Newton's Third Law of Physics, which says that for every action there is always an equal and opposite reaction. In aeronautics this law is exploited by attaching the aircraft wings to the aircraft body at a slight angle so that the tail is tilted downwards. This angle between the wing and the horizontal airflow is called the *angle of attack* and deflects the air that is passing the wing due to the dynamic movement downward, causing an opposite upward force on the wing. By further increasing the angle of attack more

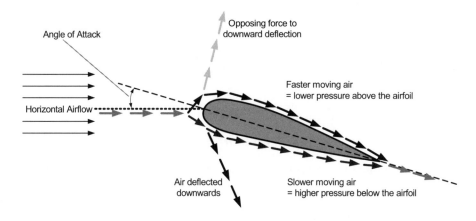

Figure 2.4 Airfoil and lifting force

lift can be generated, but only up to a point where the smooth airflow is broken and the lift cannot be sustained.

Besides the aerodynamic lift there are further three aerodynamic forces that act on the aerodyne in flight, as depicted in Figure 2.5, i.e. thrust, drag and gravity. In order to make the aerodyne fly an adequate forward thrust (F_T) must be provided by the craft's engine to overcome the resistance of the air or drag (F_D), and provide the sufficient lifting force (F_L) to compensate for the gravity (F_G). Expressions for the aerodynamic forces are given by Equations (2.4)–(2.7), where v is the velocity of the air ahead of the airfoil, ρ_{air} is the density of the air, S is wing area, C_L is the lift coefficient, A is the area on which the drag acts, C_D is the drag coefficient, m is mass, g is the gravitational acceleration and a is acceleration obtained by thrust.

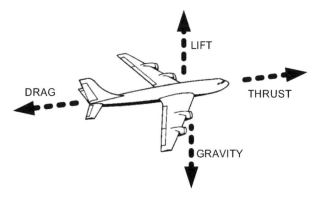

Figure 2.5 Aerodynamic forces

$$F_L = \frac{1}{2} \cdot \rho_{\text{air}} \cdot v^2 \cdot S \cdot C_L \tag{2.4}$$

$$F_D = \frac{1}{2} \cdot \rho_{\text{air}} \cdot v^2 \cdot A \cdot C_D \tag{2.5}$$

$$F_G = m \cdot g \tag{2.6}$$

$$F_T = m \cdot a \tag{2.7}$$

The lift coefficient is a measure of the amount of lift obtained from a particular airfoil shape, whereas the drag coefficient is a measure of the amount of dynamic pressure (expressed as $\frac{1}{2} \cdot \rho_{\text{air}} \cdot v^2$) that gets converted into drag. In the process of designing an aircraft the engineers tend to obtain as high a lift coefficient and as low a drag coefficient as possible.

When a craft is in a level flight the opposing forces of lift and gravity are balanced, while thrust always needs to be greater than drag for the craft to remain in the air. This means that the aerodynes cannot stay in the air unless they move forward, which is why aerodynamic platforms have to circle above the coverage area and thus maintain a quasi stationary position. Considering the large size of wings required to obtain sufficient lift at low speed in the thin atmosphere at an altitude around 20 km, the radius of the circular flight is in the range of a few kilometres. Such circular flight, however, requires some compensation for antenna pointing.

Compared with aerostatic platforms, aerodynes are less expensive, can be manned or unmanned, require less specific ground infrastructure and generally, if powered, also provide easier station keeping in windy and modest turbulent conditions. Given the operating conditions in the lower stratosphere and the requirement for station keeping, unpowered aerodynes are clearly not suitable candidates for HAPs. As to the powered aerodynes, the most suitable candidates for HAPs are fixed wing flying manned jet planes or unmanned planes with electric motors. Obviously, the power budget for aerodynes is more constrained than for aerostats; power is also needed for continuous circular flight, while there is significantly less surface area suitable for energy harvesting from solar panels. Moreover, aerodynes also pose restrictions with respect to the weight and volume of the payload, since they are subject to aerodynamic design requirements in order to obtain high lift coefficient C_L.

2.3   Power Subsystem Alternatives

Following the basic classification of aircraft to aerostats and aerodynes, two types of platforms are currently proposed for HAPs, the lighter than air airship and the fixed wing flying plane (or aircraft). One of the basic requirements for HAPs concerns the capability of uninterrupted provision of communication services on the timescale of weeks or months which includes station keeping above the targeted coverage area as well as operation of communications payload. This requirement clearly has a crucial impact on the design of the power subsystem as well as on the overall operation of the HAP system.

So far there have been three alternative energy sources proposed and investigated for the operation of HAPs in the stratosphere [5–7]:

- conventional energy sources onboard HAPs (fuel tanks, electrical batteries);
- renewable sources of energy onboard HAPs (electric energy from solar panels in combination with electrical batteries or hydrogen fuel cells for energy storage);
- microwave power transmission beamed to HAPs from the ground.

2.3.1   Conventional Energy Sources for HAPs

The limited space and load capacity for fuel tanks and/or electrical batteries represent a big limitation on the endurance of conventionally powered HAPs, so even if unmanned, the missions of such platforms typically last up to 48 h. In the case of manned platforms, of course, frequent landing is necessary in any case for changing the onboard crew and filling the energy. Due to the need for frequent landing and take-off, conventionally powered platforms are seen as a temporary solution, be it for emergency situations or the relatively short term phasing in of a new service. Another reason for considering such platforms as a temporary solution, are concerns related to the yet unknown long-term consequences of such platforms on the chemical/thermodynamic balance of the stratosphere due to the gases exhausted by the turbo propeller or jet engines.

The use of conventional energy sources has an adverse impact also from the point of uninterrupted service provision, requiring a seamless handover of all on-going connections from the currently serving to another in-flying aerial platform, before being landed for maintenance, refilling or change of the onboard crew.

2.3.2   Renewable Energy Sources for HAPs

In the case of unmanned platforms the uninterrupted provision of services typically calls for a power subsystem using renewable sources of energy, most notably solar energy collected from panels on the airship envelope or aircraft wings coupled by regenerative fuel cells (RFCs) or electrical batteries, to avoid the need for frequent

landing. In fact, most HAP programmes worldwide investigating unlimited operation endurance were considering the use of renewable sources onboard HAPs.

HAPs are particularly well suited for using solar energy. On one hand they have
large surfaces, be it wings or gas-filled envelopes, for flying at stratospheric
altitudes, and these surfaces can be efficiently covered by solar panels. On the
other hand, HAPs are operating well above the cloud ceiling, so they can produce
electricity regardless of the weather and seasonal changes on the earth. With
significantly improved efficiency of photovoltaic cells in recent years, the amount
of electrical energy obtained during the day cycle at least in equatorial and moderate
latitudes suffices for steady flight and for storing the energy required for the
night cycle.

As mentioned above, the surplus of energy captured by solar cells during the day
can be stored in electrical batteries or in RFCs. In the case of RFC energy storage,
the excess energy is used in a hydrolyser to obtain hydrogen and oxygen from
water in a closed circuit electrolytic process. Subsequently, hydrogen and
oxygen are stored in two separate tanks, available to supply the fuel cells when
the solar energy from the photovoltaic array is insufficient. In a fuel cell the reverse
chemical reaction results in electricity and water with no side products and no
pollution. The schematic of a power subsystem using solar panels and RFCs is
shown in Figure 2.6.

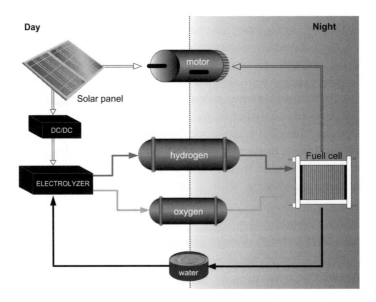

Figure 2.6   Schematic of a power subsystem based on renewable energy sources

2.3.3 Remotely Beamed Energy for HAPs

The third alternative energy source for the operation of unmanned aircraft, proposed for instance in the early 1980s within the SHARP (Stationary High Altitude Relay Platform) programme [6,7] and also in an early design of pilotless planes from Jet Propulsion Laboratories (JPL) [5], assumes microwave beams from the ground.

The basic concept is depicted in Figure 2.7 [7]. Microwave power required for the operation of the HAP would be transmitted from a ground-based antenna system with large diameter of microwave beam and antennas to keep the power flux density low. On the HAP the microwaves would be captured by an appropriate collector consisting of a large number of rectifier antennas and converted to DC for regular

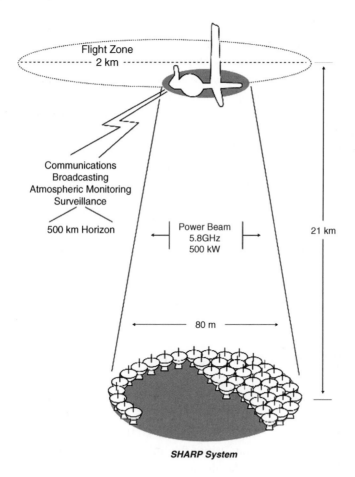

Figure 2.7 Microwave power beaming system proposed by the SHARP programme [7] (CRC Photo 81-3391). Reproduced by permission of © Communications Research Centre

operation as well as for charging the standby energy storage. Instead of microwaves, laser beams can also be used as power transmission medium in a similar manner, and there have been experiments with microwave-powered and laser-powered platforms undertaken in Japan, USA, and Israel. In spite of the remarkable advantages in terms of the HAP mission endurance and environmental friendliness due to absence of any polluting gases, this technology is not really taking off. The reason lies in high power irradiation risks by both microwave-powered and laser-powered platforms, so they are not considered as safe and environmentally friendly solutions.

2.4 Flight/Altitude Control

In terms of flight and attitude control it is necessary to distinguish between the two basic types of aerial vehicles used, the aircraft and the airship, although both are characterised by relatively loose station keeping. In normal operating conditions the airship may only need to compensate for movement from the nominal position by flying against the wind, whereas the aircraft has to fly along some kind of quasi circular path to have sufficient aerodynamic lift to stay in the air and to remain above the coverage area. However, although the stratosphere is a layer of relatively mild turbulence [2], the platform will inevitably encounter sudden wind gusts. As a result, the platform could move in any direction, requiring further compensation of the platform location, using suitable propulsion mechanisms, and/or payload stabilisation, using mechanical or electronic antenna steering mechanisms to compensate for the HAP movement. For nonstabilised or partially stabilised payloads there is a requirement to hand over communications from one antenna footprint (cell) to another [8]. Of course, handoffs at very high bit rates coupled with the stringent quality of service and latency requirements demanded by multimedia traffic require advanced handoff schemes.

2.4.1 HAP Station Keeping

Regardless of the type of aerial platform and the reasons for the displacement from the nominal position, the ITU-R has recommended that a platform should be geostationary within a location sphere of 500 m radius [9]. This may prove easier for airships than for aircraft. Angel Technologies [10], the developer of the original manned plane-based solution using the manned stratospheric plane Proteus, specified that their HALO plane should fly within a toroidal volume of radius of 2.5–4 nautical miles (approximately 4–6 km). The HeliNet project, one of the first projects examining an unmanned plane option, chose to specify two location cylinders, as shown in Figure 2.8. The platform should be located in the inner cylinder with radius of 2.5 km and altitude of 1 km (+/− 0.5 km from nominal position) for 99% of the time, and in the outer cylinder with radius of 4 km and altitude of 3 km (+/− 1.5 km from nominal position) for 99.9% of the time [11,12]. This is somewhat more relaxed than the ITU-R specification, but is intended to be realistic for an unmanned plane-based application.

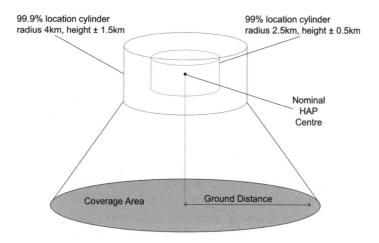

Figure 2.8 Unmanned aircraft location cylinders for 99% and 99.9% of time as specified in the HeliNet project

As to payload stabilisation, various antenna steering mechanism techniques have been proposed that can be employed to practically compensate for the HAP movement. These mechanisms can be used for fixed users on Customer Premises Equipment (CPE) as well as on the platform itself. The antenna steering mechanisms [4,13] on the platform have several constraints associated with them. For example, they must be powered, adding up to the power demands on the platform, while they also add extra weight and cost to the payload. Furthermore, even if they are employed it is impossible to guarantee a stationary position for the HAP during its service hours and so the patterns of received power and interference on the ground (footprints) will still be moving.

In the case of nonstabilised or partially stabilised payloads, the uninterrupted connections can only be maintained by employing an appropriate handoff technique. However, in a HAP system the handoff procedure is just the opposite of the one in terrestrial systems, where the user moves and the Base Station (BS) is fixed. Namely, the platform in a HAP system is moving whereas the users can be fixed or mobile. In reality, however, it is advisable to employ a combination of handoff techniques and a steering mechanism and thus possibly eliminate unnecessary handoffs and related signalling, and therefore unnecessary power consumption that could be vital for the continuous operation of the HAP [14].

2.4.2 HAP Mobility Models

Due to stratospheric wind and sudden wind gusts HAPs may move in any direction at a varying speed. In order to be able to design and analyse a HAP system capable of coping with these movements it is necessary to develop appropriate mobility models. These models have to cover all six degrees of freedom that a flying object such as a HAP, be it in the form of a plane or airship, can be subjected to. These are horizontal/vertical displacements with respect to the x-, y- and z-axis as well as yaw, pitch and roll. The six degrees of freedom can be defined in terms of vectors x, y, z, $x_\theta, y_\theta, z_\theta$ where the first three denote drift-based movement and the latter three rotational-based movement with the subscript denoting the axis of rotation. In practice the HAP can perform any (or a combination) of the six degrees of freedom as far as it remains within certain boundaries, as explained in Section 2.4.1.

In the following example of a HAP mobility model the platform, be it a plane or an airship, is assumed to be able to serve the nominal coverage area at all times. Therefore, regarding the boundary area it is most appropriate to adopt the location cylinder model proposed by the HeliNet Project [11,12] and depicted in Figure 2.8.

2.4.2.1 Drift with Respect to the x-, y- and z-Axis

Drifts with respect to the x- or y-axis of the platform have a similar effect on the ground. This is due to the similar movement, because the x- and y-axis are lying on the same plane. Results for one direction can therefore represent both cases. Assuming that the payload is stabilised against rotation but not against drift, the HAP is allowed to drift up to a distance R_{cyl} away from the centre of the coverage area, as shown in Figure 2.9.

We also assume that the current position vector r_t is dependent on the vector $r_{t-\Delta t}$ generated Δt seconds earlier, which can be represented by Equation (2.8), where v is the HAP velocity, a_f represents the vector corresponding to a destination point A_F and a_{f-1} corresponds to the vector of the point of the previous point update [15]. A_F in this example represents the edge of the cylinder towards which the HAP will move. An example of this is shown in Figure 2.10.

$$r_t = r_{t-\Delta t} + v \cdot \Delta t \cdot \hat{a}'$$

where

$$\hat{a}' = \frac{a_f - a_{f-1}}{|a_f - a_{f-1}|}$$

(2.8)

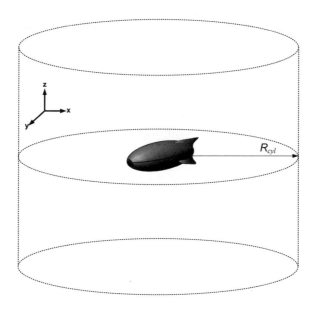

Figure 2.9   Example of y-axis drift

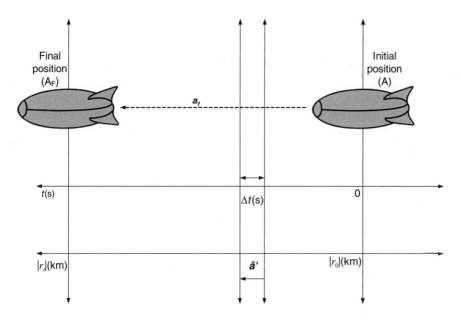

Figure 2.10   Graphical representation of x- or y-axis drift

Drift along the z-axis, graphically depicted in Figure 2.11, causes expansion and contraction of the size of every cell on the ground as well as the entire coverage area. Assuming that there is no steering mechanism on the HAP (apart from the rotation stabilisation mechanism mentioned before) and depending on the chosen cellular structure, the HAP is allowed to drift up to a half of distance H_{cyl} [12] away from its initial central point to ensure that all users remain located within the coverage area. In both cases where the HAP moves upwards or downwards, the footprints must be able to serve the nominal coverage area.

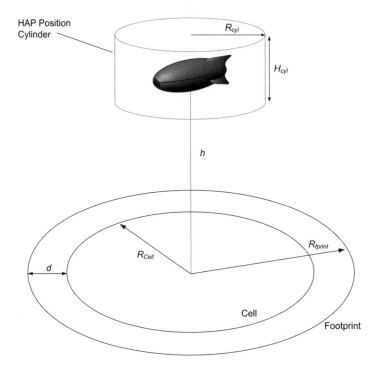

Figure 2.11   Example of z-axis drift

More specifically, the footprint must cover the cell area where this is the area on the ground where the loss to one particular HAP antenna is minimum when the HAP is at the centre of its coverage and it is not moving (i.e. initial position). In this case, the size of the footprint will change whilst the HAP moves upwards or downwards. However, the effect depends on the antenna mask employed. In the case of a highly directional antenna profile the footprint gets smaller if the HAP moves downwards, whereas if it moves upwards the footprint gets bigger and the user, initially out of coverage area, can establish a link to the HAP. These conditions are depicted in Figure 2.12. On the other hand, if the antenna profile used was less directional, such as the one in Figure 2.13, then the effective size of the footprint might become bigger, when the platform drifts closer to the ground, and smaller, when the platform drifts upwards, since the power density on the ground decreases. Consequently, the user initially within the coverage area maintains the connection if the HAP drifts downwards, whereas the user is out of coverage if the HAP drifts in the opposite direction.

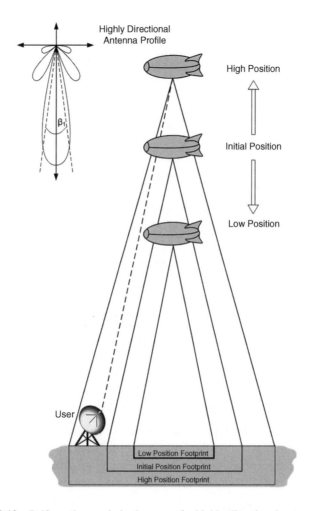

Figure 2.12   Drift on the z-axis in the case of a highly directional antenna profile

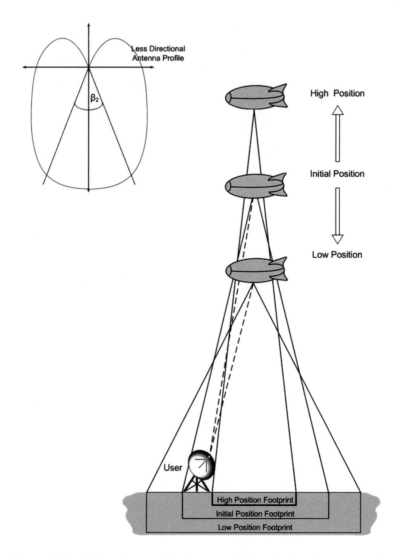

Figure 2.13   Drift on the z-axis in the case of a less directional antenna profile

2.4.2.2 Rotation with Respect to the x-, y- and z-Axis

x- and y-axis rotation (roll and pitch) have similar effect on the ground. This is again because the rotation is performed on two axes both lying on the same plane. In both cases the HAP rotates $\pm\theta$ degrees such that there are no users left without coverage (see Figure 2.14). For this work we consider that θ varies between:

$$-\arctan\left(\frac{R_{\text{fprint}}-R_{\text{cell}}}{h}\right) \leq x_\theta \leq \arctan\left(\frac{R_{\text{fprint}}-R_{\text{cell}}}{h}\right) \qquad (2.9)$$

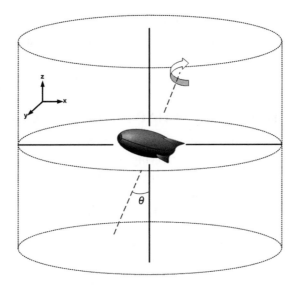

Figure 2.14 Example of y-axis rotation (pitch)

$$-\arctan\left(\frac{R_{\text{fprint}}-R_{\text{cell}}}{h}\right) \leq y_\theta \leq \arctan\left(\frac{R_{\text{fprint}}-R_{\text{cell}}}{h}\right) \qquad (2.10)$$

For the case of the rotation with respect to the z-axis, i.e. yaw, Figure 2.15 shows the HAP rotating at a constant speed anticlockwise (viewing it from top-view).

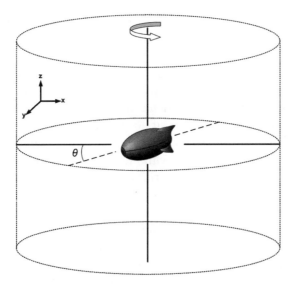

Figure 2.15 Example of z-axis rotation (yaw)

Unmanned planes are expected to be in a constant rotational type of orbit with respect to the centre of the location cylinder in order to maintain service over the coverage area.

2.5   Typical Characteristics of HAP Aircraft and Airships

Since the mid 1990s we have been witnessing intensified development of aircraft (i.e. planes) and airships capable of carrying communications payload and operating at stratospheric altitudes from several hours to days and even months. The main activities in this area were concentrated in USA, Europe and Japan, although there have been a few related projects also in some other countries including China, Malaysia and South Africa.

Considering the type of the lifting force and the source of the energy we can define four basic categories of HAPs suitable for the delivery of communication services, with different technical characteristics, maturity of technology and consequently implications to business modelling:

1. Fuel powered manned plane, e.g. Proteus [10] and Grob G520 Egrett [16,17].
2. Fuel powered unmanned plane, e.g. AV Global Observer [18].
3. Solar powered unmanned plane, e.g. Heliplat [12] and AV/NASA Pathfinder-Plus [19].
4. Solar powered unmanned airship, e.g. Lockheed Martin HAA [20,21].

Fuel powered manned planes are proven technology and the only type of HAP available for commercial use in 2010. Their main drawback is short endurance of the platform and the relatively large number of crew members required. Unmanned planes have only been tested in various experiments so far, not even necessarily carrying any communications payload. While they will provide longer endurance of the platform they will be more restricted in available payload mass and power consumption. Unmanned airships for stratospheric altitudes are yet to be realised. The capabilities of platform types increase with their decreasing maturity or availability for commercial use, i.e. with increasing time to market. The general characteristics and capabilities of platform types, estimated or adopted from various sources, are listed in Table 2.2. Unmanned airships appear to be the farthest from potential use, but once available they are promising by far the best operating conditions. Namely, in the longer term important factors for the decision on the type of the platform will not be only the availability, payload capacity and the cost for the deployment, acquisition and operation, but also environmental compatibility and reliability. And while manned and unmanned planes are well suited for early

Table 2.2 HAP operating parameters, state of maturity and indicative costs (based on [22])

	Manned	Unmanned		
	Plane	Plane (fuel)	Plane (solar)	Airship (solar)
Total development cost (€M)	0	50		225
CAPEX per unit (€M)	18	4		30
OPEX per annum continuous operation (€M)	6	1		4
Payload – mass (kg)	1000	250	150	1000
Payload – power (kW)	20	1		10
Operating altitude (km)	20	17		20
Flight endurance	12–16h	3 days	weeks	years
Staff required	4 pilots + 20 ground crew	1 pilot + 10 ground crew on launch		1 pilot + 20 ground crew on launch
Platform lifetime (years)	10 years		5 years	
No. for continuous operation per coverage area	1.3 (assuming multi area and fleet management)		1	
Redundancy factor	0.3		0.2	
Fleet/reserve multiplier	0.6		0.2	
Estimate when commercially available	Now	2010	2012	2015

Table 2.3   Example HAP capabilities in terms of fronthaul and backhaul capacities (based on [22])

	Manned	Unmanned		
	Plane	Plane (fuel)	Plane (solar)	Airship (solar)
Fronthaul				
Availability at min. data rate (%)		99.9		
Max. clear air cell capacity (Mbps)		120		
No. of cells	121	7	19	121
Total fronthaul capacity (Mbps)	14520	840	2280	14520
Diameter of cell (km)	6	15		6
Diameter of coverage area (km)	60	36	59	60
Required redundancy (%)	0	5		20
Backhaul				
Gateway capacity (Mbps)		960		
Capacity per 50 MHz backhaul carrier		320		
No. of carriers per gateway station		3		
Required redundancy (%)	0	5		30

demonstrations and trials, airships will be more suitable for actual operation scenarios.

Clearly, link capacities are closely related to the size of the HAP, weight available for the communications payload and power that can be used for its operation. This in turn determines the size of the single HAP coverage area, whereas the service area is constrained by the architectural and HAP payload configurations. Examples of typical HAP fronthaul and backhaul link capacities are given in Table 2.3.

References

1. Understanding the stratosphere – layers, http://www.atmosphere.mpg.de/enid/1__Understanding_the_ stratosphere/-_layers_lz.html, January 2010.
2. L. J. Ehernberger, *Stratospheric Turbulence Measurements and Models for Aerospace Plane Design*, NASA Technical Memorandum 104262, Edwards, CA, USA, December 1992.
3. A. K. Widiawan and R. Tafazolli, *High Altitude Platform Station (HAPS): A Review of New Infrastructure Development for Future Wireless Communications*, Wireless Personal Communications, 2007, Vol. **42**, pp. 387–404.
4. D. I. Axiotis, M. E. Theologou and E. D. Sykas, *The Effect of Platform Instability on the System Level Performance on HAPS UMTS*, IEEE Commun. Lett., 2004, Vol. **8**, pp. 111–113.
5. G. M. Djuknic, J. Freidenfelds and Y. Okunev, *Establishing Wireless Communications Services via High-Altitude Aeronautical Platforms: A Concept Whose Time Has Come?*, IEEE Commun. Mag., September 1997, pp. 128–135.
6. R. Struzak, *Mobile Telecommunications via Stratosphere*, International Communications Project, No. 1, http://www.intercomms.net/AUG03/content/struzak1.php, August 2003.

7. Stationary High Altitude Relay Platform (SHARP) on The Friends of CRC Association Website, http://www.friendsofcrc.ca/Projects/SHARP/sharp.html, February 2009.

8. K. Katzis, L. Dong, D. D. Luong, D. Grace and P. Mitchell, *Resource Allocation and Handoff Techniques for High Altitude Platforms*, FP6 CAPANINA Project, Doc Ref: CAP-D23a-WP24-UOY-PUB-01, http://www.capanina.org/documents/CAP-D23a-WP24-UOY-PUB-01.pdf, August 2006.

9. ITU Recommendation ITU-R F.1500, *Preferred Characteristics of Systems in the Fixed Service Using High Altitude Platforms Operating in the Bands 47.2-47.5 GHz and 47.9-48.2 GHz*, International Telecommunications Union, Vol. ITU-R S.672, Geneva, Switzerland, 2000.

10. N. J. Colella, J. N. Martin and I. F Akyildiz, The HALO Network, *IEEE Commun. Mag.*, June 2000, pp. 142–148.

11. M. Bobbio Pallavicini, *Systems Requirements/Specifications*, FP5 HeliNet Project, Doc Ref: HE-RQ-CGS-001-WP4, July 2000.

12. D. Grace, J. Thornton, T. Konefal, C. Spillard and T. C. Tozer, *Broadband Communications from High Altitude Platforms - The HeliNet Solution*, Proc. 4th International Symposium on Wireless Personal Multimedia Communications (WPMC 2001), Aalborg, Denmark, 9–12 September 2001.

13. M. H. Capstick and D. Grace, *High Altitude Platform mm-Wave Aperture Antenna Steering Solutions*, Int. J. Wireless Personal Commun., February 2005, Vol. **32**, pp. 215–236.

14. D. Akerberg, *On Channel Definitions and Rules for Continuous Dynamic Channel Selection in Coexistence Etiquettes for Radio Systems*, Proc. 44th IEEE Vehicular Technology Conference (VTC \lsquo94), Stockholm, Sweden, June 1994, pp. 809–813.

15. D. Grace, K. Katzis, D.A.J. Pearce and P.D. Mitchell, Low-Latency MAC-Layer Handoff for a High Altitude Platform Delivering Broadband Communications, *URSI Radio Sci. Bull. – HAPs Special Issue*, March 2010.

16. Grob G520T Egrett Platform, www.globalsecurity.org/military/world/europe/egrett.htm, January 2008.

17. Grob G520T Egrett Platform on Grob Website, www.grob-aerospace.net/index.php?id=137, January 2008.

18. AeroVironment Launches Global Observer Program for the US Special Operations Command, www.defense-update.com/newscast/0907/news/260907_hale_av.htm, September 2007.

19. NASA Pathfinder-Plus Platform, www.dfrc.nasa.gov/gallery/photo/Pathfinder-Plus/index.html, January 2008.

20. Lockheed Martin High Altitude Airship, www.globalsecurity.org/intell/systems/haa.htm, January 2008.

21. Lockheed Martin HAA Program, www.lockheedmartin.com/products/HighAltitudeAirship/index.html, January 2008.

22. D. Grace and P. Likitthanasate, A Business Modelling Approach for Broadband Services from High Altitude Platforms, Invited paper at International Conference on Telecommunications (ICT\rsquo06), Madeira, Portugal, May 2006.

3

Operating Scenarios and Reference Architectures

3.1 Operating Scenarios

High altitude platforms (HAPs) are establishing themselves as an alternative communication infrastructure that can integrate with or supplement terrestrial and satellite communications for provision of broadband wireless access both to fixed and mobile users [1]. As discussed in Section 1.3, HAPs combine some of the best characteristics of terrestrial and satellite communication systems, while avoiding some of their drawbacks. Relatively large quasi-stationary coverage area, low propagation delays, straightforward cellular architecture enabling broadband capability, little ground-based infrastructure, and easy maintenance and upgrading make HAPs particularly well suited not only to serve as a complement to existing wireless technologies predominantly for rural and remote areas but also for the provision of temporary and ad-hoc networks.

Except in stand-alone deployments and for some specific missions such as disaster relief, the traffic generated in a HAP network will be predominantly long distance. This means that connections will mostly extend beyond the single HAP coverage area, requiring terrestrial or airborne backbone network and in most cases a transition to other networks via appropriate gateways. Airborne backbone network can either consist of backhaul link via satellite or, more appropriately, interplatform backbone link to other HAPs. This means that HAPs can play an important role not only in the access part of the network but also in the backbone.

The fundamental point is that HAPs can and should exploit the advantages of both terrestrial and satellite systems. In this respect and as particularly flexible

Broadband Communications via High Altitude Platforms   David Grace and Mihael Mohorčič
© 2011 John Wiley & Sons, Ltd

infrastructure, HAPs are especially well suited for the following set of operating scenarios:

1. Establishment of a regional/national network during a service roll-out phase.
2. Provision of long-term broadband wireless access for remote and rural areas.
3. Servicing short-term large-scale events.
4. Establishing ad-hoc networks for disaster relief.
5. Providing backhaul interconnection of remote networks such as corporate local area networks (LANs) or backhaul gateway into a backbone network.

The first scenario can be seen as a fast and relatively low cost deployment of a pilot network for new technologies, services and applications including for instance fixed/mobile WiMAX or 3G pilot networks. In this scenario temporary networks should typically be implemented with cellular structure on the ground to guarantee sufficient system capacity and with narrow beam antennas to avoid or minimise potential interference into pre-existing networks.

With respect to the second scenario, it has been proven in the CAPANINA project [2], that HAPs are also an attractive solution for providing the Internet access and broadband multimedia services to passengers travelling on high-speed trains. This can be seen as an extension of the second scenario towards the mobile segment.

The third and the fourth scenarios capitalise on HAP characteristics in the best possible way, addressing a niche without real competition from existing terrestrial and satellite systems. These scenarios are even more important as they are driven by totally different business models than other scenarios. On one hand they build on offsetting large investments of the longer-term economically irrational deployments, such as servicing one-time or very rare events of high visibility and large interest with enormous requirements for communication services (e.g. Olympics, world cups, summits, etc.), and on the other hand on the fact that in the case of large scale natural or manmade disasters economic laws (should) play no role in life saving, in providing safety to rescue teams in the field, and as seen in many recent disasters, even in enforcing law and order and supporting post-disaster recovery.

The fifth scenario is again mostly building on the capability of short term or temporary service provisioning, be it because of the need to compensate for yet nonexistent or damaged ground infrastructure or because of many dislocated offices/facilities of a temporary nature requiring common access to a firewalled company Intranet, e.g. construction sites in the area with no or insufficient pre-existing ground infrastructure. This scenario also encompasses potential provision of backhaul gateways for LAN hot spots.

From the discussion above it is clear that the operating scenario directly impacts the requirements for the communication payload on HAPs. In this respect and considering the operating scenarios the HAP needs to be seen as a vehicle capable of:

- relatively fast take off and deployment of service in the targeted coverage area in accordance with mission planning;
- carrying different payloads that best meet the requirements of the mission;
- landing for payload replacement, maintenance, upgrade.

These capabilities are clearly best exploited by adopting the modular payload concept with a wide range of payload modules differentiated in terms of system capacity they can provide, network topology they can establish, mix of services and communication standards they can support, types of missions they can serve, etc. This means there will have to be separate communication payload modules for backhaul links and for fronthaul access links, as well as further application specific payload modules such as remote sensing modules, sensor network gateway modules, etc. All these modules can be mixed and matched on a HAP to support requirements of a specific mission.

3.1.1   HAPs User Scenarios

A broad range of user scenarios can be considered for a HAP-based communication system, with different sets of services and applications that they are likely to require. A deeper treatment of the services and applications themselves is given in Chapter 4, whereas here only some of the most representative examples of user scenarios are identified:

- Residential household: In order to successfully serve the domestic market, the cost of the HAP user terminal equipment must be low, services must be carefully tailored to the market and service charges must be competitive with terrestrial broadband service charges. Typical services will include broadband Internet access, voice, TV and video on demand.
- Corporate/SME office: Broadband Internet access is a strong candidate service for HAPs, especially in areas of poor terrestrial infrastructure and without the immediate likelihood of DSL. Business is more likely to pay a premium rate for a broadband service than domestic customers. SME/corporate applications tend to require a more symmetrical data flow, and it may be that HAP is better suited to this than ADSL. An office might be regarded as a single subscriber, but within that office there could be many employees all requiring broadband access. So, the total bandwidth requirement may be higher than could be accommodated by a single

ADSL line. HAPs could thus be a solution to businesses in urban or rural locations, in advance or possibly in place of VDSL.

- Remote offices/facilities: There is a common requirement to link a corporate headquarters with remote offices and facilities via a broadband Intranet. This opens up a series of common applications throughout the company, including the ability to download large data files from company servers. Broadband access to the Internet can be easily provided via the Intranet. Security is an issue, so VPN and IPSec must be considered. Other applications such as business TV, distance learning/training, videoconferencing, etc. are also possibilities. Another consideration is where DNS, DHCP is performed, since the location of these services affects the backhaul traffic/remote terminal complexity trade area.
- Teleworkers: Requirements for teleworkers are broadband Intranet/Internet and other services applicable to remote offices. Like for remote offices, security is an issue, so VPN and IPSec must be considered. In general, by providing a teleworking service where terrestrial infrastructure cannot meet this demand, HAPs could improve the quality of life, help maintain a work–life balance and enhance business efficiency. Of course, quality communications, including audio-visual applications such as videoconferencing, are a prerequisite.
- Hospitals: Distance learning applications can be augmented by live two-way audio-video assistance to remote clinics, emergency units in the field or even disaster areas. In this respect, specialised audio-video medical applications typically requiring excessive throughputs well beyond the DSL range of bit rates, should be supported, and HAPs could well meet such requirements on a municipal/regional level.
- Schools, colleges, universities and campuses: Services and applications will include broadband Internet/Intranet and distance learning. Due to a typically very high density of users, cumulative throughput requirements are very high.
- Government departments: Government departments require both broadband and secure intra-communication, as well as the ability to provide information to the citizen. In most requirements, though, they are very similar to corporate offices.

3.1.2   HAPs Network Scenarios

HAPs can be used to provide broadcasting and multicasting services as well as point-to-point services. Due to similarities between HAPs and satellites, the broadband satellite multimedia scenario studied by ETSI in [3] is also relevant for HAPs. In particular, HAPs can provide local and backbone networks and access networks, including Internet access, as well as direct connectivity of corporate Intranet/Internet, provided that dislocated corporate offices/facilities are within the coverage area.

The ETSI model divides the global IP network into three parts: core network; distribution network; and access network. The distribution network is described as an intermediate IP sub-network that is used to connect an access network into the core network. The ETSI model adapted to HAPs is summarised in Table 3.1.

Table 3.1 HAPs broadband multimedia network scenarios

		Point-to-point	Multicast	Broadcast
Access network scenarios	Corporate intranet	Corporate HAPs network, i.e. site interconnections	Corporate Multicast (e.g. Data distri- bution, video conferencing)	Datacasting TV broadcast (private)
	Corporate internet	Internet access via corporate ISP or via third party ISP	IP multicast RT streaming ISP caching	ISP caching
	SME intranet	Small HAP network	SME multicast	
	SME internet	Internet access via third party ISP	IP multicast RT streaming ISP caching	ISP caching
	SoHo	Internet access via ISP Company access via VPN	IP multicast RT streaming ISP caching	ISP caching
	Residential	Internet Access via ISP	IP multicast RT streaming ISP caching	ISP caching
Distribution network scenarios	Content to edge	ISP to backbone	IP multicast RT streaming Caching at ISP/ Edge	TV broadcast (public)
Core network scenarios	ISP inter- connect	Trunk interconnect	N/A	N/A

It can be seen from the discussion on HAPs user scenarios in Section 3.1.1 that most users require direct broadband connectivity from their terminal into a core network, from where they can obtain a defined broadband service or run an application. This link forms the access network. When used in this way, HAPs can connect user network interfaces, located at customer premises (residential households, SMEs or corporates), to a service node interface of a broadband core network.

Some general characteristics required for identified scenarios include the following:

- All HAP systems are expected to support IP.
- HAP systems may support both symmetric and asymmetric data flows.
- In general, the system will use multiple access to the radio bandwidth in order to optimise the efficiency with which multiple users utilise the spectrum.
- HAP systems may be deployed by Telcos, who provide network services to the public, or by private operators, who use the network for their own purposes.
- HAP systems should allow both terminal interoperability and service inter-operability.
- HAP systems should allow for network management functions consistent with those used to manage terrestrial networks.

The advantages and disadvantages of HAPs relative to alternative transmission systems such as terrestrial copper/fibre, terrestrial radio and satellite must be kept in mind in developing the services, applications and business cases for HAPs. The terrestrial cable/fibre infrastructure does not offer broadband connectivity to most rural and remote locations even in developed countries, let alone the developing countries worldwide. In fact, for most rural and remote locations even the DSL range of bit-rates is still an unavailable commodity. Satellites, on the other hand, can offer a broadband solution also to remote areas, but it is better suited to broadcast applications. So compared with terrestrial and satellite infrastructures HAPs with their superior link budget and cellular structure are better positioned for cost competitive point-to-point links to fixed and mobile users, whereas multicast and broadcast services, unless required only on the local/regional level within the coverage area of a single HAP or a network of HAPs, are better provided by satellites.

3.2   Antenna Requirements and Related Challenges

3.2.1   Introduction

Fundamental to the performance of HAP communications systems are the antennas both on the HAP and the user end of the link. Typically, such antennas are highly directional, a requirement of the long link lengths and high data rates being proposed for broadband communications. Normally, antennas can be categorised by a model or mask, enabling system engineers to predict the best case or worst case performance from a system. If we consider an antenna mask, both the main lobe (often specified by the half-power beam width) and side lobes (often defined by a side lobe floor) are critical to system performance. The main lobe can be used to specify the maximum expected performance from a link, with the side lobe often being used to control the worst case interference to and from other systems. Thus, care needs to be taken with both.

The configuration of the HAP antennas needs to be carefully considered, as these will illuminate an area on the ground, the so called footprint. Cellular structures can be readily formed on the ground by arranging them in appropriate clusters. Interference between cells can be controlled by adjusting the channel reuse number, with angular separation between two antennas controlling performance, unlike with a terrestrial situation, where it is the physical reuse distance that controls performance. For the most part it is the performance of the user antenna's mainlobe that is of interest in conventional deployments. However, in Chapter 9, we look at how we can use the sidelobe performance to allow configurations of multiple HAPs to share a common set of resources. The purpose of this section is to introduce the basic design tradeoffs for mm-wave broadband deployments, while also referring the reader to a plethora of other sources on the subject.

3.2.2 Types of Antennas for the Delivery of Broadband Services in the mm-Wave Bands

In deciding on the type of antenna to be employed on the HAP the following aspects must be considered [4]:

- number of cells;
- antenna gain;
- required total aperture area.

Here we consider a scenario where a HAP is required to provide broadband access at mm-wave frequencies over a 60 km diameter footprint on the ground from an altitude between 17 km and 20 km (so each user has a minimum 30° elevation angle).

First, the area of aperture of the antennas will increase as the number of cells increases, assuming that each cell is serviced by a single antenna. This is approximately an N^2 relationship where N is the number of cells [4]. Thus, as the cells get smaller for a fixed service area the beam width must decrease, hence causing the aperture to increase. Therefore, the combined effect of more antennas and larger antennas leads to a rapid increase in aperture area with cell number [4].

Here we will now provide a very brief overview of the types of antennas that are suitable for this mm-wave broadband application by group – aperture antennas, reflector antennas, and printed antennas. A brief discussion of array-based antennas is left until Chapter 8. Antenna types not applicable to the application include low gain antennas and wire antennas (dipole, Yagi, helix).

3.2.2.1 Aperture Antennas

Aperture antennas are common at microwave and millimetre wave frequencies, and are typically fed by waveguide and thus lend themselves to integration with waveguide components, thus avoiding the material losses associated with co-axial or planar transmission lines [4]. The radiation patterns from aperture antennas are a function of the field distribution in the aperture. The far field pattern may be derived from a spatial Fourier transform from the aperture into the far field. Such characteristics make them well suited for HAP-based applications, owing to their high aperture efficiency and low sidelobe level.

Conical Feedhorns
In its most basic form, the conical feedhorn is a pure mode horn where the aperture field distribution is that of the fundamental TE_{11} mode [4]. Such a horn can be fabricated fairly easily at microwave and millimetre wave frequencies, being smooth walled and with uniform flare.

Corrugated Conical Horn
In this variant of the multi-mode horn antenna, corrugations (annular grooves) are machined in the walls of a conical horn to suppress axial current flow and hence azimuthal magnetic field [4].

Pyramidal Feed Horn
The pyramidal horn is a very common antenna, as it will integrate well with a rectangular waveguide. The E- and H-plane waveguide walls are usually flared with a linear taper to produce a rectangular aperture [4]. They are fairly difficult to fabricate at millimetre wave frequencies when compared with the conical horns. However, despite this they may prove useful for this application as it is possible to readily produce elliptic beams by modifying the relative lengths of the rectangular aperture. It also has the added advantage that a transition between a rectangular and circular waveguide is not needed.

Lens/Horn Hybrid
For practical horn dimensions, the finite flare angle presents a disadvantage through the phase distribution across the aperture [4]. In such a case, a lens may be used to taper the aperture distribution more accurately. The curve of the lens may be given by [4]:

$$r = f \frac{n-1}{n \cos \theta - 1} \tag{3.1}$$

where f is the focal length of the lens and n is the refractive index of its material. This is an equation for a hyperbola of eccentricity n. This approach assumes that the theory of geometric optics is valid, which is a good approximation despite the lens being in the near field of the antenna. Examples of lenses use for this type of application can be found in [5].

3.2.2.2 Reflector Antennas

Reflector antennas are an important and common class of antenna, where a primary feed such as a waveguide horn is used to illuminate a shaped, reflective surface. There are a number of common types [4].

Front Fed Parabolic Reflector
Here the reflector has a parabolic cross-section which results in a single focus, which is where the primary feed is situated. In this configuration, the divergent beam from the primary feed is approximately collimated by the reflector, resulting in a parallel beam. A weakness of this geometric optical treatment is that diffraction at the reflector edge is not taken into account and hence side lobe patterns can be difficult to predict. For this reason, reflectors are often made over-size so as to reduce illumination of their edges. Also, the primary feed antenna, with its associated transmission line feed and mechanical supports, inevitably lead to some blockage of the reflector aperture. Reflector antennas tend to have lower aperture efficiencies than for the aperture antennas discussed above.

Sub-Reflector (Cassegrain) Antenna
A two-reflector system may be used to improve the performance of reflector antennas. The primary reflector is a paraboloid with the sub-reflector being a hyperboloid, thereby allowing for a more convenient location of the primary feed. To achieve good radiation characteristics, the sub-reflector needs to be several wavelengths in diameter, which inevitably leads to some blockage of the much larger primary reflector.

Multiple-Feed Reflector
The possibility of producing multiple beams by use of multiple primary feeds can be achieved with this type of configuration. However, where the primary reflector is parabolic, deviation of the feed point from the focus leads to aberration of the beam, which limits the scan angles which can be achieved.

3.2.2.3 Printed Antennas

Printed antennas are also commonly known as patch antennas, and are popular in a wide range of applications [4]. They can be fabricated using standard photolithographic techniques, so they can be readily integrated with other printed circuit components. Printed antennas are by nature planar, lightweight and mechanically robust structures. Antennas that will operate in the mm-wave bands can be made highly directional, and are small in size. A disadvantage with these types of antennas is their relatively poor aperture efficiency.

3.2.3   Antenna Model Example

Conventional aperture antennas of medium and high directivity (D) have main lobe patterns which can be approximated by [6]:

$$D = D_{max} \left(\cos \theta\right)^n \tag{3.2}$$

For very low sidelobe levels peak directivity can be approximated by [7]:

$$D_{max} = \frac{32\ln 2}{\theta_{3dB}^2 + \phi_{3dB}^2} \tag{3.3}$$

where θ_{3dB} and ϕ_{3dB} are the 3 dB beam widths in two orthogonal planes. For a circularly symmetric beam, this can be rewritten as:

$$D = \left(\cos \theta\right)^n \frac{32\ln 2}{2\theta_{3dB}^2} \tag{3.4}$$

The 3 dB beam width is also a function of n [8]:

$$\theta_{3dB} = 2 \arccos\left(\sqrt[n]{\tfrac{1}{2}}\right) \tag{3.5}$$

enabling the directivity to be expressed as a function of only θ and n [8]:

$$D = \left(\cos\theta\right)^n \frac{32\ln 2}{2[2\arccos\left(\sqrt[n]{\tfrac{1}{2}}\right)]^2} \tag{3.6}$$

For a cellular scenario on a HAP results show that it is possible to maximise the power at the cell edge when the edge of cell power roll-off is approximately 4.5 dB below the boresight gain. This is an important design rule, and the reader is referred to [8] for more information. Also in [8] are detailed derivations of how to design a cellular reuse design for a broadband system.

This can then be converted into a general model for the antenna by taking into account a sidelobe floor (s_f) and aperture efficiency η:

$$A(\theta) = D_{\max}\eta\{\max[\cos^n(\theta), s_f]\} \tag{3.7}$$

where $D_{\max}$ is the maximum directivity at boresight.

Figure 3.1 shows the effect of antenna roll-off in terms n, and its relationship to the beam width. In this example here it is assumed that the sidelobe floor is 30 dB below boresight (peak) gain, with an aperture electrical efficiency of 90%.

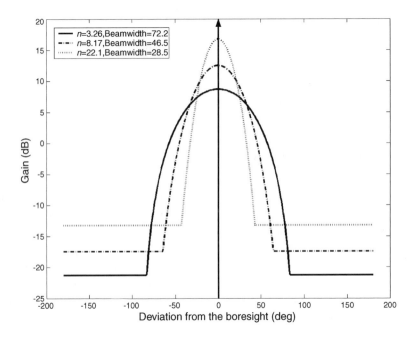

Figure 3.1 The effect of the antenna roll-off on antenna profile, with a sidelobe floor of -30 dB relative to boresight gain [9]

A key issue when the HAP is using a circular beam pattern is the fact that at low elevation angles the beams will spread by different amounts, resulting in a noncircular almost elliptical footprint, as shown in Figure 3.2. This can be compensated for using an antenna with an elliptical pattern, with the eccentricity in the opposite plane to the spread [8,10].

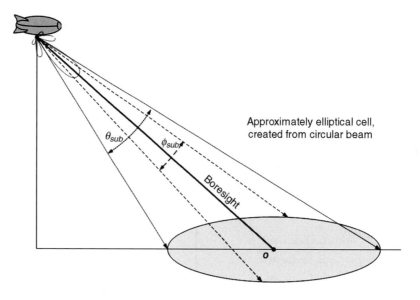

Figure 3.2 HAP and cell geometry

3.2.3.1 Channel Reuse Schemes and the Impact of Antenna Beam Patterns

In the case of a hexagonal cellular layout the coverage is affected by the number of cells and the reuse angle each cell is away from another sharing the same channel. It is assumed that interference is largely caused by the sidelobes and not the mainlobe.

For each beam it is possible to generate a matrix of the form $\{x, y, power\}$, where the power can be derived from the antenna directivity seen at a point $\{x, y\}$ on the ground, minus the excess free space loss with respect to the sub-platform point [8]. This excess path loss is included here because the link length variation across the coverage (and hence path loss) varies significantly for typical sizes of coverage area. Note that this is different from satellite scenarios where link length is largely the same.

Hence the carrier-to-interference ratio (CIR) may be derived for the group of co-channel cells from [8]:

$$CIR(x, y) = \frac{P_{\max}(x, y)}{-P_{\max}(x, y) + \sum_{i=1}^{n_{cc}} P_i(x, y)} \tag{3.8}$$

where n_{cc} is the number of co-channel cells. $P_{\max}(x, y)$ is the maximum power from one of the beams and is therefore the signal. The interference component of the denominator in Equation (3.8) is the sum of powers of all the other beams, at point $\{x,y\}$.

It is shown in [8] that when conventional reuse schemes of 3, 4 and 7 are applied, the reuse scheme of 7 has a higher CIR for a particular level of coverage, with the reuse scheme of 3 having the worst. This is exactly the same as with a conventional terrestrial scenario with the exception that now the level of interference will increase with the number of cells (antennas) in the deployment. However, even the reuse of 3 has adequate CIR for at least the lower order schemes, e.g. QPSK.

3.2.3.2 Antennas for Backhaul Links

Dedicated backhaul antennas may be used on the platform. Alternatively, to avoid this requirement, it may be possible for backhaul data to be transmitted via the user antenna array, albeit with a loss of overall system capacity. The antenna groups sharing a channel will have their own CIR characteristics on the ground.

Diversity studies indicate that backhaul ground stations sharing a common data signal (for site diversity/redundancy) should be separated by at least 10 km [11], and in this case co-channel interference will be from sidelobes only, given the narrow beam width antennas selected. More discussion of antenna beam width selection is given in later sections of this chapter.

The backhaul links are not required to provide continuous cellular coverage so there will be no advantage in selecting elliptic beams. If antennas are used which have narrower beamwidths than those used to generate the cellular footprints, several beams may be packed in close proximity without main lobe overlap.

For example, using 5° circular symmetric beams, 5 beams at 5 km ground distance and 10 beams at 10 km ground distance could be used for a sidelobe floor of -40 dB [12].

3.3 System and Network Architecture of HAP-Based Communication Systems

3.3.1 Overview

The configuration of the payload on board a HAP is particularly dependent on available space, maximum weight and available power [12]. This in turn has an effect on the possible architectures. We examine these implications here and also consider methods of reducing the link capacity required through the use of broadcasting and caching. We finally suggest three options for the payload configuration. Much of this work in Section 3.3 was originally done as part of the FP5 HeliNet project so here we rely heavily on a deliverable from the project [13], whose content has not been made public until now.

A range of possible system architectures is available, depending on scenario and circumstances. The HAP in most cases will be at the edge of a network. It is very likely that it will form the final/first hop for transmissions to and from anywhere in the world. Initially there will be a few HAPs. The advantage of stratospheric platform-based services are that they can be rolled out to provide pockets of coverage, most likely connected into a wider network, using either terrestrial or satellite infrastructure. If local traffic only is to be supplied there may also be a case for a standalone HAP with no network connection. Network connections will normally be provided by a backhaul link (either terrestrial or satellite). One possible service would be that of a LAN interconnect between sites, so in this case no backhaul link may be required; instead a network connection is provided at one of the office sites. A second family of architectures requires multiple HAPs that are networked together. This can be done using interplatform links, backhaul links, or a mixture of both. Architectures will rely on a mixture of user, backhaul and interplatform links. Network issues may be similar for some of the more complicated networks, i.e. dependent more on number of hops and less on PHY and MAC attributes.

3.3.1.1 The Impact of Platform Station-Keeping on HAP System Architecture

It is worth reiterating here that HAPs are quasi-stationary, so it is important that their movement is taken into account with the design of any system architecture. Changes in position of the HAP will give rise to movement of the cells on the ground unless it is compensated for either electrically or mechanically. Station keeping will also be a major constraint for the backhaul links.

Figure 3.3 illustrates the angular variation in position of the HAP as seen from the ground as a function of ground distance (defined as the distance away from the point immediately below the centre of the location cylinder). HAP movement can be both horizontal and vertical. This angular variation can be used to determine whether fixed or steerable ground-based antennas are required. If the angular variation is greater than the beam width of the ground-based antenna then it is necessary to use steerable antennas. The traces on the graph represent the HeliNet scenarios for the two location cylinders and the ITU-R scenario as described in Section 2.4. The greatest angular variation is immediately below the HAP. However, it may be possible to use wider antenna beam widths at these points due to the shorter path length. Changes in vertical height cause the most change to the angular variation at short and very long ground distances. At long ground distances, this change is of greater significance because changes in vertical height make up a larger proportion of the angular variation.

By comparison, Angel Technologies' HALO scheme considering a horizontal radius of 4–6 km, will obviously cause even larger angular variations, which is probably why they propose steerable ground-based antennas [10].

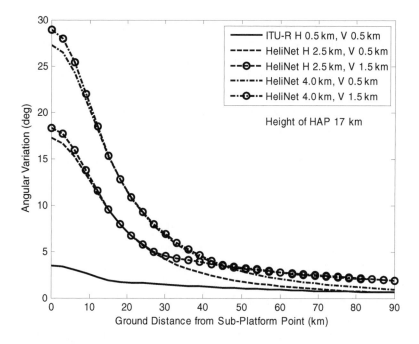

Figure 3.3 Angular variations in the position of a HAP on the ground as a function of ground distance, for worst case horizontal and vertical HAP displacements (based on [14])

3.3.1.2 Payload Configuration Options

Three options can be considered for a HAP payload, depending on the eventual size and weight available, all of which will have some impact on the overall architecture and services available. The business related implications of different types of HAP architectural configurations are discussed in Chapter 4, but the main options are:

- *Option 1 – size/mass efficient configuration.* Hardware is placed wherever possible on the ground (and not on the HAP). This effectively makes the HAP a transparent transponder. This architecture is the most difficult to design, because efficient ways must be found to aggregate and split up the traffic between cells. The backhaul requirements are also more demanding, as all traffic must be brought to ground for processing, which also results in longer time delays. It is also difficult to use interplatform links (IPLs) as data must be backhauled to ground and then sent back up to the HAP for transmission over the IPL. Using IPLs with this configuration would only be of benefit when there was no other network infrastructure.

- *Option 2 – On board switching and routing.* Switching and routing hardware is placed on the HAP, allowing local traffic to be handled on the HAP without being brought to ground, thereby reducing the backhaul requirements and time delays. Traffic can also be routed to IPLs relatively easily. A network of HAPs is less reliant on ground-based infrastructure and networks.
- *Option 3 – On board switching, routing and caching.* Effectively as much hardware as possible is placed on the HAP, reducing its dependency on existing networks. The addition of caching considerably reduces the backhaul requirements, allowing large libraries of information to be stored, for easy and high speed access by the user.

3.3.1.3 Broadcasting, Caching and Diversity Options

Broadcast (or multicast) information can be used to reduce the data rate requirements on all link types. The requirements effectively reduce the capacity to $1/n$ over independent transmissions (where n is the number of users receiving the broadcast). Broadcast traffic can be both real and nonreal time. Broadcasting can be used with all payload configuration options.

Caching, the temporary storage of nonreal time information (e.g. web pages and video-on-demand), has two obvious benefits:

- It allows the amount of traffic to be reduced over a communications link. As with broadcast information the capacity requirement reduces by $1/n$ over independent downloads (where n is the number of users that download the page from the cache). This is particularly useful in reducing the information on the backhaul links.
- It provides a means of time diversity for the system. Providing the information is already stored in the cache an end user need not know that an intermediate link has failed (due to rain), allowing the system availability to be much higher than any link availability. Information may be already in the cache because another user has requested it, or it may be possible to predict which information is required and fill the cache during quiet periods. Prediction methods could include the use of web page statistics or the top x most popular films. Pricing of films/information can also encourage the use of cached information.

Caching is best exploited with payload Option 3, although it would also have a more limited use with the other two options, where the user could have a local cache. A typical caching example is shown in Figure 3.4 – Option 3 is assumed. In the case of Options 1 and 2 the on board cache is removed. Users are increasingly able to

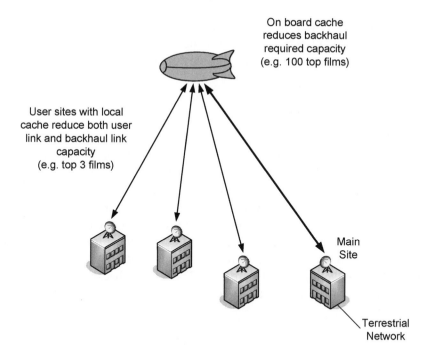

Figure 3.4 Caching architecture example

access large amounts of localised capacity. Personal video recorders (PVRs) now have many gigabytes of storage. These can be linked to broadcast schedules, examples include Sky+ in the UK [15], and TiVo in the USA [16]. Both the CPE and HAP caches can be filled during a quiet period. The local cache not only saves transmitting traffic over the link at an inconvenient moment, but can also be filled using a broadcast transmission. That is, each film in the local cache is transferred only once over the backhaul and only once over the user link, where it is transmitted simultaneously to multiple user caches. Other films not in the top 3 would be downloaded on an individual basis from the HAP. A film may also be buffered by a cache (i.e. watched with a small delay) to cope with any rain outage. In this case only traffic on the backhaul link is reduced.

Space diversity can also be used to improve the overall system availability. Rain outages tend to be localised and providing antennas can be sited a few kilometres apart it is possible to decrease the probability that both will suffer an outage simultaneously, even if the overall data rate will reduce. Such a technique is only applicable for backhaul traffic, because of this large geographical separation requirement. We look at its impact here, and additional information on the technique can be found in [11].

It may also be possible to use polarisation diversity during clear air conditions to increase capacity or provide a greater degree of isolation between the shared channel. This technique would be applicable for all link types.

3.3.2   HAP Architectures

3.3.2.1 Stand-alone HAPs

An example of a stand-alone HAP architecture is shown in Figure 3.5. This illustrates how the applications are provided bandwidth using a cellular array of spot beams. The capacity per cell is dependent on antenna spot beam design, available bandwidth and power. As a rough rule of thumb 'what goes up must come down', so it is necessary to provide sufficient capacity for the backhaul links. However, each scenario will have a certain amount of local traffic, i.e. both users are served from the same HAP, so the backhaul requirements may be reduced assuming payload Options 2 and 3 are used. The satellite backhaul scenario can be used when there is insufficient terrestrial infrastructure available on the ground.

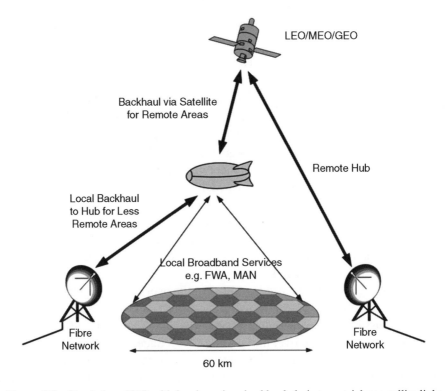

Figure 3.5   Stand-alone HAP with local services backhauled via terrestrial or satellite links

Satellite backhaul will severely constrain what is offered from the HAP as the capacity will be severely limited, unless there was a large amount of local traffic.

Figure 3.6 illustrates the case when there is no backhaul link. Now connection to a network is achieved at one of the customer sites, with HAP being used as an interconnect service (e.g. a corporate LAN, where the two office sites are connected together).

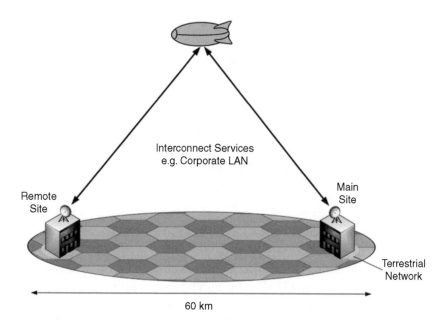

Figure 3.6　Stand-alone HAP with no backhaul and a ground base network connection

3.3.2.2 Interconnection of HAPs via IPLs

The scenario with platforms connected via IPLs is depicted in Figure 3.7. In this case the system comprises two or more platforms. Each HAP payload includes a switching device and one or more IPL transceivers, which enable communication between adjacent platforms without any ground network elements included. In this scenario ground stations can be used as gateways to other public and/or private networks, or to provide a backup interconnection between platforms in case of IPL failure (the previous scenario applies in this special case).

In this scenario only platforms with onboard switching payload are envisaged, in order to take advantage of IPL implementation. Thus, the shortest path between users A and B within a single platform comprises of uplink from user A to the

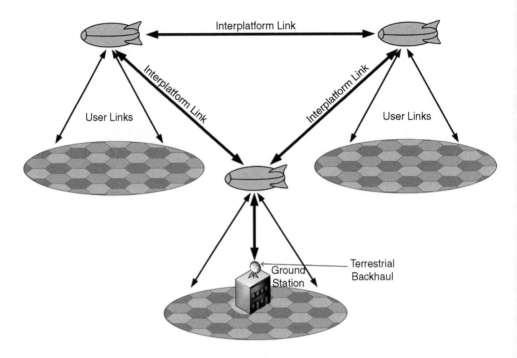

Figure 3.7   Interconnection of HAPs via IPLs

platform, where switching is performed, and downlink to user B. The shortest path between two users in distinct HAP coverages takes user uplink from user A to the platform, where switching payload chooses the most suitable sequence of IPLs towards the platform, which is serving user B. The last part of the path between users represents user downlink to user B.

Table 3.2 summarises advantages and disadvantages of this topology scenario with respect to scenarios not utilising IPLs.

Table 3.2   Advantages and disadvantages of topology scenario using IPLs

Advantages of topology scenario using IPLs	Disadvantages of topology scenario using IPLs
System operation independence from terrestrial networks	IPL transceivers represent additional weight and power consumption of platform payload
Reduced requirements for terrestrial network and up/down link segments	Expected difficulties in realisation of IPLs according to weather conditions on HAP operational altitudes
Highly flexible system coverage	Additional research and development in IPL transceivers is needed
Lower signal delays	

In the scope of HAP system design the above listed (dis)advantages need to be carefully examined and weighted against alternative topology scenarios. However, the most critical issues in IPL design are first to investigate high altitude flying conditions with all accuracy available, and, secondly, to obtain technical solutions in IPL design to overcome difficulties caused by low stratospheric winds. Thus, exact knowledge about platform stability due to wind statistics, unscheduled gusts, turbulence, flight mechanics (motor vibrations, other moving parts of the structure) is needed for a more detailed investigation on the feasibility of the topology scenario utilising IPLs. Based on such knowledge and also more detailed system specifications regarding services to be offered and quality of service required, the choice between optical (OIPL) and radio (RF IPL) interplatform links is to be made. We specifically examine options for OIPL in Chapter 7.

3.3.2.3 Scenario with Platform to Satellite Links

HAPs, placed above areas with underdeveloped (rural and remote areas) or completely without (ocean, desert,etc.) terrestrial infrastructure can be connected to other public or private networks using platform to satellite links (PSLs). In addition, on the platform utilising PSLs these could also be used as a backup solution when the connection with the rest of the network via IPLs or ground station is disabled due to a failure or extreme rain fading on UDL segment.

Due to longer communication paths and thus higher attenuation PSL transceivers will be heavier than IPL transceivers and will have higher power consumption. Interference with other satellite communication systems operating in adjacent frequency bands is expected to be another problem associated with this scenario. Taking this and also specifications in [4] into account, the use of PSLs is expected to be possible on stand-alone platforms only, since the HAP is not likely to be able to host any IPL transceivers in addition to PSL transceivers.

3.3.2.4 Interconnected HAPs using Common or Terrestrially Connected Backhaul Ground Stations

The final unique type of architecture is shown in Figure 3.8 and uses ground stations that connect to more than one HAP. Here each ground station connects with two HAP – up to n HAPs can be connected to a single ground station if required. The main problem with this architecture is that the ground stations must be located on the edge of coverage, i.e. they will have long path lengths. Connection to a wider network and backhaul capacity requirements will be the same as in the single HAP case as shown in Figure 3.5, but there would be less

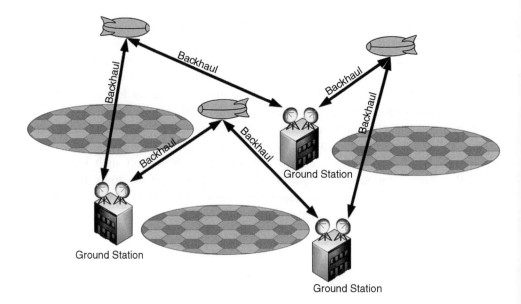

Figure 3.8 HAPs interconnected using common backhaul ground stations

reliance on an external network. An alternative to common ground stations would be the use of individual ground stations connected by terrestrial network links (TNLs), e.g. leased lines, as shown in Figure 3.9. Obviously there are many other combinations that could be adopted.

3.3.3   Broadband Communications Links

3.3.3.1 Overview

The architectures described earlier require a mixture of user, backhaul and interplatform links. In this section we evaluate the maximum data rates that can be transferred for the scenarios described earlier. Link budgets are used to determine the data rates available taking into account features such as propagation losses and link lengths. It is also assumed that there is sufficient bandwidth available to support the link budgets.

Further definition of the radiocommunication regulations governing allocated frequency bands for HAPs needs to be carried out. The ITU-R has reserved bands at 48 GHz for high altitude platform stations (HAPSs) but they are defined as flying above 20 km; HAPs are flying below this altitude. It may therefore be possible to treat the HAP as a tall terrestrial mast, and use terrestrial frequencies, e.g. 28 GHz.

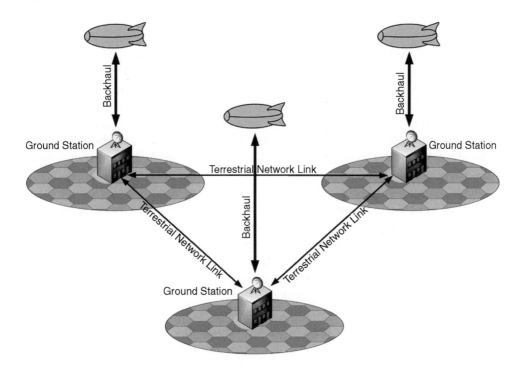

Figure 3.9 HAPs interconnected using backhaul ground stations interconnected using TNLs

Such an approach was adopted by Angel Technologies in the USA [10], and in this case their jet-powered plane operated below 20 km. In addition, bands around 28 GHz have been adopted by the ITU for HAPSs in a number of countries [9], so here we examine both possibilities, with the techniques being generally applicable to other bands which may become available at a later date. Radio regulation was discussed in more detail in Section 1.3.2.

One of the major causes of signal attenuation at these frequencies is rain attenuation. Rain attenuation statistics are available from organisations such as the ITU-R. These can be used to derive the required link margin that must be included in order for the link to operate with a particular availability (the maximum rain rate that the link must be able to handle). We are assuming each backhaul link need only operate with an availability of 99.0–99.9%, depending on spatial diversity configurations, and in any case it is unlikely that it will be possible to guarantee the position on the HAP with a higher availability than this as specified by the HeliNet position cylinder discussed earlier. Also the user will probably not be aware as to whether the unavailability is due to the user or backhaul link.

3.3.3.2 User Links

The user downlink will be shared amongst all users within a cell. The exact way in which the sharing will be done, along with handoff and frequency reuse issues will be discussed later in this book (see Chapters 8 and 9 in particular). The aggregate capacity (effectively we are assuming we have a TDM channel) is determined by a link budget. Table 3.3 describes the terms used in the user link budgets. An example link budget is shown in Table 3.4. These link budgets have been typically calculated for the worst case position within the area of coverage. This tends to be at the edge of coverage. Additionally, edge of cell losses, due to not being at the antennas boresight, are also taken into account. Although the narrower beam width HAP antennas (see Section 3.2) used for cells at the edge of coverage compensate for the increase in path loss due to the longer link length, they do not compensate for the increased margins needed to overcome rain attenuation, caused by more of the path travelling through the atmosphere.

The possible data rates available depend on the availability required and size of receive antennas, amongst other things.

Using the link budget above as an example it is possible to look at a range of scenarios and systems. These have been summarised in Tables 3.5 and 3.6. The link budget shows that it is not possible to deliver 99.99% link availability (even if HAP could be maintained on station) with any scenario. It can be seen that the nonsteerable antenna scenarios tend to have low data rates (per cell), primarily because the antenna gain on the ground is low and will suffer from higher edge of cell losses due to the greater pointing inaccuracies. Also the increased rain margins required at 48 GHz compared with 28 GHz, coupled with the increased path loss have the effect of reducing the link availability available for a given data rate.

3.3.3.3 Backhaul Terrestrial Links

The backhaul link is used to connect the HAP into a wider network. More than one backhaul link will be required in order to support the capacity involved and it is also useful to have alternative links as a backup should one of the backhaul ground stations fail. As mentioned previously, three factors can reduce the backhaul requirements:

- *Local services* – e.g. for LAN interconnect, where two users are served from the same HAP.
- *Caching* - This will be useful for two applications video-on-demand and for downloading of Internet web pages.

Table 3.3 Explanation of the terms used in the user link budgets

Line	Explanation of terms used in the downlink user link budget	Typical values
2	HAP power per carrier (dBm), P_{TXH}, chosen for the particular budget	30.0
3–4	HAP antenna elevation and azimuth 3dB beam widths (deg)	3.0, 8.0
5	HAP antenna electrical efficiency. A horn antenna is assumed	0.95
6	HAP antenna gain, G_H. This is determined from the directivity and antenna efficiency	32.6
	The directivity (in dB) is determined from the following:	

$$D = 10\log_{10}\left[\frac{32\ln(2)}{\theta^2 + \phi^2}\right]$$

where θ and ϕ are the elevation and azimuth beam widths, respectively (in rad)

$$G_H = 10\log_{10}(\eta) + D$$

where η is the electrical efficiency of the antenna and D is the diameter of the antenna given above

7	HAP antenna feed loss (dB), A_L	1.0
8	HAP EIRP (dB) $= P_{TXH} + G_H - A_L$	
11	The Boltzmann constant (dBJ/K)	
12	Noise temperature T (K)	300
13	Noise density N_o (dBm/Hz). This is calculated using $N_o = kT$	−174
14	Receiver noise figure NF (dB)	5
15	HAP receiver noise density (dBm/Hz) $= N_o + $ NF	
16	Receiver interference noise density (dBm/Hz). The maximum level of interference. Worst case for a noise limited situation is assumed to be when equal to the noise power. This is because if the interference level is higher than noise the transmit power can be reduced because the system is now interference limited. If the transmit power is reduced so that the interference is well below the noise floor then the available $C/(I + N)$ is unnecessarily reduced	
17	Total effective noise $+$ interference NI_{TOT} (dBm)	
19	Backhaul ground station antenna beam width (deg)	5.0
20	Backhaul ground station aperture efficiency. A horn antenna is assumed	0.95
21	Backhaul ground station antenna gain G_G (dB), determined on separate sheet	
22	Cable loss at the backhaul ground station (dB)	2
23	Maximum $C/(I_o + N_o)$ (dBHz)	
25	Modulation scheme. A range of schemes has been chosen which have differing E_b/N_o requirements. We expect that these will be switched depending on the level of signal attenuation	
26	Required E_b/N_o (dB). We have chosen figures for a BER 10^{-9}	7.8–25

(*continued*)

Table 3.3 (*Continued*)

Line	Explanation of terms used in the downlink user link budget	Typical values
27	Bit/symbol	2–6
29	System bandwidth (MHz), B	50
30	Code rate, CR: Either no coding or $^1/_2$ rate convolutional coding has been used	0.78–1
31	Data rate. Taking lines 29 and 30, and using a 50% roll-off filter (γ) yields a data rate of $\left(\dfrac{BC_R}{1+\gamma}\right)$	
32	Line 30 in log form: $R = 10\log_{10}\left(\dfrac{BC_R}{1+\gamma}\right)$	
33	Required $C/I_o + N_o$ (dBHz) $= E_b/N_o +$ line 31	
35	Maximum allowed losses. This takes into account the maximum $C/I_o + N_o$ and data rate $=$ line 22–line 32	
38	Frequency (GHz)	28/48
39	Corresponding wavelength (m) determined from $c = f\lambda$	0.011–0.013
40	Ground distance GD (km).	10
41	Platform height H (dB)	17
43	The line of sight (LOS) distance (L) (in km) is determined by Pythagoras, taking into account horizontal ground distance (GD) from the sub-platform point (SPP) to the ground based transmitter and the height (H): $L = \sqrt{GD^2 + H^2}$	19.72
44	FSPL (free space path loss) given (in dB) by: $\text{FSPL} = 20\log_{10}\left(\dfrac{4\pi L}{\lambda}\right)$	
45	Miscellaneous atmospheric losses, P_L (dB). This takes into account oxygen and water vapour	0.7–1.5
46	Edge of cell and antenna beam losses (dB). Only antennas pointing at each other on boresight will have no losses. In practice stations may be on the edge of a cell and some way off boresight	1–6
47	Clear air losses (dB) $=$ FSPL $+ P_L$	
49	Received margin at boresight in clear air (dB)	
50	Minimum transmit power required (dBm). This takes into account the level of margin given by the typical transmit power. $=P_{TX}$-margin	
52–54	Rain attenuation figures for 99.9% and 99% (dB). They use the latest ITU Recommendations and figures assume climate zone K (Italy)	
56–58	Received margin for 99.9% and 99% availability, assuming rain attenuation at boresight (dB)	
60–62	Minimums transmit power required (dBm) for 99.9% and 99% availability. This takes into account the level of margin given by the typical transmit power. $=P_{TX}$-margin	

Table 3.4   Example user link budget

User Link for High Rate Broadband Services at 28GHz

#		64QAM	16QAM		8AMPM	QPSK	
1	**Transmitter (HAP)**						
2	Power per backhaul carrier (dBm)	30.0					a
3	Antenna beamwidth - theta (degrees)	3.8					
4	Antenna beamwidth - phi (degrees)	8.2					
5	Antenna electrical efficiency	0.95					
6	Antenna gain (dBi)	29.3					b
7	Antenna feed loss (dB)	1.0					c
8	HAP EIRP (dBm)	58.3					d=a+b-c
9							
10	**Receiver (Ground Station)**						
11	The Boltzmann Constant (dBJ/K)	-228.6					
12	Noise Temperature (K)	300.0					
13	Thermal noise density (dBm/Hz)	-173.8					e
14	Receiver noise figure (dB)	5.0					f
15	Receiver noise density (dBm/Hz)	-168.8					g=e+f
16	Receiver interference noise density (dBm/Hz)	-168.8					j = g
17	Total effective noise density (dBm/Hz)	-165.8					k= 10*log(10^(g/10)+10^(j/10))
18							
19	Antenna beamwidth (degrees)	2.0					
20	Antenna electrical efficiency	0.95					
21	Antenna gain (dBi)	39.4					l
22	Cable loss at ground station	2.0					m
23	Maximum C/(Io+No) (dBHz)	261.5					o=d-k+l-m
24							
25	Modulation Scheme	64QAM	16QAM		8AMPM	QPSK	
26	Required Eb/No (BER 10-6)	20	16		14	7.8	ab
27	Bit/symbol	6	4		3	2	ac
28							
29	Bandwidth (MHz)	25.0	25.0		25.0	25.0	
30	Code Rate	1.00	1.00		1.00	0.50	aj
31	Data Rate (Mbit/s) (25% rolloff)	120	80		60	20	ad
32	Data Rate (dBbit/s)	80.8	79.0		77.8	73.0	p
33	Required C/(Io+No) (dBHz)	100.8	95.0		91.8	80.8	ae=ab+p
34							
35	Maximum allowed losses (dB)	160.7	166.4		169.7	180.7	q=o-ae
36							
37	**Link Parameters**						
38	Frequency (GHz)	28.0					
39	Wavelength (m)	0.011					
40	Ground Distance (km)	30.0					
41	Platform Height (km)	17.0					
42							
43	LOS Distance (km)	34.48					
44	FSPL (dB)	152.1					r
45	Misc Atmospheric Losses (dB)	0.7					s
46	Edge of cell and antenna beam losses	5.0					sa
47	Clear air losses (dB)	157.8					t=r+s+sa
48							
49	Received margin clear air (dB)	2.9	8.6		11.9	22.9	u=q-t
50	Minimum required transmit power clear air (dBm)	27.1	21.4		18.1	7.1	
51							
52	Rain Attenuation Zone K 99.99% (dB)	32.5					w
53	Rain Attenuation Zone K 99.90% (dB)	12.3					x
54	Rain Attenuation Zone K 99.00% (dB)	3.3					y
55							
56	Received margin 99.99% (dB)	-29.6	-23.8		-20.6	-9.6	z=u-w
57	Received margin 99.90% (dB)	-9.4	-3.6		-0.4	10.6	aa=u-x
58	Received margin 99.00% (dB)	-0.4	5.4		8.6	19.6	ba=u-y
59							
60	Minimum required transmit power 99.99% (dBm)	59.6	53.8		50.6	39.6	ca=a-z
61	Minimum required transmit power 99.90% (dBm)	39.4	33.6		30.4	19.4	da=a-aa
62	Minimum required transmit power 99.00% (dBm)	30.4	24.6		21.4	10.4	ea=a-ba

Table 3.5 Downlink data rates per cell for different scenarios using the 28 GHz band

Scenario (28 GHz)	Data rate per cell (Mbit/s)		
	99.90%	99.00%	Clear air
Highly portable/vehicle terminal, 60° ground antenna		2	2
Fixed antenna at edge of coverage (30 km GD), 8° ground antenna		3	10
Fixed antenna at sub-platform point, 28° ground antenna		3	3
Low data rate (12.5 MHz BW) 2° steerable antenna at 30 km GD	30[a]	60	60
High data rate (25.0 MHz BW) 2° steerable antenna at 30 km GD	20[b]	80	120

[a] This data rate assumes 8 AMPM modulation (3 bits per symbol) and no coding with a margin of 1.1 dB.
[b] This data rate assumes QPSK (2 bits per symbol) and half rate coding and as such is lower than the low bandwidth case. The margin for 8 AMPM is –0.9 dB, which would yield a data rate of 60 Mbit/s. Link budgets for the final outline specification will ensure that anomalies such as this do not occur – it is not worth doing this at this stage until the modulation schemes have been fully examined.

- **Broadcast services** – If the same transmission is sent simultaneously to n users the backhaul requirement reduces to $1/n$ of that required from individual transmissions.

Backhaul links will be asymmetric, as will be the user links. We shall assume that the ratio of traffic will be $10:1$. This figure is based on the fact that telephony, videoconference traffic will be approximately symmetric, but video-on-demand and Internet traffic will be highly asymmetric. In the case of the user links the downlinks on average will carry more traffic, e.g. videos and Internet traffic,

Table 3.6 Downlink data rates per cell for different scenarios using the 48 GHz band

Scenario (48 GHz)	Data rate per cell (Mbit/s)		
	99.90%	99.00%	Clear air
Highly portable/vehicle terminal, 60° ground antenna			
Fixed antenna at edge of coverage (30 km GD), 8° ground antenna		1	1
Fixed antenna at sub-platform point, 28° ground antenna		1	1
Low data rate (12.5 MHz BW) 2° steerable antenna at 30 km GD		30	60
High data rate (25.0 MHz BW) 2° steerable antenna at 30 km GD		20	80

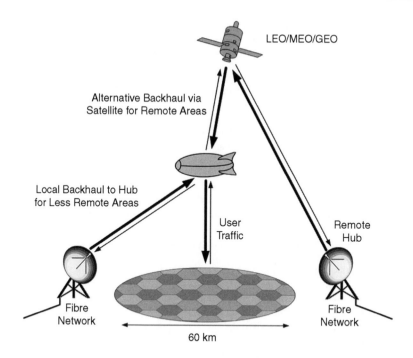

Figure 3.10   Backhaul link configurations

meaning that the backhaul uplink will carry more traffic. Such a representation is shown in Figure 3.10. One constraint on the user downlink traffic is the amount of traffic that can be sent to the HAP. In practice the number of ground stations available for backhaul will be limited (say, to <20). Later we calculate how many cells can be served by a terrestrial backhaul link. One advantage of the architecture is that multiple uplinks can be served from each backhaul station because power is not constrained for ground transmissions (assuming that there is sufficient bandwidth available). This extra capacity can be used to:

- reduce the number of backhaul ground stations required; or
- increase the choice of multicast transmissions available to users; and/or
- increase the material available in the on-board caches.

Backhaul traffic sent over satellite will be further limited by both power and bandwidth constraints.

The backhaul link budgets are similar to the user link budgets, except that we assume that more bandwidth will be available, and the ground-based equipment can be more expensive, allowing for narrower beam width ground-based antennas. We also assume that separate backhaul antennas are used on the HAP. Table 3.7 explains

Table 3.7 Explanation of the terms used in the backhaul link budgets

Line	Explanation of terms used in the downlink backhaul link budget	Typical values
2	HAP power per carrier (dBm), P_{TXH}, chosen for the particular budget	30.0
3	HAP antenna beam width (deg)	5.0
4	HAP antenna efficiency. A horn antenna is assumed	0.95
5	HAP antenna gain, G_H from the beam width we need first to determine the diameter of the antenna (in m)	32.6

$$D \approx \frac{70\lambda}{BW}$$

where 70 is a rule of thumb constant, depending on various factors including the aperture illumination function, λ is the wavelength (in m) and BW is the half power beam width (in deg)

To calculate the gain of the antenna (in dB)

$$G = 20\log_{10}\left(\frac{\sqrt{\eta}\pi D}{\lambda}\right)$$

where η is the aperture efficiency of the antenna and D is the diameter of the antenna given above

Line	Explanation	Typical
6	HAP antenna feed loss (dB), A_L	1.0
7	HAP EIRP (dB) $= P_{TXH} + G_H - A_L$	
10	The Boltzmann constant (dBJ/K).	
11	Noise temperature T (K)	300
12	Noise density N_o (dBm/Hz). This is calculated using $N_o = kT$	−174
13	Receiver noise figure NF (dB)	5
14	HAP receiver noise density (dBm/Hz) $= N_o + $ NF	
15	Receiver interference noise density (dBm/Hz). The maximum level of interference. Worse case for a noise limited situation is assumed to be when equal to noise power. This is because if interference level is higher than noise the transmit power can be reduced because the system is now interference limited. If the transmit power is reduced so that the interference is well below the noise floor then the available C/(I + N) is unnecessarily reduced	
16	Total effective noise + interference NI_{TOT} (dBm)	
18	Backhaul ground station antenna beam width (deg)	5.0
19	Backhaul ground station aperture efficiency. A horn antenna is assumed	0.95
20	Backhaul ground station antenna gain G_G (dB), determined on separate sheet	
21	Cable loss at the backhaul ground station (dB)	2
22	Maximum $C/(I_o + N_o)$ (dBHz)	
24	Modulation scheme. A range of schemes has been chosen which have differing E_b/N_o requirements. We expect that these will be switched depending on the level of signal attenuation	
25	Required E_b/N_o (dB). We have chosen figures for a BER 10^{-9}.	7.8–25

(continued)

Table 3.7 (*Continued*)

Line	Explanation of terms used in the downlink backhaul link budget	Typical values
26	Bit/symbol	2–6
28	System bandwidth (MHz), B	50
29	Code rate, CR: Either no coding or $1/2$ rate convolutional coding has been used	0.78–1
30	Data rate. Taking lines 28 and 29, and using a 50 % roll-off filter (γ) yields a data rate of $\left(\dfrac{BC_R}{1+\gamma}\right)$	
31	Line 30 in log form: $R = 10\log_{10}\left(\dfrac{BC_R}{1+\gamma}\right)$	
32	Required $C/I_o + N_o$ (dBHz) $= E_b/N_o + $ line 31	
34	Maximum allowed losses. This takes into account the maximum $C/I_o + N_o$ and data rate = line 22 − line 32	
37	Frequency (GHz)	28/48
38	Corresponding wavelength (m) determined from $c = f\lambda$	0.011–0.013
39	Ground distance GD (km)	10
40	Platform height H (km)	17
42	The line of sight (LOS) distance (L) (in km) is determined by Pythagoras, taking into account horizontal ground distance (GD) from the sub-platform point (SPP) to the ground based transmitter and the height (H): $L = \sqrt{\text{GD}^2 + H^2}$	19.72
43	FSPL (free space path loss) given (in dB) by: $\text{FSPL} = 20\log_{10}\left(\dfrac{4\pi L}{\lambda}\right).$	
44	Miscellaneous atmospheric losses, P_L (dB). Takes into account oxygen and water vapour	0.7–1.5
46	Clear air losses (dB) $=$ FSPL $+ P_L$	
48	Received margin at boresight in clear air (dB)	
49	Minimum transmit power required (dBm). This takes into account the level of margin given by the typical transmit power. $= P_{TX}$- margin	
51–52	Rain attenuation figures for 99.9% and 99% (dB). They use the latest ITU Recommendations and figures assume climate zone K (Italy)	
54–55	Received margin for 99.9% and 99% availability, assuming rain attenuation at boresight (dB)	
57–58	Minimum transmit power required (dBm) for 99.9% and 99% availability. This takes into account the level of margin given by the typical transmit power. $=P_{TX}$-margin	

the terms and parameter values used in the link budget with Tables 3.8 and 3.9 illustrating the link budgets themselves. As with the user links the data rate available depends on link availability. The 48 GHz band has a lower availability for a given data rate than the 28 GHz band.

Table 3.8 Example link budget for the backhaul link at 28 GHz

Backhaul Link at 28GHz

1 **Transmitter (HAP)**						
2 Power per backhaul carrier (dBm)	30.0		a			
3 Antenna beamwidth (degrees)	5.0					
4 Antenna efficiency	0.95					
5 Antenna gain (dBi)	32.6		b			
6 Antenna feed loss (dB)	1.0		c			
7 HAP EIRP (dBm)	61.6		d=a+b-c			
8						
9 **Receiver (Ground Station)**						
10 The Boltzmann Constant (dBJ/K)	-228.6					
11 Noise Temperature (K)	300.0					
12 Thermal noise density (dBm/Hz)	-173.8		e			
13 Receiver noise figure (dB)	5.0		f			
14 Receiver noise density (dBm/Hz)	-168.8		g=e+f			
15 Receiver interference noise density (dBm/Hz)	-168.8		j = g			
16 Total effective noise density (dBm/Hz)	-165.8		k= 10*log(10^(g/10)+10^(j/10))			
17						
18 Antenna beamwidth (degrees)	5.0					
19 Antenna efficiency	0.95					
20 Antenna gain (dBi)	32.6		l			
21 Cable loss at ground station	2.0		m			
22 Maximum C/(Io+No) (dBHz)	258.1		o=d-k+l-m			
23						
24 Modulation Scheme	256QAM	64QAM	16QAM	8AMPM	QPSK	
25 Required Eb/No (BER 10-6)	25	20	16	14	7.8	ab
26 Bit/symbol	8	6	4	3	2	ac
27						
28 Bandwidth (MHz)	50.0	50.0	50.0	50.0	50.0	
29 Code Rate	1.00	1.00	1.00	1.00	0.50	aj
30 Data Rate (Mbit/s) (25% rolloff)	320	240	160	120	40	ad
31 Data Rate (dBbit/s)	85.1	83.8	82.0	80.8	76.0	p
32 Required C/(Io+No) (dBHz)	110.1	103.8	98.0	94.8	83.8	ae=ab+p
33						
34 Maximum allowed losses (dB)	148.1	154.3	160.1	163.3	174.3	q=o-ae
35						
36 **Link Parameters**						
37 Frequency (GHz)	28.0					
38 Wavelength (m)	0.011					
39 Ground Distance (km)	10.0					
40 Platform Height (km)	17.0					
41						
42 LOS Distance (km)	19.72					
43 FSPL (dB)	147.3		r			
44 Misc Atmospheric Losses (dB)	0.7		s			
45						
46 Clear air losses (dB)	147.9		t=r+s			
47						
48 Received margin clear air (dB)	0.1	6.4	12.1	15.4	26.3	u=q-t
49 Minimum required transmit power clear air (dBm)	29.9	23.6	17.9	14.6	3.7	
50						
51 Rain Attenuation Zone K 99.90% (dB)	10.0					w
52 Rain Attenuation Zone K 99.00% (dB)	2.6					x
53						
54 Received margin 99.90% (dB)	-9.9	-3.7	2.1	5.4	16.3	z=u-w
55 Received margin 99.00% (dB)	-2.5	3.7	9.5	12.8	23.7	aa=u-x
56						
57 Minimum required transmit power 99.90% (dBm)	39.9	33.7	27.9	24.6	13.7	
58 Minimum required transmit power 99.00% (dBm)	32.5	26.3	20.5	17.2	6.3	

Table 3.9 Example link budget for the backhaul link at 48 GHz

Backhaul Link at 48GHz

1 Transmitter (HAP)

2 Power per backhaul carrier (dBm)	30.0	a				
3 Antenna beamwidth (degrees)	5.0					
4 Antenna efficiency	0.95					
5 Antenna gain (dBi)	32.6	b				
6 Antenna feed loss (dB)	1.0	c				
7 HAP EIRP (dBm)	61.6	d=a+b-c				
8						

9 Receiver (Ground Station)

10 The Boltzmann Constant (dBJ/K)	-228.6	
11 Noise Temperature (K)	300.0	
12 Thermal noise density (dBm/Hz)	-173.8	e
13 Receiver noise figure (dB)	5.0	f
14 Receiver noise density (dBm/Hz)	-168.8	g=e+f
15 Receiver interference noise density (dBm/Hz)	-168.8	j = g
16 Total effective noise density (dBm/Hz)	-165.8	k= 10*log(10^(g/10)+10^(j/10))
17		
18 Antenna beamwidth (degrees)	5.0	
19 Antenna efficiency	0.95	
20 Antenna gain (dBi)	32.6	l
21 Cable loss at ground station	2.0	m
22 Maximum C/(Io+No) (dBHz)	258.1	o=d-k+l-m
23		

	256QAM	64QAM	16QAM	8AMPM	QPSK	
24 Modulation Scheme	256QAM	64QAM	16QAM	8AMPM	QPSK	
25 Required Eb/No (BER 10-6)	25	20	16	14	7.8	ab
26 Bit/symbol	8	6	4	3	2	ac
27						
28 Bandwidth (MHz)	50.0	50.0	50.0	50.0	50.0	
29 Code Rate	1.00	1.00	1.00	1.00	0.50	aj
30 Data Rate (Mbit/s) (25% rolloff)	320	240	160	120	40	ad
31 Data Rate (dBbit/s)	85.1	83.8	82.0	80.8	76.0	p
32 Required C/(Io+No) (dBHz)	110.1	103.8	98.0	94.8	83.8	ae=ab+p
33						
34 Maximum allowed losses (dB)	148.1	154.3	160.1	163.3	174.3	q=o-ae
35						

36 Link Parameters

37 Frequency (GHz)	48.0	
38 Wavelength (m)	0.006	
39 Ground Distance (km)	10.0	
40 Platform Height (km)	17.0	
41		
42 LOS Distance (km)	19.72	
43 FSPL (dB)	152.0	r
44 Misc Atmospheric Losses (dB)	0.7	s
45		
46 Clear air losses (dB)	152.6	t=r+s
47		

	256QAM	64QAM	16QAM	8AMPM	QPSK	
48 Received margin clear air (dB)	-4.6	1.7	7.4	10.7	21.7	u=q-t
49 Minimum required transmit power clear air (dBm)	34.6	28.3	22.6	19.3	8.3	
50						
51 Rain Attenuation Zone K 99.90% (dB)	22.0					w
52 Rain Attenuation Zone K 99.00% (dB)	6.2					x
53						
54 Received margin 99.90% (dB)	-26.6	-20.3	-14.5	-11.3	-0.3	z=u-w
55 Received margin 99.00% (dB)	-10.8	-4.5	1.3	4.5	15.5	aa=u-x
56						
57 Minimum required transmit power 99.90% (dBm)	56.6	50.3	44.5	41.3	30.3	
58 Minimum required transmit power 99.00% (dBm)	40.8	34.5	28.7	25.5	14.5	

The link budgets in Tables 3.8 and 3.9 are summarised in Tables 3.10 and 3.11, respectively, and show that with a bandwidth of 50 MHz, and using variable rate modulation, the links should be able to support data rates of 40–320 Mbit/s. The 28 GHz links should be able to supply 320 Mbit/s during clear air conditions, with the data rate decreasing to 240 Mbit/s for 99.9% of the time, whereas the 48 GHz links can only supply 320 Mbit/s during clear air conditions and 240 Mbit/s for 99% of the time owing to the increased margin required at those frequencies to overcome rain attenuation.

Table 3.10 Downlink data rates per backhaul link for different scenarios using the 28 GHz band

Scenario (28 GHz)	Data rate per link (Mbit/s)			
	99.99%	99.90%	99.00%	Clear air
Backhaul 5° HAP antenna, 1° ground antenna at 10 km GD, 50 MHz BW	40	240	320	320
Backhaul with diversity 5° HAP antenna, 1° ground antenna at 10 km GD, 50 MHz BW	320	320	No extra benefit	

Table 3.11 Downlink data rates per backhaul link for different scenarios using the 48 GHz band

Scenario (48 GHz)	Data rate per link (Mbit/s)			
	99.99%	99.90%	99.00%	Clear air
Backhaul 5° HAP antenna, 1° ground antenna at 10 km GD, 50 MHz BW			240	320
Backhaul with diversity 5°HAP antenna, 1° ground antenna at 10 km GD, 50 MHz BW	120	240	No extra benefit	

Using the spatial diversity configuration discussed earlier by pairing two backhaul stations situated say 10 km apart it is possible to achieve an availability to 99.99%. This is achieved because it is possible to significantly reduce the rain margin, because there is a low probability that there will be heavy rain at *both* ground stations at once [11]. No extra benefit is seen through ground station pairing at 99% and clear availabilities, because they are already transferring the maximum amount of traffic, assuming a 256 QAM modulation scheme and the 50 MHz BW. The use of spatial diversity allows the backhaul link to be extremely reliable, albeit by reducing the capacity by 50% compared with independent stations. However, such stations need only be paired during rainfall conditions. This will also mean that the backhaul link is likely to be the most available part of the system, exceeding that of the user links and the platform station-keeping.

Backhaul Link Requirements

We identified earlier on that it may be necessary to adopt a different configuration depending on the payload power, weight, and volume available. Option 1 will be most reliant on backhaul capacity, neglecting broadcast traffic all user traffic must be channelled up and down the backhaul links. For Option 2 it will be possible to reduce the backhaul requirements by allowing local traffic. Such traffic can be switched or routed on the HAP without being brought to ground. Option 3 will additionally allow traffic to be cached on board further reducing the requirements.

The previous tabulated results can be used to determine the number of cells that can be served from a single backhaul link and these figures are illustrated in Tables 3.12 and 3.13. Expressing the requirements in this way allows the total number of cells to be varied depending on the platform constraints.

Table 3.12 Number of cells served by an individual backhaul link for the 28 GHz band

Scenario (28 GHz links)		Number of cells/backhaul link			
		99.99%	99.90%	99.00%	Clear air
Highly portable/vehicle terminal, 60° ground antenna (with conventional backhaul)				160	160
Fixed antenna at edge of coverage (30 km GD), 8° ground antenna (with conventional backhaul)				106	32
Fixed antenna at sub-platform point, 28° ground antenna (with conventional backhaul)				106	106
Low data rate (12.5 MHz BW) 2° steerable antenna at 30 km GD	Conventional backhaul	8.0	5.3		5.3
	Backhaul with diversity	5.3	No extra benefit		
High data rate (25.0 MHz BW) 2° steerable antenna at 30 km GD	Conventional backhaul	12.0	4.0		2.7
	Backhaul with diversity	8.0	No extra benefit		

These figures indicate that the number of backhaul links required could be large, especially if the clear air rates are desired. Caching and broadcasting can reduce the number of backhaul links. We assume that video-on-demand and Internet traffic on the users links will contain a large proportion of either cached or broadcast traffic, thereby substantially reducing the backhaul requirements. Assuming that Internet/video-on-demand traffic on the user downlink makes up f_i of the total user traffic (T_u).

Table 3.13 Number of cells served by an individual backhaul link for the 48 GHz band

Scenario (48 GHz links)	Number of cells/backhaul link			
	99.99%	99.90%	99.00%	Clear air
Highly portable/vehicle terminal, 60° ground antenna (with conventional backhaul)				
Fixed antenna at edge of coverage (30 km GD), 8° ground antenna (with conventional backhaul)			240	320
Fixed antenna at sub-platform point, 28° ground antenna (with conventional backhaul)			240	320
Low data rate (12.5 MHz BW) 2° steerable antenna at 30 km GD	Conventional backhaul link		8.0	5.3
	Backhaul link with diversity	No extra benefit		
High data rate (25.0MHz BW) 2° steerable antenna at 30km GD	Conventional backhaul link		12.0	4.0
	Backhaul link with diversity	No extra benefit		

Such real time traffic can be restricted to the low availability, with the high availability rates being made available for other real time services, such as telephony and videoconferencing.

We can assume that a fraction c_i of the Internet/video-on-demand traffic on the user downlink originates from a cache (with the cache being filled during periods of quiet activity) then the total backhaul requirement can be calculated using:

$$T_B = [(1-f_i) + f_i(1-c_i)]T_u \qquad (3.9)$$

Using the example traffic loadings, i.e. Internet/video-on-demand traffic makes up 80% of the traffic (i.e. $f_i = 0.8$) and 80% originates from a cache (with the cache being filled during periods of quiet activity), 36% of user traffic needs to be transferred over the backhaul links during period of demand.

$$T_B = 0.36T_u \qquad (3.10)$$

Table 3.14 Number of cells served by an individual backhaul link with broadcast/caching for the 28 GHz band

Scenario (28 GHz links)	Number of cells/backhaul link	
	99.00%	Clear air
Low data rate (12.5 MHz BW) 2° steerable antenna at 30 km GD Conventional backhaul	14.7	14.7
High data rate (25.0 MHz BW) 2° steerable antenna at 30 km GD Conventional backhaul	11.1	7.5

Table 3.15 Number of cells served by an individual backhaul link with broadcast/caching for the 48 GHz band

Scenario (48 GHz links)	Number of cells/backhaul link	
	99.00%	Clear air
Low data rate (12.5 MHz BW) 2° steerable antenna at 30 km GD Conventional backhaul	8.0	5.3
High data rate (25.0 MHz BW) 2° steerable antenna at 30 km GD Conventional backhaul	12.0	4.0

Tables 3.14 and 3.15 illustrate the adjusted number of cells per backhaul link during 99.00% and clear air availabilities and they indicate that the number of backhaul links required for a 121 cell system would be between 16 and 30, assuming the full available data rate was to be transferred during clear air conditions. The number of backhaul links required at higher availabilities decreases because the user data rates decrease faster than the data rates on the backhaul links. If we assume that we can provide two backhaul uplinks per ground station, the number of ground stations required ranges from 8 to 15, which will be feasible.

Delivery of the Backhaul Links

Section 3.2 examines the backhaul antenna configurations and channel reuse issues in detail and this significantly affects how the backhaul links can be delivered. We assume (unless explicitly stated) that the backhaul traffic will be carried on a separate 50 MHz carrier from the user traffic. Three approaches are possible for the HAP backhaul links:

- *Separate narrow beamwidth antennas.* Using separate antennas has the advantage that they can have narrow beams (less than a cell) and can be steered to ensure that they point at the ground stations. It was shown in [11] that the optimum distance to site backhaul ground stations is 5–10 km radius of the sub-platform point (SPP). However, due to frequency reuse considerations it is only possible to site a few stations at this distance. It may therefore be most appropriate to configure the backhaul links in two rings. The inner ring should be used to maintain the high availability traffic in the presence of rain and the outer ring for the lower availability traffic. Therefore the outer ring would not need to overcome rain outages, so it would not matter that any rain margin was large. This approach also does not require the use of transmit/receive diplexors, which reduces complexity. The major drawback with this approach is that continuous coverage could not be maintained, unless the antennas on the HAP tracked the ground stations and vice versa.

- *Use the user spot beam antennas.* This is the most likely configuration as space on the HAP in many situations will be very restricted. Again only a limited number of backhaul ground stations can be placed around 5–10 km from the SPP. However, because the spot-beams compensate for the increase in path loss through higher gain the optimum distance from the SPP to site the ground stations from a spatial diversity perspective is much less critical – the rain margin required varies only very slightly beyond 5 km [11]. If spatial diversity is used it is therefore much less critical where the backhaul ground stations are sited. However, if it is required that all stations operate (i.e. without spatial diversity) then stations far from the SPP will be most affected. Therefore to maximise flexibility it will be best to assume that stations are again placed in rings. From a frequency reuse perspective this is also beneficial because it is possible to ensure that the main lobe beams do not overlap. A significant advantage of this approach is that as the array provides continuous coverage over the coverage area it is possible that the backhaul traffic can be handed-off to an alternative beam in the same way as the user traffic. A major problem with this configuration is that diplexors must be used, if two distinct carriers are to be transmitted/received simultaneously. These radio frequency devices are difficult to make, introduce losses, and take up additional space.

- *Use both sets of antennas.* The fact that the traffic will be highly asymmetric on both user and backhaul links means that it may be possible to carry downlink backhaul traffic (on a user carrier) along with the user downlink traffic, using the user spot beam antenna array, and vice versa. This will eliminate the need for diplexors to be used because the user antennas are only carrying a single carrier. It is unlikely that there will be sufficient capacity using this approach alone, so additional capacity could be provided by the separate antennas. The tracking requirements could be much less restrictive, because continuous transmission could be provided by the user antenna array and handoffs, whilst beams from the separate antennas are redirected and switched.

References

1. G. M. Djuknic, J. Freidenfelds and Y. Okunev, *Establishing Wireless Communications Services via High-Altitude Aeronautical Platforms: A Concept Whose Time Has Come?*, IEEE Commun. Mag., September 1997, pp. 128–135.
2. D. Grace, M. Mohorčič, M. H. Capstick, M. Bobbio Pallavicini and M. Fitch, *Integrating Users into the Wider Broadband Network via High Altitude Platforms*, IEEE Wireless Commun., October 2005, Vol. **12**, No. 5, pp. 98–105.
3. ETSI TR 101 984 V1.1.1 (2002-11), Satellite Earth Stations and Systems. (SES); Broadband Satellite Multimedia; Services and Architectures.
4. M. Capstick and J. Thornton, *Antenna and mm-Wave System Requirements*, HeliNet Document: HE-026-T5-UNY-RP-01, 15 December 2000.
5. Q. Xu, J. Thornton, *Report on Steerable Antenna Architectures and Critical RF Circuits Performance*, CAPANINA Doc Ref: CAP-D24-WP32-UOY-PUB-01, http://www.capanina.org/documents/CAP-D24-WP32-UOY-PUB-01.pdf, 29 September 2006.
6. C.A. Balanis, *Antenna Theory, Analysis and Design*, 2nd Edn, John Wiley & Sons, Ltd, 1997, pp. 812–813.
7. C.A. Balanis, *Antenna Theory, Analysis and Design*, 2nd Edn, John Wiley & Sons, Ltd, 1997, pp. 48–49.
8. J. Thornton, D. Grace, M.H. Capstick and T.C. Tozer, *Optimising an Array of Antennas for Cellular Coverage from a High Altitude Platform*, IEEE Trans. Wireless Commun., May 2003, Vol. **2**, No. 3, pp. 484–492.
9. G. Chen, Y. Liu, D. Grace, P.D. Mitchell, T. Celcer, T. Javornik and M. Mohorčič, *Evaluation of the Performance Enhancements Offered by Multiple Platform Architectures*, CAPANINA Doc Ref: CAP-D23b-WP24-UOY-PUB-01, http://www.capanina.org/documents/CAP-D23b-WP24-UOY-PUB-01.pdf, 18 December 2006.
10. J. Thornton, D.A.J. Pearce, D. Grace, M. Oodo, K. Katzis and T.C. Tozer, *Effect of Antenna Beam Pattern and Layout on Cellular Performance in High Altitude Platform Communications*, Int. J. Wireless Personal Commun., October 2005, Vol. **35**, No. 1–2, pp. 35–51.
11. T. Konefal, C. Spillard and D. Grace, *Site Diversity for High Altitude Platforms: A Method for the Prediction of Joint Site Attenuation Statistics*, IEE Proc. Ant. Prop., April 2002, Vol. **149**, No. 2, pp. 124–128.
12. M. Bobbio Pallavicini *et al.*, *Systems Requirements/Specifications*, HeliNet Doc: HE-RQ-CGS-001-WP4, 28 July 2000.
13. D. Grace *et al.*, *System Design Level Aspects for the Delivery of Broadband Services over HeliNet*, HeliNet Doc Ref: HE-032-T1-UNY-RP-01, 29 March 2001.
14. T. C. Tozer and D. Grace, Broadband Service Delivery from High Altitude Platforms, Communicate 2000 (Online conference), 2–13 Oct 2000.
15. The Sky Website, www.sky.com/portal/site/skycom/skyproducts/skytv/skyplus, December 2009.
16. The TiVo Website, www.tivo.com, December 2009.

4

Applications and Business Modelling

4.1 Introduction

High altitude platforms (HAPs) have the potential to deliver a range of communications services and other applications cost effectively, e.g. broadband, 3G mobile, and disaster relief/event servicing. It has been discussed previously that they are essentially airships or aircraft operating in the stratosphere, and due to their altitude (17–22 km) have the potential to integrate hard-to-reach users located over a wide coverage area into existing broadband networks [1]. A key point is that they can offer a step-change in performance and availability in such circumstances, as they are able to deliver the high capacity similar to that available from terrestrial systems and wide area type coverage similar to that available from satellites [2]. From a techno-economic perspective it is not intended that HAPs will replace these existing technologies, but instead work with these in a complementary and integrated fashion.

The purpose of this chapter is to discuss possible ways that a variety of applications may be delivered from HAPs, and how from a techno-economic perspective some of the possible business models for broadband services and applications should be developed. Central to the analysis is the adoption of a business modelling approach from two perspectives:

- The HAP operator.
- The service providers.

Broadband Communications via High Altitude Platforms   David Grace and Mihael Mohorčič
© 2011 John Wiley & Sons, Ltd

These two perspectives are shown in Figure 4.1.

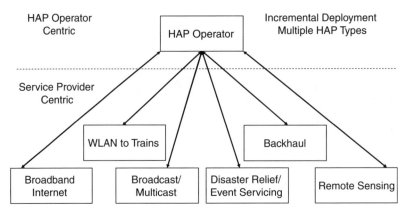

Figure 4.1 HAP operator and service provider centric business model linkages (based on [3])

It is assumed that the HAP operator will be responsible for providing and maintaining the communications payload and connecting this payload into the wider network. The service providers will then be responsible for the ground-user segment of the communications system, and billing the end customer. Capacity can be leased from the HAP operator under appropriate service level agreements (SLAs). A key point is to ensure that each link in the value chain is able to make profits over the appropriate timescale. Figure 4.2 illustrates the technology owner-ship divisions between service provider and HAP operator used here.

This chapter is organised as follows, first the potential applications delivered from HAPs will be discussed. This is followed by the business models and operations are placed in the context of a HAP broadband roadmap. Several service provider centric business models are then described, along with their operating constraints. This is followed by a HAP operator centric business model, which includes a discussion on incremental deployment options and HAP types. The final section looks at a technology and commercial assessment of the business sector.

4.2 Applications and Services

To be able to realise the potential of HAPs it is important that the techno-economic development of this new technology is linked to the development timetable of the

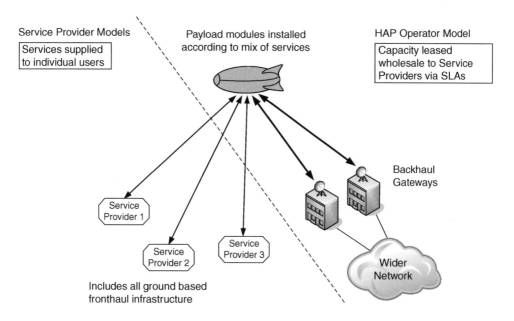

Service Provider Models

Services supplied
to individual users

Payload modules installed
according to mix of services

HAP Operator Model

Capacity leased
wholesale to Service
Providers via SLAs

Backhaul
Gateways

Service
Provider 1

Service
Provider 2

Service
Provider 3

Wider
Network

Includes all ground based
fronthaul infrastructure

Figure 4.2 Technology divisions between service provider and HAP operator (based on [3])

HAPs themselves. This section describes a possible development roadmap for the delivery of broadband from HAPs. A roadmap for other services could be developed in a similar way. The roadmap links the business strategy and market data with the product and technology decisions.

There are several potential markets identified for HAPs including, governments and military private networks for niche applications (e.g. secure VLANs, surveillance, disaster recovery and special events, etc.) and also competitive applications based on broadcast applications (e.g. HDTV, IP Multicast, etc.). In the long term, however, it is envisaged that HAPs technology will play an important part in the global fully integrated, multi-platform telecommunication network of the future. A fundamental point about HAPs broadband is the capacity (up to 121 cells in the medium term), coverage (up to 60 km diameter) and peak data rate (up to 120 Mbps) combination that is unavailable from competing technologies. A potential evolutionary path for HAPs services is illustrated in Figure 4.3.

The ideal HAP in the long term is a vehicle with large payload (>1000 kg, 20 kW of power) that is quasi-stationary – probably an airship. However, this type of platform is also the most ambitious, and possibly even unachievable in the current economic and political climate. In the short and medium terms it will

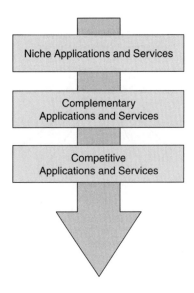

Figure 4.3   HAPs roadmap

be possible to make use of existing or other platforms currently in the development phase. Use of these platforms for less complex missions will help generate confidence, develop the technology, and perhaps more importantly provide revenue streams. A key feature is to address the investment–confidence cycle, as illustrated in Figure 4.4.

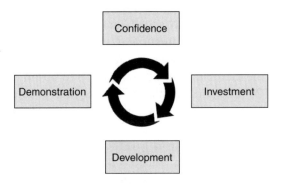

Figure 4.4   Confidence, investment, development and demonstration cycle

Given that the state of the HAP development is a key factor influencing the business models and types of business that can be offered we split HAPs into four categories:

1. Fuel powered manned plane HAP.
2. Fuel powered unmanned plane HAP.
3. Solar powered unmanned plane HAP.
4. Solar powered unmanned airship HAP.

To date only the first type of HAP is available for commercial use, although categories 2 and 3 HAPs have been tested experimentally. The fourth category, the unmanned solar powered airship, still has to be realised. Examples of each type of HAP are given in Figure 4.5 and discussed more fully in Chapter 1 and Chapter 2.

Manned Planes
—e.g. Grob G520T Egrett
(a)

Unmanned Hydrogen Powered Planes
– e.g. Global Observer
(b)

Unmanned Solar Powered Planes
—e.g. NASA/AV Pathfinder-Plus
(c)

Unmanned Solar Powered Airships
—e.g. Lockheed Martin HAA
(d)

Figure 4.5 Examples of the main types of high altitude platforms: (a) manned plane. Reproduced by permission of @ Grob Aircraft AG; (b) unmanned plane (fuel). Reproduced courtesy of AeroViron-ment Inc. www.avinc.com; (c) unmanned plane (solar); Reproduced from NASA - http://www.dfrc .nasa.gov/gallery/photo/index.html; (d) unmanned airship (solar). Reproduced by permission of @ Lockheed Martin

We place at the core of the roadmap and subsequent business models the notion that payloads and payload characteristics should be described in a modular way suitable (ideally) for all platform types, reinforcing commonality of requirements and specifications, and making the technology more accessible to nonspecialists.

We expect that one or more platform modules will be incrementally deployed to serve a common coverage area, with each platform serving one or more payload modules (telecom or other). Such platforms will be networked, with the detailed operations transparent to the end user.

The roadmap links the business strategy and market data with the product and technology decisions by highlighting potential gaps in product and technology plans and by prioritising investments based on drivers such as:

1. Product families (e.g. wireless standards, WiMAX, etc.).
2. HAPs platform manufacturing and supply requirements.
3. Linkage with appropriate bodies and organisation for research, standards, intellectual property, etc.

The speed of this evolution of the technology is ultimately controlled by the platform availability and characteristics. HAPs and lower altitude aerial platforms are available now, but in general, mission life is short. Hence, we divide the roadmap here into the **short** (2–3 years), **medium** (3–6 years) and **long term** (>6 years). Technology aspects of the roadmap are discussed in more detail in Section 5.2.

4.2.1 Short Term

In the **short term,** the first missions will be centred on delivery of broadband services for event servicing and disaster relief, providing high data rate services, where currently none are available. An example mission is given in Table 4.1.

Table 4.1 Example mission possible in the short term and required capabilities

Example mission	Event servicing/disaster relief
Telecom payload	Broadband (e.g. WiMAX) either mm-wave, or ideally below 11 GHz
	Possible applications:
	• High speed Internet
	• Broadcast/multicast
	• WLAN on high speed vehicles
	• 3G
Mission duration	1 day to 2 weeks – availability 95.0 %
	To maintain continuity of service, platforms could be replaced on a regular basis. The replacement interval would be set by platform type
Types of platform	Manned plane HAPs, tethered aerostats (not stratospheric)

4.2.2 Medium Term

Over the **medium term,** as platform capabilities increase, mission life can increase as well. The first longer range missions are likely to be in developing countries, where there is an absence of existing infrastructure, with continuous (or near continuous coverage provided) by platforms that may be changed on a regular basis. Details of an example mission are given in Table 4.2.

Table 4.2 Example mission possible in the medium term and required capabilities

Example mission	Broadband communications for fixed users in developing countries
Telecom payload	Broadband (e.g. WiMAX) either mm-wave, or ideally below 11 GHz, limited number of cells (or additional platforms) to increase capacity
Mission duration	Continuous – availability 95.0 %
	To maintain continuity of service, platforms could be replaced on a regular basis. The replacement interval would be set by platform type
Types of platform	Unmanned planes (solar) operating in fleets, airship HAPs (if available)

4.2.3   Long Term

Over the **longer term** – the 'true' HAPs – the ones of the artist's impressions as illustrated in Figure 4.5, will provide the broadband services to the hard-to-reach regions, the mobile users or fixed users in remote locations. They will initially be used as a complementary technology, but as equipment improves, costs reduce and confidence grows, they could increasingly be seen as a competing technology. An example mission is given in Table 4.3.

Table 4.3 Example mission possible in the long term and required capabilities

Example mission	Mobile and broadband communications for both high speed mobile and fixed users (developed countries, with established infrastructures)
Telecom payload	High capacity broadband (e.g. WiMAX) either mm-wave or ideally below 11 GHz, high number of cells (and/or additional platforms) to increase capacity
	Possible applications: • High speed Internet • Broadcast/Multicast • WLAN on high speed vehicles • 3G
Mission duration	Continuous – availability 99.9 %
	To maintain continuity of service, platforms could be replaced on a regular basis. The replacement interval would be set by platform type.
Types of platform	Airship HAPs
	Unmanned planes (solar) operating in fleets

4.3 Business Model Introduction

Development of suitable business modelling methodologies for broadband communications is of critical importance to the future of the HAPs business sector. Many models that have been developed to date have failed to address the fundamental technical constraints or opportunities of HAPs technology. To date many models assume that a HAP can serve many more customers than is readily practical and/or are based on airship HAPs only. Such models do not take into account the fact that payloads (and capacity) are constrained by the power, weight and volume available and station-keeping must be taken into account. Additionally, it is important that the models also can integrate well with existing value chain-based modelling ensuring rapid integration with existing businesses and infrastructure.

In this section we try to address these fundamental constraints in order to illustrate a possible business modelling strategy that will deliver a profitable business, given the appropriate number of customers. We first describe the operating scenario assumed in the models, we then explain the financial assessment parameters used in both service provider and HAP operator centric models. The detailed discussion and revenue projections of six service provider centric models and HAP operator centric model will be presented in later sections.

4.3.1 Operating Scenario

In this chapter it is assumed that one or more HAPs will provide services to a coverage area of 60 km in diameter. This size of coverage area has been chosen to fit in with potential payload constraints likely to be found on early platforms.

The mm-wave band communications will be used to supply broadband communications fronthaul and backhaul. Bands at 48/47 GHz are licensed for worldwide use [4], with 31/28 GHz permitted for use in 40 countries worldwide [5]. Optical communications could additionally be used to provide interplatform links and additional backhaul to ground, but are not directly included in these models.

To supply an ever increasing requirement for capacity we exploit a multiple platform overlapping coverage area technique, as shown in Figure 4.6, which exploits the directional antenna characteristics of the ground-based antenna to enable frequency reuse from a number of platforms sharing the same frequency band [6] (the technical aspects of this are discussed in Chapter 9). In this way capacity can be incrementally deployed from an increasing number of HAPs. Early HAP deployments need only carry a small payload and later deployed HAPs can carry larger payloads using the latest technology. Each HAP type also has a varying number of cells depending on its capabilities, providing further frequency reuse. Key communications parameters, mentioned earlier, are shown in Table 4.4.

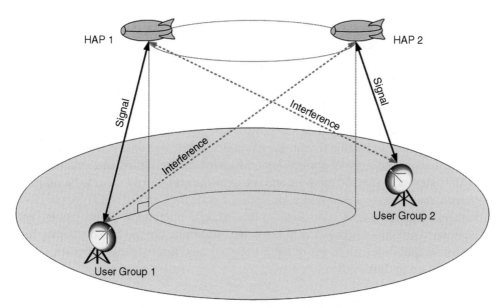

Service area, served by overlapping HAP coverage

Figure 4.6 Multiple HAP with overlapping coverage area incremental deployment model

Table 4.4 Communications operating parameters (based on [3])

	Manned	Unmanned		
	Plane	Plane (fuel)	Plane (solar)	Airship (solar)
Fronthaul				
Availability at min. data rate (%)	99.9			
Max. clear air cell capacity (Mbps)	120			
Number of cells	121	7	19	121
Total fronthaul capacity (Mbps)	14 520	840	22 80	14 520
Cell diameter (km)	6	15		6
Coverage area diameter (km)	60	36	59	60
Redundancy (%)	0		5	20
Backhaul				
Gateway capacity (Mbps)	960			
Capacity per 50 MHz backhaul carrier	320			
No. of carriers per gateway station	3			
Link redundancy (%)	0		5	30

As discussed previously, the second important concept is that of the modular payload. The vehicle is one of the most costly items in the business and it is important it is fully and widely utilised, enabling volume production to create economies of scale, minimising costs. HAPs with a generic multi-module payload bay, with a standard set of operating constraints suitable for a wide range of payload modules, should be developed. This will make the HAP simpler for nonspecialists to use boosting the take-up of the technology. A payload module should be thought of as a fundamental functional unit, for example a backhaul link, a fronthaul cell, a sensor for remote sensing applications. Modules can be mixed and matched on a HAP to support a specific mission, and it will be the norm for services to be supplied to more than one service provider. Here we focus on the four different types of HAP, namely, manned plane, unmanned plane (fuel), unmanned plane (solar) and unmanned airship (solar). Each HAP has different operating constraints and payload capabilities (in terms of the number of modules they can support), which are outlined in Table 4.5.

Table 4.5 HAP operating parameters and costs (based on [3])

	Manned	Unmanned		
	Plane	Plane (fuel)	Plane (solar)	Airship (solar)
Total development cost (€M)	0	50		225
CAPEX per unit (€M)	18	4		30
OPEX per annum continuous operation (€M)	6	1		4
Payload – mass (kg)	1000	250	150	1000
Payload – power (kW)	10	1		10
Operating altitude (km)	20	17		20
Flight duration	12 h	3 days	weeks	years
Staff required	4 pilots + 20 ground crew	1 pilot + 10 ground crew on launch		1 pilot + 20 ground crew on launch
Platform lifetime (years)	10	5		
No. for continuous operation per coverage area	2		1	
Redundancy factor	0.3		0.2	
Fleet/reserve multiplier	0.6		0.2	

Built in redundancy is required for each HAP type and payload and will depend on HAP characteristics and mission length. HAPs with longer flight durations will require higher payload redundancy. HAPs with short flight durations will need to operate in fleets.

4.3.2   Business Model Assumption

Both the service provider and HAP centric models use the same assessment methods. These are:

- **Operating profit** (EBITDA) – Earnings before interest, tax, depreciation, and amortization.
- **Net free cash flow** (FCF) – This takes into account income and outgoings (CAPEX and OPEX) on a yearly basis.
- **Discounted free cash flow** – This is similar to FCF, except that it is discounted (by up to 18 % per annum), which allows different business types to be assessed on a comparable basis in terms of risk and reward.
- **Accumulated discounted FCF** – This is used to calculate the payback period, the time it takes to achieve a net positive return on investment.
- **Internal rate of return** (IRR) – The return on the original CAPEX investment, normalised on an annual basis.
- **Net present value** (NPV) – The accumulated annual discounted FCF of the business after a specified number of years related to the present year.

We divide our HAP business model into two different models, which are the service provider centric model and HAP operator centric model. The service provider centric model deals with service providers who provide their services directly to users. The HAP operator centric model deals with the HAP operator who provides a wholesale capacity basis to service providers.

4.3.2.1 Service Provider Centric Model

We consider the service provider centric models are mainly based on the capacity supplied by the long-term solar power HAP airship with 121 cells per HAP coverage area. However we consider the short-term fuel HAP aircraft to provide services in the event servicing model.

The cost items in the service provider centric models are categorised into common subgroups as shown in Table 4.6.

Each group contains both common and specific cost elements depending on the different types of service. The revenue stream of the service provider centric models mainly comes from the price charged to their customers. We set the price of each service based on the competitive price of the same or similar services in the current

Table 4.6   Operating expense assumptions of service provider centric models

Manpower costs	Order handling
	Repairs
	Training
	Design costs
	Engineering
Equipment costs	Depends on type of service
Other operating costs	Marketing costs (% of revenue)
	Distribution costs (€/sub/year)
	Billing system (€/sub/year)
	Customer care system (€/sub/year)
Network costs	OSS
	Network operation and maintenance
	Software
HAPs capacity costs	HAPs capacity (per Mbps)

EU market. However, the number of customers in each model is not based on any specific market situation but assumed in order to evaluate the feasibility of the business based on a required market demand to ensure the business will be financially viable and the HAP technology can deliver the required capacity. Therefore we take all the constraints on both business and technical sides into account to set up the market demand. Moreover, the objective of these business models is only for demonstration rather than developing the actual business plan. The intention is that these models can be used as a skeleton and developed further for an actual business plan, which requires market information from a targeted market and relevant financial assumptions.

4.3.2.2 HAP Operator Centric Model

The HAP operator centric model deals with the supply of multiple services on a wholesale basis to one or more service providers. The cost side of the HAP operator centric business model is more complicated than the service provider centric business model, since this model contains the cost of the payload, which varies and depends on services provided to service providers. In addition, it also contains the HAP vehicle development cost and the HAP operating cost. However, there are still common costs as discussed in the service provider centric model, such as the manpower cost and the network cost. The costs are shown in Table 4.7.

Table 4.7 Operation expense assumptions of HAP operator centric model

Manpower costs	Order handling
	Repairs
	Training
	Design costs
	Engineering
HAP development costs	HAP vehicle development cost
Vehicle costs per HAP	HAP vehicle cost
	Additional HAP operating cost
Payload costs per HAP	Fronthaul equipment
	Common elements
	Backhaul equipment
Network costs per gateway	Gateway
	OSS
	Network operation and maintenance
	Software

4.4 Service Provider Centric Models

Broadband applications have been selected that have the potential to generate the most revenue from the HAPs deployment [7]. The selection process has taken into account potential competing services and the unique advantages of HAPs architecture – the regional high capacity coverage and unobstructed look angle. Key factors that impact on all models are the costs of the equipment owned by the customer or service provider and the capacity required from the HAP system. The models assume that each service provider pays a leasing charge for capacity (per Mbps), with the price determined by the competitive advantage that HAPs-based capacity delivers to that business – ultimately being based on the amount the market will bear.

4.4.1 Bandwidth Utilisation and Contention Ratio

One of the important factors affecting the service provider centric models is the issue of contention ratio and bandwidth utilisation. For broadcast, streaming and IP Multicast services the bandwidth utilisation is very high which means a cheaper service charge per customer (i.e. the system cost is independent of the number of end users). However, for TCP-based services, such as broadband Internet browsing, e-mail, downloads, etc., the bandwidth utilisation is determined by the contention ratio offered.

The contention ratio is defined as the ratio of the total bit rate that would be required by all users of the system if they connected simultaneously at peak data rate to the maximum achievable system data rate. As an example, if 10 000 users taking a service of 500 kbps peak were supported by capacity of 10 Mbps, the contention ratio would be 500 : 1. A contended service uses packet-mode transmission to share capacity in such a way as to provide both the required peak data rate, and the required average traffic per user. For success it depends on an individual user traffic pattern with a high peak-to-mean ratio, the peak occurring when receiving high-volume content, but for only a short proportion of the time.

The telecoms industry uses two typical contention ratios [7]. One corresponds with the ADSL residential figure of 50 : 1, the other with a typical low-cost satellite competitor at 150 : 1, and thus the higher the contention ratio, the lower the achievable quality of service. In our analysis we have chosen to use 50 : 1 for residential, and 2 : 1–5 : 1 for backhaul depending on the type of service to be backhauled. When developing a real business from HAPs, the operator may wish to use the contention ratio offered to their customers as a market differentiator.

In practice actual contention at peak times as seen by the user (note this is a slightly different definition of contention ratio from that above) anecdotally could be much lower, e.g. 10 : 1 for business users and 15 : 1 for residential users [8,9], but service providers in general like to keep these figures secret.

4.4.2   WLAN to Trains

With this model it is assumed that the service provider installs conventional WLAN access points and an on-board server on trains, which are then connected via mm-wave transceivers on top of the train to the HAP. It is expected that the cost per train will be around €125k CAPEX. The service provider then sells WLAN capacity in a conventional way on the train. HAPs have the potential to deliver significant capacity and throughput improvements compared with alternative means but the price a customer is willing to pay is very sensitive. A similar pricing model to that used on GNER trains in the UK has been adopted in the models [10]. Key drivers for this model are the number of passengers that will use the service on each train, the number of trains that need to be equipped to provide a sensible marketable service, and from a technical viewpoint the availability of usable alternative network connections, e.g. GPRS and satellite to provide a more limited service when out of the HAP coverage area.

4.4.2.1 Financial Model Assumptions

As discussed previously, we divided the cost items into subgroups which are the network cost, the manpower cost, the cost of multiple equipment installed on the train and the other operating costs such as marketing cost and distribution cost. Table 4.8 shows all the cost items for these financial model assumptions.

Table 4.8   The cost items of WLAN on trains

Manpower costs	Order handling
	Repairs
	Training
	Design costs
	Engineering
Equipment costs	Trials/ref. model
	APs
	Wiring and installation
	Box on train
	Cellular modems
	Cache
	Server + antennas
	Software
Other operating costs	Marketing costs (% of revenue)
	Distribution costs (€/sub/year)
	Billing system (€/sub/year)
	Customer care system (€/sub/year)
Network costs per n trains	OSS
	Cellular capacity
	Network capacity
	Software
HAPs capacity costs	HAPs capacity (per Mbps)

For the revenue side, the Wireless LAN-on-train business model gains its revenue only from users using the Internet on trains. By applying the price structure of the train company in the UK, we have four different price levels for different periods of time. In addition we assume a distribution of users on different price levels in the model as shown in Table 4.9.

Table 4.9   Price and customer distribution assumptions of WLAN on trains

Type (min)	Price (€)	Distribution (%)		
		Year 0	Year 4	Year 9
30	5.00	50	25	25
60	8.00	30	30	30
120	12.00	10	30	30
180	15.00	10	15	15

We assume that the initial number of travellers per train is 400 with 2 % growth rate each year and 5 % of these travellers are using the Internet service on the train in the first year and the numbers grow at a rate of 30 % each year.

4.4.2.2 Financial Results

At the end of the projection period, we will have 478 travellers per journey per train with253 travellers being Internet users spending €9.50 on average per person. Revenues are projected to reach €39.7M by the end of the projection period.

The main operating expenses, as shown in Figure 4.7, are the network and train operation costs which are dominant and around 70 % of the total annual OPEX for the first half of the projection period. However, as the number of customers increases towards the end of the projection period, the marketing cost and the HAP capacity cost are dominant instead, accounting for 50 % of the total annual OPEX at the end of the projection period.

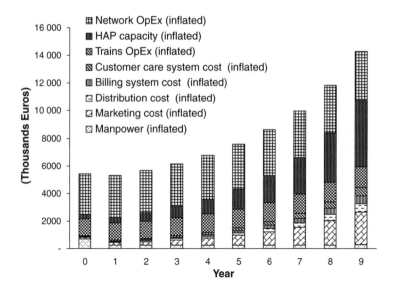

Figure 4.7   Operational expenditures of the WLAN on trains service

The on board train communications equipment is the main CAPEX of this model, which is incurred at the start of the business. The cash flow model can be seen in Figure 4.8. The Wireless LAN on train business has its payback period in 6 years and 9 months with NPV €17.4M and IRR 40 %.

The financial results are summarised and shown in Table 4.10.

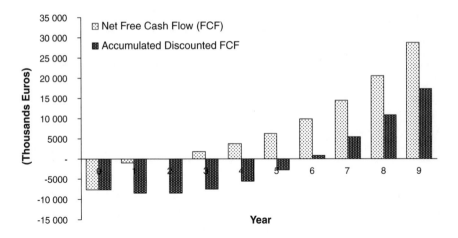

Figure 4.8   Cash flow of the WLAN on trains service

Table 4.10   Financial summary of the WLAN on trains service

	Year 0	Year 4	Year 9
No. of travellers per train	400	433	478
No. of customers per train	20	62	253
Service price (average) (€)	7.60	9.50	9.50
Total service revenue (k€)	2508	9719	39 658
EBITDA (k€)	−2915	3709	28 769
EBITDA margin (%)	−107	38	73
CAPEX (k€)	4977		
NPV (k€)	17 379		

4.4.2.3 Sensitivity Analysis

The Change of assumptions can affect the net present value (NPV) of the business. We have considered varying the key parameters in our financial and market assumptions and examined the change on NPV as shown in Table 4.11. An important aspect to note is that the NPV is not particularly sensitive to the price of the HAP, which means that it would be possible for the HAP operator to raise the cost of this component if required. The costs of HAP development is a cause of significant uncertainty as will be discussed in more detail in Section 4.5.

4.4.3   Backhaul for Terrestrial Base Stations/Access Points

The purpose of this application is to provide a backhaul communications link from ground-based terrestrial 3G base stations and WLAN access points on a wholesale

Table 4.11   Sensitivity analysis results of WLAN on trains

Changing of assumptions	Effect on NPV
Growth rate of number of travellers decreases by 10 % (from 2 to 1.8 %)	NPV decreases by € 0.7M (4 %)
Growth rate of percentage of WLAN users decreases by 10 % (from 30 to 27 %)	NPV decreases by € 5.7M (33 %)
WLAN price decreases by 10 %	NPV decreases by € 5.2M (30 %)
HAP capacity cost increases by 10 %	NPV decreases by € 0.2M (1 %)

basis, with the customer being a retail service provider. It should not be confused with HAP backhaul that connects the HAP with the wider network. Conventionally over 60 % [11] of links are provided by point-to-point microwave links, but this becomes very costly in remote locations, with an extra factor that there are ever-increasing capacity demands by users. We assume a basic €520 CAPEX [7] cost for each ground terminal (although this aspect is not too price sensitive for this application). Given that the customer will pay fully for the cost of this equipment, this application is much more lucrative than models where the CPE has to be subsidised.

4.4.3.1 Financial Model Assumptions

Our cost assumptions for the backhaul service are grouped into five areas as shown in Table 4.12.

Table 4.12   The cost items of the backhaul service

Manpower costs	Order handling
	Repairs
	Training
	Design costs
	Engineering
Equipment costs	Customer premise equipment (CPE)
Other operating costs	Marketing costs (% of revenue)
	Distribution costs (€/sub/year)
	Billing system (€/sub/year)
	Customer care system (€/sub/year)
Network costs	OSS
	Network operation and maintenance
	Software
HAPs capacity costs	HAPs capacity (per Mbps)

The backhaul link business can provide a backhaul link service to WiFi/WiMAX networks or mobile telephone networks such as GSM and 3G. This business gains its income from such services. Based on the price structure of terrestrial lease line circuit in the US market, we set the cost of one backhaul link for WiFi/WiMAX at €2000 per month with 120 Mbps peak data rate and 5 : 1 contention ratio, and GSM/3G mobile backhaul link at €1000 with 20 Mbps peak data rate and 2:1 contention ratio. In addition we put the cost of CPE (€520) to customers who normally are the operators proving WiFi/WiMAX or mobile telephone services. In addition we assume the price of both WiFi/WiMAX and GSM/3G mobile backhaul link increase at a rate of 2 % each year.

We assume that we initially get two WiFi/WiMAX backhaul link customers for each HAP cell and growth at a rate of 25 % to eventually reach the maximum available capacity. In addition, we initially get one 3G backhaul link customer for each cell of HAP with a 25 % growth rate each year until it reaches its maximum available capacity.

4.4.3.2 Financial Results

Assuming that 70 % of HAP capacity is allocated to the WiFi/WiMAX backhaul link service, then the remaining 30 % is for the 3G backhaul link service. At the end of the projection period, we will achieve the maximum four WiFi/WiMAX-backhaul links and four 3G-backhaul links in each cell. Revenues are projected to reach €16.7M by the end of the projection period.

The operating expenses are shown in Figure 4.9 and the main operating expenses are shown to be the network operation costs, which are dominant and stable around 60 % of the total annual OPEX until the end of the projection period. The marketing cost and the HAP capacity cost play a more important role in the later years, as the number of customers grows and peaks at the end of the projection period.

Since the CPE cost is charged to the end customers, only the network CAPEX features in the backhaul link business model. The backhaul link business has a payback period in its first year with NPV of €44.2M, as can be seen from Figure 4.10, which shows the cash flow of the backhaul service.

The financial results are summarised and shown in Table 4.13

4.4.3.3 Sensitivity Analysis

We have varied the key parameters in the financial and market assumptions on the backhaul link business model. These are the percentage of cell utilisation per HAP coverage area, the growth rate in the number of customers per cell, the price of the backhaul link and the HAP capacity cost. The absolute and percentage change in NPV is shown in Table 4.14. It is shown that the system is most sensitive to the growth rate in the number of customers.

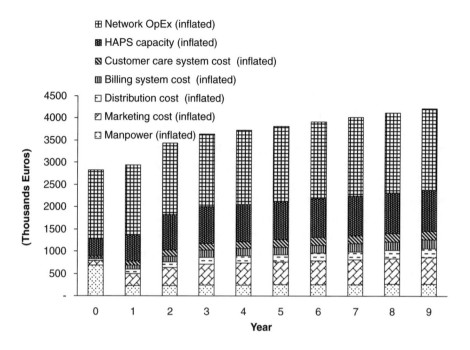

Figure 4.9   Operational expenditures of the backhaul service

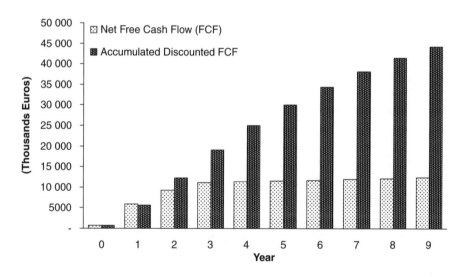

Figure 4.10   Cash flow of the backhaul service

Table 4.13 Financial summary of the backhaul service

	Year 0	Year 4	Year 9
No. of WiFi/WiMAX-backhaul link per HAP coverage	194	387	387
No. of 3G-backhaul link per HAP coverage	194	387	387
WiFi/WiMAX backhaul service price (€)	2000	2164	2390
3G backhaul service price (€)	1000	1082	1195
Total service revenue (k€)	3492	15 080	16 650
EBITDA (k€)	668	11 354	12 423
EBITDA margin (%)	19	75	75
CAPEX (k€)	15		
NPV (k€)	44 203		

Table 4.14 Sensitivity analysis results of backhaul service

Changing of assumptions	Effect on NPV
Percentage of cell utilisation per HAP coverage area decreases by 10 % (from 80 to 72 %)	NPV decreases by € 5.5M (12 %)
Growth rate of no. of users per cell decreases by 100 % (from 30 to 0 %)	NPV decreases by € 23.2M (52 %)
Backhaul link price decreases by 10 %	NPV decreases by € 5.9M (13 %)
HAP capacity cost increases by 10 %	NPV decreases by € 0.4M (1 %)

4.4.4 Broadband Internet

This model considers the delivery of a broadband Internet-based service directly to the end user via a dish antenna, with capacity delivered of 10 Mbps per customer with a contention ratio of 50 : 1. We assume that the CPE terminal will still cost €520 [7]. Getting a foothold into the already established broadband Internet market is potentially very difficult owing to the ever-increasing speeds offered by xDSL technologies. In order to compete, the market segment where the user is too remote from the exchange could be exploited (the more rural user). Alternatively, the mm-wave CPE needs to be subsidised heavily, say to down €100, which eats significantly into the profits, given that income per user is only likely to be €300 per annum. Use of lower frequency bands may provide a lower cost CPE, but that is beyond the scope of this book. Another significant techno-economic factor is that the overall capacity requirements are likely to be far in excess of that needed for the other applications. Even though a HAP system has significantly more capacity than say a satellite system, the other applications are much more lucrative in terms of the revenue they can raise per Mbyte. A HAP system should not rely on this model alone.

4.4.4.1 Financial Model Assumptions

Table 4.15 shows all the cost items for this financial model.

Table 4.15   The cost items of the broadband Internet service

Manpower costs	Order handling
	Repairs
	Training
	Design costs
	Engineering
Equipment costs	Customer premise equipment (CPE)
Other operating costs	Marketing costs (% of revenue)
	Distribution costs (€/sub/year)
	Billing system (€/sub/year)
	Customer care system (€/sub/year)
Network costs	OSS
	Network operation and maintenance
	Software
HAPs capacity costs	HAPs capacity (per Mbps)

The broadband Internet business provides Internet access service direct to users. The prices charged to customers are based on the current broadband Internet access service in the UK market, which is €40 per month and constant for the whole projection period. We assume the number of users will initially start at 200 users with a growth rate of 15 % each year.

4.4.4.2 Financial Results

The operating expenses for the broadband Internet business model are shown in Figure 4.11. The HAP capacity cost is the main operating expense of the broadband Internet business, which is 41 % of the total annual OPEX in the first year and increases to 54 % at the end of the projection. The network operation cost is still significant whereas the marketing cost and the HAP capacity cost play a more important role in later years, as the number of customers increases and reaches a peak at the end of the projection period.

The cash flow for the broadband Internet model is shown in Figure 4.12. The broadband Internet business gains its investment back by the end of year 7 with NPV of €5.5M and 29 % IRR.

The financial results are summarised and shown in Table 4.16.

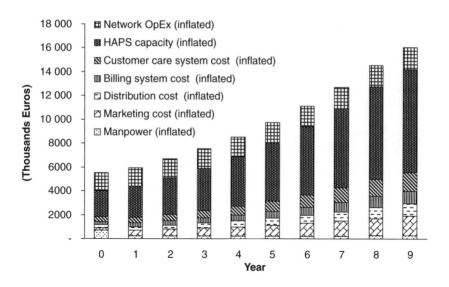

Figure 4.11   Operational expenditures of the broadband Internet service

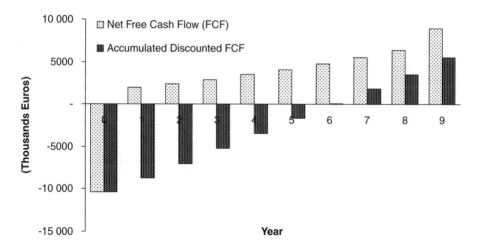

Figure 4.12   Cash flow of broadband Internet

4.4.4.3 Sensitivity Analysis

Some key parameters in the financial and market assumptions of the broadband Internet business model are varied to understand the effect on the NPV. These are the initial number of customers per cell, the growth rate of the number of customers per

Table 4.16 Financial summary of the broadband Internet service

	Year 0	Year 4	Year 9
No. of users per cell	180	315	600
Broadband Internet service price (€)	40	40	40
Total service revenue (k€)	4312	14 111	27 552
EBITDA (k€)	−1222	5580	11 477
EBITDA margin (%)	−28	40	42
CAPEX (k€)	9185	2089	2547
NPV (k€)	5543		

cell, the price of the broadband Internet service and the HAP capacity cost. The results are shown in Table 4.17, where it is seen that the model is particularly sensitive to a decrease in the price charged to the customer for broadband Internet. Unlike the wireless LAN on trains service this business model is also sensitive to changes in the cost of HAP capacity. This illustrates the benefits of having a multi-strand (e.g. triple play) business, so that the costs of providing a basic broadband Internet package can be supplemented by other more highly valued services.

Table 4.17 Sensitivity analysis results of the broadband Internet service

Changing of assumptions	Effect on NPV
Initial no. of customers per cell decreases by 10 % (from 200 to 180)	NPV decreases by € 2.6M (47 %)
Growth rate of no. of user per cell decreases by 10 % (from 15 to 13.5 %)	NPV decreases by € 1.2M (22 %)
Broadband Internet price decreases by 10 %	NPV decreases by € 7.1M (128 %)
HAP capacity cost increases by 10 %	NPV decreases by € 4.7M (84 %)

4.4.5 Broadcast/Multicast

This model deals with providing the next generation of broadcast/multicast services to the end user. Satellite competitors are obvious, but the degree of local control (particularly important in many non-European countries) and the ability to package these services along with others offered on the HAP, make it particularly attractive. The application has the potential to be highly profitable. It is assumed that the CPE cost will be €520, which will require a subsidy, but this is more than compensated for by the significantly reduced capacity requirements (and costs), due to the same information being supplied to multiple users.

4.4.5.1 Financial Model Assumptions

The cost items are divided into six groups which are the network costs, the manpower costs, the equipment costs, the content costs, the HAP capacity costs and the other operating costs such as marketing cost and distribution cost. Table 4.18 shows all the cost items for this financial model.

Table 4.18   The cost items of the Broadcast/Multicast service

Manpower costs	Order handling
	Repairs
	Training
	Design costs
	Engineering
Equipment costs	Customer premise equipment (CPE)
Other operating costs	Marketing costs (% of revenue)
	Distribution costs (€/sub/year)
	Billing system (€/sub/year)
	Customer care system (€/sub/year)
Content costs	Domestic TV content
Network costs	OSS
	Network operation and maintenance
	Software
HAPs capacity costs	HAPs capacity (per Mbps)

The Broadcast/Multicast service provides broadcasting content such as TV programmes. We set the prices based on the SKY subscribed TV programmes, in the UK market, which is €25 per month and constant for the whole projection period. We assume the number of users initially starts at 30 000 users with a growth rate of 10 % each year.

4.4.5.2 Financial Results

The operating expenses for the Broadcast/Multicast business are shown in Figure 4.13. The TV content cost and the network operation cost are the main operating expenses of the Broadcast/Multicast business, which is more than 50 % of the total annual OPEX. The marketing cost adds a significant amount to the annual OPEX in the later years, as the number of customers grows. On the other hand, the HAP capacity cost for this type of business is constant, as we discussed previously.

The Broadcast/Multicast business has its payback period within 6 years and 6 months with NPV of €7.2M and 30 % IRR, as shown in Figure 4.14.

The financial results are summarised and shown in Table 4.19.

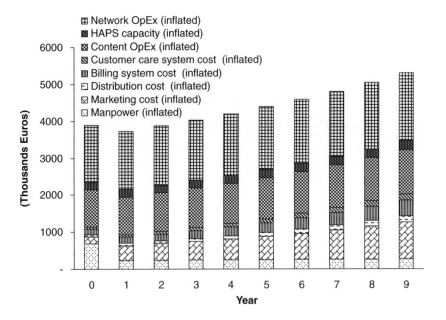

Figure 4.13   Operational expenditures of the Broadcast/Multicast service

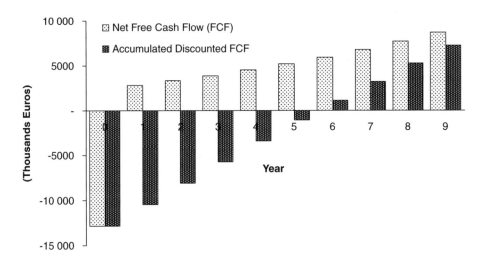

Figure 4.14   Cash flow of the Broadcast/Multicast service

Table 4.19   Financial summary of the Broadcast/Multicast service

	Year 0	Year 4	Year 9
No. of subscribers	30 000	43 923	70 739
Broadcast/Multicast service price (€)	25	25	25
Total service revenue (k€)	3713	10 377	16 712
EBITDA (k€)	−169	6179	11 410
EBITDA margin (%)	−5	60	68
CAPEX (k€)	12 645	1681	11 410
NPV (k€)	7228		

4.4.5.3 Sensitivity Analysis

Again, key parameters in the financial and market assumptions are varied. In the case of the Broadcast/Multicast business model these are: the initial number of customers per cell, the growth rate of number of customers per cell, the price of content and TV broadcast service and the HAP capacity cost. The effect on the NPV of this business is shown in Table 4.20. Just as in the case of broadband Internet, the model is most sensitive to the price charged to the customer for the service.

Table 4.20   Sensitivity analysis results of the Broadcast/Multicast service

Changing of assumptions	Effect on NPV
Initial no. of customers per cell decrease by 10 % (from 30 000 to 27 000)	NPV decreases by € 2.5M (3 5%)
Growth rate of no. of user per cell decrease by 10 % (from 10 to 9 %)	NPV decreases by € 0.5M (7 %)
Content price increases by 10 %	NPV decreases by € 0.6M (8 %)
TV Broadcast price decreases by 10 %	NPV decreases by € 4.7M (65 %)
HAP capacity cost increases by 10 %	NPV decreases by € 0.1M (1 %)

4.4.6   Event Servicing and Disaster Relief

An event servicing/disaster relief type application operates in a quite different way to the longer deployments discussed previously. In both circumstances the requirement will be to supply services and applications for a short period of time, and will need to be deployed and removed rapidly when no longer needed.

There is also likely to be potentially more vertical integration of the business, with the HAP operator potentially supplying the end service in some circumstances. We side step this issue here by focussing on the HAP operator centric model for the business, leaving the specific service provider situation aside. It is important that service providers are aware of the HAP operator centric side of the business, especially the costs and charging model, so that they develop their own (potentially entrepreneurial service-provider centric models). There are two potential competitors:

- **Helicopters/planes** – anecdotal evidence has suggested these were used to provide emergency 3G coverage in New Orleans after Hurricane Katrina.
- **Satellites** – These could provide backhaul services for some types of users but it is certain that HAP-based technologies will ultimately deliver high-link capacities and be more portable.

The most important aspect here is that it is assumed that the HAP will need to be deployed and removed rapidly, and that the mission will have limited duration.

It is assumed that multiple modular payloads will be required to serve several types of application, with the aim to replace or augment infrastructure. Two modes of operation are foreseen:

- provide backhaul services (to temporary infrastructure);
- services direct to the user or operator.

The working assumption for the scenario is assumed not to be capacity intensive, with applications provided from a single overlay cell. Such an assumption may not fit every scenario, particularly in the case of natural disasters but it would be possible to adapt the model for multiple cell applications.

4.4.6.1 Possible Operating Scenarios

To gain a better understanding of the business, it is helpful to consider real examples where such short-term HAP communications deployments would have been beneficial, in addition to the Hurricane Katrina example already given:

- Events
  - World Cup
  - Olympic Games – anecdotal evidence shows that HAP start-up companies are already looking at providing HAP coverage for the London 2012 Games.
- After disasters
  - Earthquakes. – e.g. in Pakistan in October 2005; Java, Indonesia in May 2006; Port-au-Prince, Haiti in January 2010.
  - Tsunami. – e.g. Pacific Rim, December 2004.

○ Hurricanes. – e.g. Hurricane Katrina, New Orleans, USA, 2005.
○ Volcanic eruptions. – e.g. Guatemala 2010, Java, Indonesia 2006 and Mt St Helens, USA, 1980.
○ Floods. – e.g. Bangladesh, October 2004.
• After civil unrest.

In all examples it is anticipated that there will also be a remote sensing application linked to the communications application. A key drawback of all disaster management, and to a lesser extent the event scenarios, is the ability to have high resolution continuous real time images. In most situations today, it is just not feasible; LEO satellites can provide reasonable resolution images but the dwell time is short and the repeat frequency is long. Conventional planes can provide an alternative but their coverage area is relatively small, and their costs are high, because it is often a dedicated remote sensing mission – costs are not shared with other applications. Continuous monitoring is still difficult.

4.4.6.2 Target Applications

It is anticipated that the HAP operator will provide a core set of applications. These would be provided in a modular format, which could be mixed and matched depending on the customer's requirements. Ultimately it will be the end customer who will drive the requirements and these could include:

• displaced citizens or general public;
• NGOs, emergency services;
• Military.

Currently it is assumed that these customers would require a core set of wireless broadband technologies, including WiFi, 3G, WiMAX, DVB-H, and HDTV, plus in some cases a specialist communications package. These technologies would either be delivered direct to the user from the HAP or alternatively the HAP can be used as backhaul for temporary ground-based infrastructure.

In developing the business model we have developed a generic set of applications to be provided via the business. The relative importance of each will depend on the scenario and end customer. The applications are:

1. Temporary broadband Internet service.
2. Increased capacity backhaul service.
3. Temporary backhaul service.
4. Special event Broadcast/Multicast service.
5. Emergency coordination service.

6. Real time remote sensing service.
7. 3G service (cell fill-in or extra capacity).

Items 1–3 would be provided via a broadband payload and appropriate capacity would need to be supplied by the HAP. Charging would be based on a per-Mbit-per-second-per-day basis, with each application being charged, based on the importance to the end customer and market alternatives.

Item 4 may require an alternative payload depending on the nature of the broadcast/service (e.g. HDTV). Again it could be priced on a per-Mbit-per-second-per-day basis, and is likely to be much higher than the broadband applications, given its revenue generation possibilities. Clearly, with convergence, the division between broadcast and broadband technologies is becoming blurred, so a separate charging model may become increasingly difficult to maintain as technologies change.

Item 5 will require a specialist communications payload, and will currently depend on the emergency service and region. Standardisation is underway to provide a common standard for such equipment [12], but we have assumed for the moment that this is bespoke. Again a capacity charge (per-Mbit-per-second-per-day) is assumed.

Item 6 will also require a specialist payload consisting of appropriate camera technologies linked to the wider network by the broadband backhaul. It is assumed that a lease price per day will cover the costs of providing the service, which would include communications network costs.

Item 7, the 3G service, will require a payload of comparable size with the broadband payload. Again a capacity charge per-Mbit-per-second-per-day will be charged. Depending on the type of HAP used and the amount of capacity required, it may not be possible to fly this payload along with all the others, so as to provide a service direct to the user. In such circumstances it may prove sensible to use item 2 and provide backhaul to infrastructure on the ground.

4.4.6.3 Financial Model Assumptions

The revenue/charging assumption has been modified compared with the long endurance HAP-centric model to align it better to the costs involved in providing the HAP and the service. It is also linked to the perceived willingness to pay of the service provider customer, or other stakeholder (e.g. NGOs or governments, insurance companies), for the given scenario. Three revenue streams are assumed:

• **Rental of the HAP** (in terms of a daily charge) – this aims to cover the daily OPEX costs of the HAP deployment.

- **Retainer fee** (in the form of an annual charge) for operators who require an emergency response – this, in effect, can be seen as a form of insurance policy for the service provider. It aims to cover the costs for HAPs that are 'idle'. Clearly, it is assumed that there will be a fleet of HAPs managed by the HAP operator and it will be possible to use them for a mix of events and disaster management, and minimise the idle time for HAPs, while maintaining the service level agreements (SLAs), thus maximising profits.
- **Communications and other applications** – These would be provided to the service provider. It is anticipated that the end costs could be comparable with existing long term communications provision, i.e. the premium for the specialist nature is charged for the first two items. The charging structure for these applications will be discussed in more detail later.

We set the assumption of key market demand parameters such as number of customers, number of event days per year, retainer charge per year and the HAP rental which are applied to our event servicing/disaster relief business model as shown in Table 4.21.

Table 4.21   Key market demand assumptions in event servicing/disaster relief model

	Year 0	Year 4	Year 9
No. of customers	3	4	4
Retainer charge per year (k€)	90	90	90
No. of event days per year	60	100	150
HAP lease price per day (k€)	90	90	90
No. of ES/DR per year	20	25	30
Average no. of days per event	3	4	5
Total no. of event days in service	60	100	150

In addition we also assume the traffic distribution of different types of services that are shown in detail in Table 4.22.

Table 4.22   Traffic distribution assumptions of event servicing/disaster relief model

	Year 1	Year 4	Year 9
Temporary broadband Internet service (capacity) (%)	40	40	40
Increased capacity backhaul service (capacity) (%)	27	27	27
Temporary backhaul service (capacity) (%)	33	33	33
Temporary 3G mobile service (capacity) (%)	0	0	0
Special event Broadcast service (Mbps)	120	120	120
Emergency coordination service (Mbps)	120	120	120
Remote sensing service	1	1	1

On the other hand, the operational expenditure assumption is similar to the long term HAP operator centric model that contains the costs as shown in Table 4.23.

Table 4.23   The cost items of the event servicing/disaster relief model

Manpower costs	Order handling
	Repairs
	Training
	Design costs
	Engineering
HAP development costs	HAP vehicle development cost
Vehicle costs per HAP	Fixed HAP vehicle cost
	Variable HAP vehicle cost per event day
Fronthaul payload costs per HAP	Broadband base station
	Broadcast/Multicast processor
	Event servicing special payload
	Remote sensor equipment
	3G mobile base station
	Fronthaul payload software
Common payload costs per HAP	Antenna farm/RF element
	Wiring and installation
Backhaul payload costs per HAP	Data aggregation unit
	Backhaul base station
	Antenna farm/RF element
	Backhaul payload software
Network costs per gateway	Gateway
	OSS
	Network operation and maintenance
	Software

4.4.6.4 Financial Results

The HAP vehicle operating cost is the major expense of the event servicing/disaster relief model, which is around 90 %, as is shown in Figure 4.15. However, this is because we include all support costs to operate the HAP vehicle, such as manpower to operate the HAP vehicle, maintenance, fuel and all the HAP vehicle consumables costs into this HAP vehicle operating cost. The remaining operating expenditures belong to the manpower cost, the HAP payload operating cost and the network operating cost, which are less significant to this model and only around 10 % of the total operating expense.

Figure 4.16 shows the cash flow for the event servicing/disaster relief business. It has its payback period in 7 years and 5 months with NPV of €22.5M and 28 % IRR.

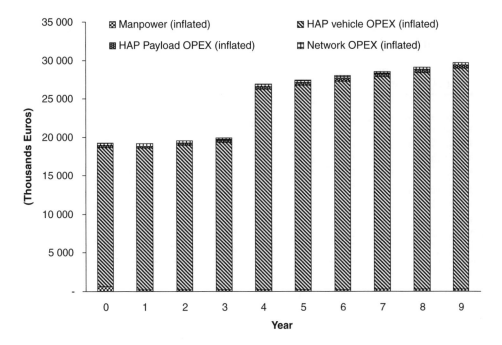

Figure 4.15   Operational expenditures of the event servicing/disaster relief model

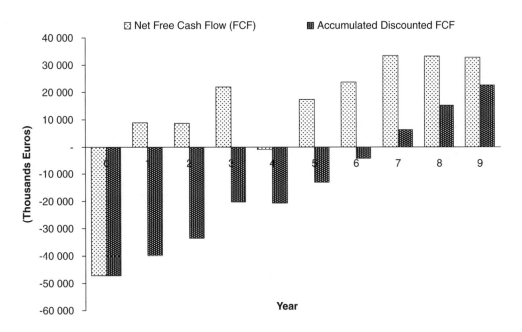

Figure 4.16   Cash flow of the event servicing/disaster relief model

The financial results of the event servicing/disaster relief model are summarised and shown in Table 4.24.

Table 4.24 Financial summary of the event servicing/disaster relief model

	Year 0	Year 4	Year 9
No. of customers	3	4	4
Total no. of event days in service	60	100	150
Retainer charge per year (k€)	90	90	90
HAP lease price per day (k€)	90	90	90
Total service revenue (k€)	29 433	48 965	73 268
EBITDA (k€)	−10 162	22 048	43 583
EBITDA margin (%)	−35	45	59
CAPEX (k€)	55 945	18 605	—
NPV (k€)	22 470		

4.4.6.5 Sensitivity Analysis

We have varied some key parameters in the financial and market assumptions of the event servicing/disaster relief business model, which are the initial number of customers, growth rate of the number of users per cell, service price and the HAP capacity cost. The effect on the NPV of this business is shown in Table 4.25.

Table 4.25 Sensitivity analysis of event servicing/disaster relief model

Changing of assumptions	Effect on NPV
No. of events/disasters per year decreases by 10 %	NPV decreases by € 17.2M (77 %)
Retainer charge per year decreases by 10 %	NPV decreases by € 0.1M (0.4 %)
HAP lease price per day decreases by 10 %	NPV decreases by € 0.3M (1 %)
HAP capacity price per Mbps decreases by 10 %	NPV decreases by € 1.7M (8 %)
HAP vehicle operating cost increases by 10 %	NPV decreases by € 8.1M (36 %)
HAP vehicle cost increases by 10 %	NPV decreases by € 5.4M (24 %)

4.4.7 Third Generation (3G) Mobile Telephone

In this section we examine the possible benefits of providing a long-term 3G mobile service from a HAP, from a service provider centric perspective. We assume that sufficient space is made available on the HAP in order to guarantee sufficient capacity for the service provider. We have assumed that the power, weight and volume requirements of a 3G cell are comparable with a broadband cell, so it is important for HAP operators to choose their payload mix appropriately. For example, in the case of an airship HAP capable of supporting 121 cells, this could be used to support for example 61 cells of broadband payload and 60 3G cells. The HAP operator would have to decide whether it was sensible to sacrifice a reduction in broadband capacity of 50 % in order to provide a 3G service or vice versa.

From a technical perspective the 3G service can be provided in a similar way to a broadband service, namely that cells are formed using a directional antenna on the HAP, with the roll-off of the antenna at the cell edge being a key design parameter [13]. We have assumed that each 3G cell can support up to 80 simultaneous speech users [13], with capacity being ultimately limited by the self-interference of multiple users sharing the same frequency in a cell. We assume a mobile activity factor of 1.25 mErlang, meaning that each cell has a maximum number of supportable customers of 550, assuming a 1 % blocking requirement. This means that the maximum number of customers that can be supported by each HAP is approximately 70 000, considerably lower than other forecasts.

4.4.7.1 Financial Model Assumptions

Table 4.26 shows all the cost items for this financial model.

Table 4.26 The cost items of the 3G mobile phone service

Manpower costs	Order handling
	Repairs
	Training
	Design costs
	Engineering
Other operating costs	Marketing costs (% of revenue)
	Distribution costs (€/sub/year)
	Billing system (€/sub/year)
	Customer care system (€/sub/year)
Network costs	OSS
	Network operation and maintenance
	Software
HAPs capacity costs	HAPs capacity (per Mbps)

The 3G mobile phone business provides the third generation mobile telephone service direct to users. We base the mobile service prices in this model on the average of the current mobile service price in the UK market, which is €18 per month and constant for the whole projection period. We assume the number of users initially starts at 150 users per cell and is uniformly distributed over the cells of the HAP coverage area, with a growth rate of 15 % each year.

4.4.7.2 Financial Results

Figure 4.17 shows that the network operation cost is the major operating expense at the beginning of the mobile phone business, which is around 50 % of the total annual OPEX and decreases in later years as the cost related to the number of customers increases due to an increased number of customers. Even though the costs related to the number of customers, such as the marketing cost, the customer care, the distribution and the billing system costs play a more significant role in the overall operating expense, the network operation cost is still the major operating cost at around 28 % of the total annual OPEX at the end of the projection period.

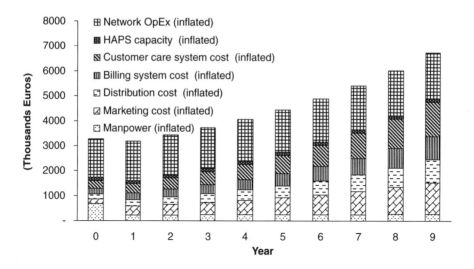

Figure 4.17 Operational expenditures of 3G mobile phone

The mobile phone business has its payback period in 4 years and 8 months with NPV of €4.3M and 52 % IRR, as can be seen from the cash flow model in Figure 4.18.

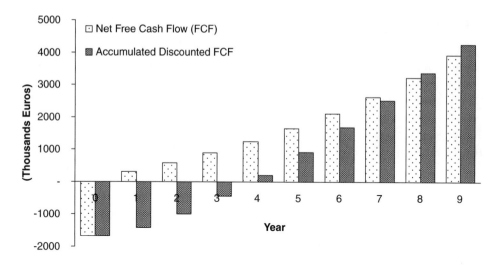

Figure 4.18 Cash flow of 3G mobile phone

The financial results are summarised and shown in Table 4.27.

Table 4.27 Financial summary of 3G mobile phone

	Year 0	Year 4	Year 9
No. of users per cell	194	387	387
3G mobile phone service price (€)	18	18	18
Total service revenue (k€)	1617	5304	10 641
EBITDA (k€)	−1653	1244	3911
EBITDA margin (%)	−102	23	37
CAPEX (k€)	15		
NPV (k€)	4255		

4.4.7.3 Sensitivity Analysis

We have varied some key parameters in the financial and market assumptions of the mobile phone business model. These are the initial number of customers, growth rate of the number of users per cell, the mobile service price and the HAP capacity cost. The effect on the NPV of this business is shown in Table 4.28. The business is shown to be most sensitive to a decrease in prices charged to the customer.

Table 4.28 Sensitivity analysis of 3G mobile phone

Changing of assumptions	Effect on NPV
Initial no. of customers per cell decreases by 10 % (from 150 to 135)	NPV decreases by € 1.6M (38 %)
Growth rate of no. of users per cell decreases by 10 % (from 15 to 13.5 %)	NPV decreases by € 0.8M (19 %)
Service price decreases by 10 %	NPV decreases by € 2.3M (54 %)
HAP capacity cost increases by 10 %	NPV decreases by € 0.1M (2 %)

4.5   HAP Operator Centric Model

The HAP operator centric model is significantly more complex than the individual service provider centric models, since it must deal with leasing of capacity to multiple types of business, as well as encompassing multiple HAP types and capacity growth.

Here we present results from a generic model that we have developed in software that is capable of flexibly coping with multiple types of HAP with the aid of pull-down menus. These fill in the unique characteristics of different HAPs (Table 4.5) and capacity requirements (Table 4.4). We have made some basic assumptions about the mix of types of service the model is supporting and the capacity growth requirements. Additional HAPs are brought into service as and when demanded by the need for capacity. The number of HAPs required and the deployment interval depends on the level of capacity and capacity growth. The model currently assumes that a single HAP type is deployed, but it would be possible to build in HAPs with improved capabilities for later years.

A few points to note are that if the capacity requirements are relatively modest, the unmanned plane (solar) is the best solution. If the requirements for capacity are significant, then airship-based HAPs look a better through life solution for the long term. The cost per Mbps is cheaper, but the initial outlay is much greater. They are also perhaps the most futuristic of all HAP types. The manned-plane-based HAP looks the most costly solution (but does have the advantage that it could be deployed today). The main disadvantage is the significant outlay for the OPEX – in terms of people to maintain and fly the HAP and the associated fuel costs. In the case of unmanned HAPs, it is very difficult to estimate the OPEX accurately. In all situations a pilot is required to at least oversee even autonomous flight operations, and some military unmanned aerial vehicles (UAVs) are flown with the support of several tens of people. Fuel costs will be largely eliminated in the case of the solar powered vehicles, and likely to be much lower in the case of the hydrogen powered unmanned aircraft, mainly due to the lower weight of the vehicle. In all cases we have amortised the HAP development costs over the first 50 vehicles. It should be noted that the development cost (and all other HAP costs) are by no means certain, given that there is virtually no concrete data on which to base the estimates. Big unknowns for example include the costs of airworthiness certification, and obtaining the necessary radio spectrum licenses.

4.5.1 Financial Model Assumptions

The HAP operator centric model contains different cost items than the service provider centric model, which are the vehicle cost and payload cost. However this model still has the other cost items as discussed in the service centric model such as the network cost and the manpower cost. Table 4.29 shows all the cost items for this financial model.

Table 4.29   The cost items of the HAP operator centric model

Manpower costs	Order handling
	Repairs
	Training
	Design costs
	Engineering
HAP development costs	HAP vehicle development cost
Vehicle costs per HAP	HAP vehicle cost
	Additional HAP operating cost
Fronthaul payload costs per HAP	Broadband base station
	Broadcast/Multicast processor
	Event servicing special payload
	Remote sensor equipment
	3G mobile base station
	Fronthaul payload software
Common payload costs per HAP	Antenna farm/RF element
	Wiring and installation
Backhaul payload costs per HAP	Data aggregation unit
	Backhaul base station
	Antenna farm/RF element
	Backhaul payload software
Network costs per gateway	Gateway
	OSS
	Network operation and maintenance
	Software

The services provided by the HAP operator are various and can be mixed. Therefore this affects both the fronthaul and backhaul HAP payload. The HAP operator business gains its revenue mainly from providing capacity to its customers. There are different pricing schemes for different types of services. However most of the pricing schemes are based on 'per Mbps'. We assume that the required HAP capacity from users initially starts at 80 % of HAP capacity and then grows by 20 % each year, with additional HAPs being deployed to serve the capacity as required.

In addition, different types of HAP vehicle affect the capital expenditure and hence the financial results of the HAP operator centric model. Therefore we consider different types of HAP vehicles and compare the financial results with each other.

4.5.2   Unmanned Solar Powered Airship

The unmanned solar powered airship is a long endurance type HAP that we use as our base type of HAP in all the service provider centric models discussed previously. The working assumptions for this type of HAP are shown in Table 4.30.

Table 4.30   Unmanned airship (solar) parameter assumptions

Total development cost ($€$)	225M
CAPEX per HAP ($€$)	30M
CAPEX per cell ($€$)	298k
OPEX per annum continuous operation per HAP ($€$)	4M
OPEX per annum continuous operation per cell ($€$)	40k

4.5.2.1 Financial Results

The cost of the HAP and its payload are the major operating expense of this business, which is above 60 % of the total annual OPEX as we can see from Figure 4.19.

The cash flow for this HAP operator centric model is shown in Figure 4.20. The HAP operator business has its payback in 5 years 1 month with NPV of $€$65M and 41 % IRR.

The financial results are summarised and shown in Table 4.31.

4.5.2.2 Sensitivity Analysis

We have varied some key parameters in the financial and the market assumptions of the HAP operator centric model. These are the initial number of customers, network cost, the HAP vehicle cost, the price of all services and the HAP payload cost. The effect on the NPV of this business is shown in Table 4.32. The model is relatively insensitive to cost or revenue changes, with the service price changes being the most critical aspect. In the case of the service provider centric models it was shown that

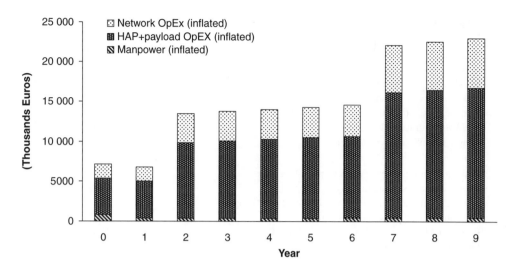

Figure 4.19   Operational expenditures of HAP operator centric model (unmanned airship)

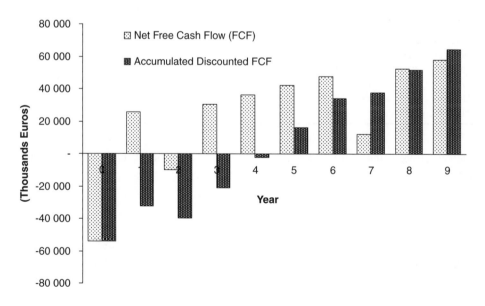

Figure 4.20   Cash flow of HAP operator centric model (unmanned airship)

Table 4.31 Financial summary of HAP operator centric model (unmanned airship)

	Year 0	Year 4	Year 9
Broadband Internet service (Mbps)	4646	9293	15 101
WLAN on train service (Mbps)	2323	4646	7550
Backhaul service (Mbps)	4066	8131	13 213
Event servicing/disaster relief service (Mbps)	581	1162	1888
Broadcast/Multicast service (Mbps)	120	120	120
Remote sensing service (module)	1	1	1
3G mobile service (Mbps)	—	—	—
Broadband Internet service price (per Mbps) (k€)	0.5	0.5	0.5
WLAN on train service price (per Mbps) (k€)	0.6	0.6	0.6
Event servicing/disaster relief service price (per Mbps) (k€)	4	4	4
Backhaul service price (per Mbps) (k€)	2	2	2
Broadcast/Multicast service price (per Mbps) (k€)	100	180	280
Remote sensing service price (per Module) (k€)	50	54	59
3G mobile service price (per Mbps) (k€)	60	60	60
Total service revenue (k€)	26 222	50 483	80 837
EBITDA (k€)	19 102	36 464	57 838
EBITDA margin (%)	73	72	72
CAPEX(k€)	73 208	—	—
NPV (k€)	64 762		

Table 4.32 Sensitivity analysis results of HAP operator centric model (unmanned airship)

Changing of assumptions	Effect on NPV
Initial no. of customers decreases by 10 %	NPV decreases by €10M (15 %)
Network cost increases by 10 %	NPV decreases by €2M (3 %)
HAP vehicle cost increases by 10 %	NPV decreases by €13M (20 %)
All service price decreases by 10 %	NPV decreases by €24M (37 %)
Payload cost increases by 10 %	NPV decreases by €2M (3 %)

they were very insensitive to capacity charges (with the exception of broadband Internet) so there would be scope to increase service prices.

4.5.3 Fuel Powered Manned Plane

The manned plane is an existing HAP type that can be used for long-term deployment providing a fleet of HAPs is available and the HAPs are rotated periodically. It is also used for the special event/disaster relief business model discussed earlier in this chapter, but all figures in this section relate to long-term deployments. Our assumptions for this type of HAP are shown in Table 4.33.

Table 4.33 Manned plane parameter assumptions

Total development cost	—
CAPEX per HAP (€)	18M
CAPEX per cell (€)	238k
OPEX per annum continuous operation per HAP (€)	6M
OPEX per annum continuous operation per cell (€)	79k

4.5.3.1 Financial Results

Figure 4.21 shows the operation expenditures for the manned plane HAP operator centric model. The cost of the HAP vehicle and its payload are still the major operating expense of this business, which is above 60 % of the total annual OPEX.

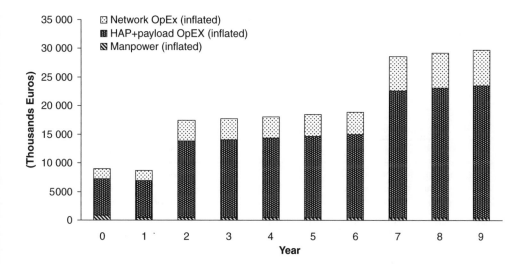

Figure 4.21 Operational expenditures of HAP operator centric model (manned plane)

The HAP operator centric business model with manned plane has its payback within 4 years and 4 months with NPV of €73M and 60 % IRR, as shown in Figure 4.22. This manned plane type offers a significant advantage for a HAP operator centric business model because there are no significant further development costs required and it has a lower CAPEX than the unmanned airship. However this type of HAP results in a higher OPEX in the model. Therefore the payback can be obtained within a shorter period with only a little higher NPV.

The financial results are summarised and shown in Table 4.34.

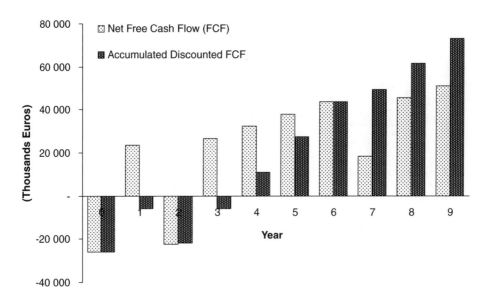

Figure 4.22   Cash flow of HAP operator centric model (manned plane)

Table 4.34   Financial summary of HAP operator centric model (manned plane)

	Year 0	Year 4	Year 9
Broadband Internet service (Mbps)	4646	9293	15 101
WLAN on train service (Mbps)	2323	4646	7550
Backhaul Service (Mbps)	4066	8131	13 213
Event servicing/disaster relief service (Mbps)	581	1162	1888
Broadcast/Multicast service (Mbps)	120	120	120
Remote Sensing Service (module)	1	1	1
3G mobile service (Mbps)	—	—	—
Broadband Internet service price (per Mbps) (k€)	0.5	0.5	0.5
WLAN on train service price (per Mbps) (k€)	0.6	0.6	0.6
Event servicing/disaster relief service price (per Mbps) (k€)	4	4	4
Backhaul service price (per Mbps) (k€)	2	2	2
Broadcast/Multicast service price (per Mbps) (k€)	100	180	280
Remote sensing service price (per Module) (k€)	50	54	59
3G mobile service price (per Mbps) (k€)	60	60	60
Total service revenue (k€)	26 222	50 483	80 837
EBITDA (k€)	19 102	36 464	57 838
EBITDA margin (%)	66	64	63
CAPEX (k€)	43 154	—	—
NPV (k€)	73 095		

4.5.3.2 Sensitivity Analysis

Key parameters in the financial and the market assumptions of the HAP operator centric model are varied once again. In the case of this model, these are the initial number of customers, the network cost, the HAP vehicle cost, the price of all services and the HAP payload cost. The effect on the NPV of this business is shown in Table 4.35.

Table 4.35   Sensitivity analysis results of HAP operator centric model (manned plane)

Changing of assumptions	Effect on NPV
Initial no. of customers decreases by 10 %	NPV decreases by € 9.5M (13 %)
Network cost increases by 10 %	NPV decreases by € 1.9M (3 %)
HAP vehicle cost increases by 10 %	NPV decreases by € 13M (18 %)
All services price decreases by 10 %	NPV decreases by € 24.1M (33 %)
Payload cost increases by 10 %	NPV decreases by € 1.7M (2 %)

4.5.4   Fuel Powered Unmanned Plane

The fuel powered unmanned plane is a developing HAP type that still requires some additional development before it can be deployed commercially. One example is the AV Global Observer, for which a scaled size prototype is already available as shown in Figure 4.5. Our assumptions for this type of HAP are shown in Table 4.36.

Table 4.36   Unmanned plane (fuel) parameter assumptions

Total development cost (€)	50M
CAPEX per HAP (€)	4M
CAPEX per cell (€)	914k
OPEX per annum continuous operation per HAP (€)	1M
OPEX per annum continuous operation per cell (€)	229k

4.5.4.1 Financial Results

Figure 4.23 shows the operation expenditures for the unmanned plane (fuel) HAP operator centric model. The cost of the HAP vehicle and its payload are still the major operating expense of this business, which is above 50 % of the total annual OPEX in the starting year and increases to 82 % at the end of the projection period.

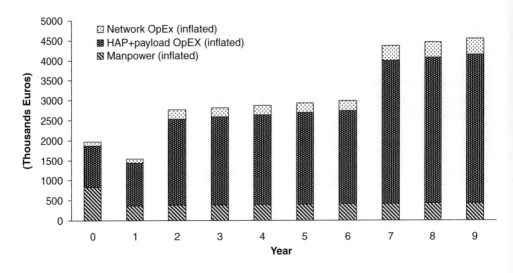

Figure 4.23 Operational expenditures of HAP operator centric model [unmanned plane (fuel)]

The HAP operator centric business model with unmanned plane (fuel) has its payback within its first year with NPV of €83M, as shown in Figure 4.24. This vehicle type still requires development costs to be able to deliver the service, and additionally its capacity is much more limited than both the airship and manned plane variants. In the case of capacity intensive broadband services, each plane can provide only 7 cells each of 15 km diameter, and hence the cost per cell (the key driver for the business models) is significantly higher than the other types of HAP. Additionally, the OPEX per annum is also much higher than the other HAP types. Therefore this can result in no profitability when it is necessary to supply capacity intensive services. The bulk of the revenues in this example are derived from noncapacity intensive services such as Broadcast/Multicast and remote sensing, which will require a single payload each, but can supply services over a wide coverage area. These will result in high profitability with high NPV.

The financial results are summarised and shown in Table 4.37.

4.5.4.2 Sensitivity Analysis

Key parameters in the financial and the market assumptions of the HAP operator centric model are varied once again. In the case of this model these are the initial number of customers, the network cost, the HAP vehicle cost, the price of all services and the HAP payload cost. The effect on the NPV of this business is shown in Table 4.38. Clearly the model is insensitive to key changes with the exception of the revenues generated from the services.

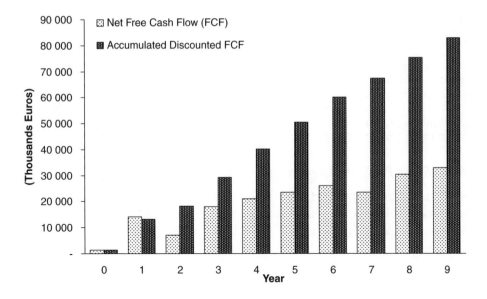

Figure 4.24 Cash flow of HAP operator centric model [unmanned plane (fuel)]

Table 4.37 Financial summary of HAP operator centric model [unmanned plane (fuel)]

	Year 0	Year 4	Year 9
Broadband Internet service (Mbps)	269	538	874
WLAN on train service (Mbps)	134	269	437
Backhaul service (Mbps)	235	470	764
Event servicing/disaster relief service (Mbps)	34	67	109
Broadcast/Multicast service (Mbps)	120	120	120
Remote sensing service (module)	1	10	20
3G mobile service (Mbps)	—	—	—
Broadband Internet service price (per Mbps) (k€)	0.5	0.5	0.5
WLAN on train service price (per Mbps) (k€)	0.6	0.6	0.6
Event servicing/disaster relief service price (per Mbps) (k€)	4	4	4
Backhaul service price (per Mbps) (k€)	2	2	2
Broadcast/Multicast service price (per Mbps) (k€)	100	180	280
Remote sensing service price (per Module) (k€)	50	54	59
3G mobile service price (per Mbps) (k€)	60	60	60
Total service revenue (k€)	12 870	23 780	37 444
EBITDA (k€)	10 895	20 900	32 892
EBITDA margin (%)	85	88	88
CAPEX (k€)	9609	—	—
NPV (k€)	82 770		

Table 4.38   Sensitivity analysis results of HAP operator centric model [unmanned plane (fuel)]

Changing of assumptions	Effect on NPV
Initial no. of customers decrease by 10 %	NPV decreases by € 0.2M (0.2 %)
Network cost increases by 10 %	NPV decreases by € 0.1M (0.1 %)
HAP vehicle cost increases by 10 %	NPV decreases by € 2.5M (3 %)
All services price decreases by 10 %	NPV decreases by € 11.4M (14 %)
Payload cost increases by 10 %	NPV decreases by € 0.2M (0.2 %)

4.5.5   Solar Powered Unmanned Plane

The solar powered unmanned plane is another HAP that is still in the development phase, and it will be some years before these HAPs can be deployed. Examples of solar powered plane HAPs include the Pathfinder Plus shown in Figure 4.5. Our assumptions for this type of HAP are shown in Table 4.39.

Table 4.39   Unmanned plane (solar) parameter assumptions

Total development cost (€)	50M
CAPEX per HAP (€)	4M
CAPEX per cell (€)	253k
OPEX per annum continuous operation per HAP (€)	1M
OPEX per annum continuous operation per cell (€)	63k

4.5.5.1 Financial Results

Figure 4.25 shows the operation expenditures for the solar powered unmanned plane HAP operator centric model. The cost of the HAP vehicle and its payload are still the major operating expense of this business, which is above 50 % of the total annual OPEX.

The HAP operator centric business model with solar powered unmanned plane also has its payback within its first year with NPV of €95M, as shown in Figure 4.26. It is important to stress though that this type of HAP still requires development cost and time, just as with the fuel powered unmanned plane in order to be used to provide services. Moreover again this type of HAP also requires a selective services and coverage area in order to make a profitable business scenario. This is because the cost per cell is higher than the unmanned airship or manned plane cases, but may be significantly cheaper than the unmanned fuel planes being proposed. However, these cost figures should be treated as indicative, and it may be possible to reduce the cost per cell if this was a significant design driver.

The financial results are summarised and shown in Table 4.40.

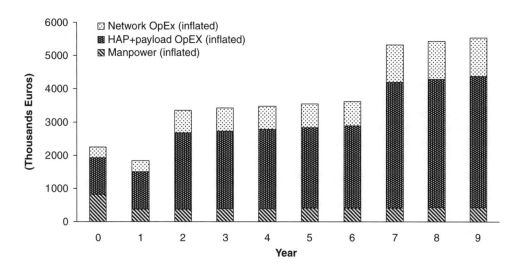

Figure 4.25   Operational expenditures of HAP operator centric model [unmanned plane (solar)]

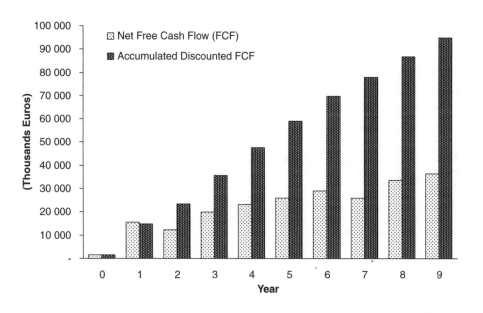

Figure 4.26   Cash flow of HAP operator centric model [unmanned plane (solar)]

4.5.5.2 Sensitivity Analysis

Key parameters in the financial and the market assumptions of the HAP operator centric model are varied once again. In the case of this model these are the initial

Table 4.40 Financial summary of HAP operator centric model [unmanned plane (solar)]

	Year 0	Year 4	Year 9
Broadband Internet service (Mbps)	730	1,459	2,371
WLAN on train service (Mbps)	365	730	1,186
Backhaul service (Mbps)	638	1,277	2,075
Event servicing/disaster relief service (Mbps)	91	182	296
Broadcast/Multicast service (Mbps)	120	120	120
Remote sensing service (module)	1	10	20
3G mobile service (Mbps)	—	—	—
Broadband Internet service price (per Mbps) (k€)	0.5	0.5	0.5
WLAN on train service price (per Mbps) (k€)	0.6	0.6	0.6
Event servicing/disaster relief service price (per Mbps) (k€)	4	4	4
Backhaul service price (per Mbps) (k€)	2	2	2
Broadcast/Multicast service price (per Mbps) (k€)	100	180	280
Remote sensing service price (per Module) (k€)	50	54	59
3G mobile service price (per Mbps) (k€)	60	60	60
Total service revenue (k€)	14 275	26 591	42 012
EBITDA (k€)	12 024	23 111	36 467
EBITDA margin (%)	84	87	87
CAPEX (k€)	10 546	—	—
NPV (k€)	94 906		

number of customers, the network cost, the HAP vehicle cost, the price of all services and the HAP payload cost. The effect on the NPV of this business is shown in Table 4.41. Again similar conclusions can be drawn as in the fuel powered unmanned plane case.

Table 4.41 Sensitivity analysis results of HAP operator centric model [unmanned plane (solar)]

Changing of assumptions	Effect on NPV
Initial no. of customers decreases by 10 %	NPV decreases by € 1.4M (1.5 %)
Network cost increases by 10 %	NPV decreases by € 0.4M (0.4 %)
HAP vehicle cost increases by 10 %	NPV decreases by € 2.3M (2.4 %)
All services price decreases by 10 %	NPV decreases by € 13M (14 %)
Payload cost increases by 10 %	NPV decreases by € 0.3M (0.3 %)

4.6 Risk Assessment

4.6.1 Technology Assessment

This section describes a number of technical risks for HAPs operators that require careful investigation. Two major technical risks are identified here: the required performance of the flying platform (e.g. station keeping, duration of the flight, antenna tracking, etc.) and lack of OSS support (e.g. integration with ISP networks, customer service, billing, etc.).

4.6.1.1 Platform Reliability and Station Keeping

The major technical risk for HAP operators remains in making a flying platform stable and reliable enough for operational use. The technologies that would enable long-flight duration (e.g. 6 months or more) are still in development. Most of the current platforms use a combination of solar powered electric and diesel engines and are designed for relatively short duration flights. It is crucial that communications projects take these factors into account when developing missions, payloads and operating scenarios. Handoff from one platform to another would enable long term service deployments, with relatively short term HAP missions. A fleet of HAPs are required to achieve this, which is discussed in more detail in Section 4.3.1. Additionally, the models work with constraints outlining opportunities available for short missions and low power applications, while following the technology road-map that is described in more detail in Section 5.2.

The station keeping characteristics of platforms proposed and available varies widely. Antenna tracking mechanisms capable of stabilising the payload gondola also pose a considerable challenge.

4.6.1.2 Propagation Characteristics in the mm-wave Bands

The ITU-R has allocated two frequency bands for HAPs broadband operations, 31/28 GHz in 40 countries worldwide (the Ka band) and 47/48 GHz worldwide (the V-band). These frequencies are planned for use by both terrestrial and satellite networks, as well as by future HAP networks. When planning link availability, engineers will therefore have to take into account not only the effects of the atmosphere on their own link, but also any ways in which radiation from one link may interfere with other links operating at the same frequency, whether on the same network or on those of other operators.

The bands have specific characteristics that make them suitable for certain services and unsuitable for others. These characteristics are:

- Reflections from, and diffraction around, most man-made and natural objects is minimal in the V-band. Thus line-of-sight propagation only must be assumed.
- Links are subject to deep fades (tens of dBs) for typically several minutes duration due to rain and other atmospheric impairments. The link is considered useless during such fades.

- The minimum elevation angle is assumed to be 20°, giving the maximum workable slant path through the atmosphere (e.g. scintillation, etc.).

4.6.1.3 Network Integration Technical Risks

The network integration requirements (i.e. backhaul to terrestrial gateways, end-user CPE equipment, inter-HAP links, etc.) may add to the complexity of the HAP broadband payload. For example, the carrier signals would need to be terminated (demodulated, decoded, decrypted, etc.), switched and routed towards the appropriate spot-beam or backbone link. A HAP platform for fixed and mobile broadband communications will need to use on-board processing (OBP); the subsequent increase in payload complexity could pose a technical risk.

4.6.1.4 Customer Premises Equipment Interoperability

HAP operators want to support all broadband Internet applications (e.g. web browsing, home shopping, on-line banking, etc). Broadcast applications for entertainment (e.g. HDTV) and education (e.g. BTV), VPN for corporate markets, etc., are also desirable.

The customer premises equipment (CPE) will need to support multiple simultaneous applications. Therefore, the Set Top Box (STB) unit is required to perform intelligent operations (e.g. IP routing, IGMP support for IP Multicast, etc.) and support interfaces for multimedia applicants including IP, TV and voice (e.g. USB).

The design of the CPE supporting multi-service broadband applications might pose a technical risk. This is particularly important when considering standard CPE interoperability with the spectrum available or dedicated for HAPs.

It is also recommended that HAP operators establish close technical and commercial ties with the CPE manufacturers at the early stages of the system design for HAP broadband provision.

4.6.1.5 Operational Support Service

The Operational Supporting System (OSS) and its subsystem components are required to support commercial aspects of HAPs service provision such as Customer Service, Network Management, Account Management, System Policy Management, Conditional Access and Security, Billing and Usage Tracking.

The general lack of an OSS in many HAPs projects is a major technical risk to commercial deployments. The OSS is often the most complex and most expensive component of the network and requires a significant amount of time to be established.

4.6.1.6 Regulatory Issues

Although the era of monopolistic practices in most of the countries around the globe is drawing to a close, there remain impediments to the operation of the free market. These include the need to obtain a license to operate HAP-based wireless broadband services, the legal uncertainty on the deployment of wireless radio technologies in public places and lack of clarity on interconnection rules with telecommunications in the local loop.

A more insidious constraint has been the willingness of large incumbent telecommunication corporations – so-called 'big telecoms' – to resort to legal action if dissatisfied; this, coupled with an imbalance in experience and funding of big telecoms relative to the local regulator in developing countries, could ensure that the wishes of big telecoms take precedence over those of either the consumer or private commercial interests (e.g. HAPs).

One of the identified benefits of HAP technology is its 'portability', or ease of deployment. Unlike satellite technology, local governments or military organisations can be offered full control over the physical platform and hence the services provided.

It has already been shown that a clear first step for HAPs will be for short missions and as a complementary technology, possibly used by the 'big telecoms' for fill-in services.

4.6.2 Market Assessment

Before developing a full business plan it will be necessary to determine the market penetration of different potentially competing technologies. This needs to be based on a specific market and geographical region, and so given the general nature of this work we have not conducted this type of competitor analysis. Below we provide a summary of the different competing technologies that could be available in a chosen market.

- ADSL
- Cable
- Fibre in the local loop (FITL)
- VDSL
- Satellite DVB-S/RCS
- Terrestrial DVB-T
- Handheld DVB-H
- 3G mobile technologies (delivered terrestrially)
- WiFi
- WiMAX
- DAB
- LTE (in the future).

References

1. G.M. Djuknic, J. Freidenfelds and Y. Okunev, Establishing Wireless Communications Services via High-Altitude Aeronautical Platforms: A Concept Whose Time Has Come?, IEEE Commun. Mag., September 1997, pp. 128–135.
2. D. Grace, J. Thornton, T. Konefal, C. Spillard and T.C. Tozer, Broadband Communications from High Altitude Platforms - The HeliNet Solution, Proc. Wireless Personal Multimedia Communications (WPMC) Conference 2001, Aalborg, Denmark, September 2001, Vol. 1, pp. 75–80.
3. D. Grace and P. Likitthanasate, *A Business Modelling Approach for Broadband Services from High Altitude Platforms*, Invited paper at International Conference on Telecommunications (ICT'06), Madeira, Portugal, May 2006.
4. Recommendation ITU-R F.1500, *Preferred Characteristics of Systems in the FS Using High Altitude Platforms Operating in the Bands 47.2-47.5 GHz and 47.9-48.2 GHz*, International Telecommunications Union, Geneva, Switzerland, 2000.
5. Footnote 5.537A and 5.543A, Article 5, Chapter II, Radio Regulations, ITU and *Potential Use of the Bands 27.5-28.35 GHz and 31-31.3 GHz by High Altitude Platform Stations (HAPS) in the Fixed Service*, Resolution 145 (WRC-03), Radio Regulations, ITU, 2003.
6. D. Grace, J. Thornton, G. Chen, G. P. White and T. C. Tozer, Improving the System Capacity of Broadband Services Using Multiple High Altitude Platforms, *IEEE Trans. Wireless Commun.*, March 2005, Vol. **4**, No. 2, pp. 700–709.
7. CAPANINA Project Deliverable, *D12, Strategy Document: Delivering Broadband for All from Aerial Platforms Including Commercial and Technical Risk Assessments*, www.capanina.org/documents/CAP-D12-WP12-BT-PUB-01.pdf.
8. DSL oversubscription, www.networkworld.com/newsletters/isp/0522isp1.html.
9. Wholesale DSL, www.adsl-service-providers.net/wholesaledsl.html.
10. GNER Website, www.gner.co.uk.
11. S. Byars, *Using Pseudo-Wires for Mobile Wireless Backhaul over Carrier Ethernet,* Blueprint Metro Ethernet, http://www.convergedigest.com/, 13 February 2006.
12. Project MESA Website, www.projectmesa.org.
13. J. Holis, D. Grace and P. Pechac, *The Optimization of Antenna Power Roll-off for High Altitude Platform WCDMA Systems*, York HAP Week, www.yorkhapweek.org/DOCS/download.php?part=hapcos&day=thursday&time=pm&wg=1&slot=9, October 2006.

5

Future Development of HAPs and HAP-Based Applications

5.1 Trends in Aeronautical Development

Previous chapters of this book have considered in general the different aspects of HAPs and their payload technology. HAPs are currently not readily available for commercial deployments, and here we will discuss some of the key aspects affecting this situation. This will be primarily from a technical perspective, building on the business viewpoint discussed in Chapter 4, by developing technical roadmaps for the short, medium and long term showing how the future roll-out of HAPs could be achieved. Discussion of how to further the development of HAP aeronautical technologies has also recently been the subject of the Specific Support Action USEHAAS, which developed a Strategic Research Agenda for the aeronautical sector [1], and many of their findings are also included here.[1]

One of the biggest drivers affecting the launch of HAP telecommunications systems are the platforms themselves. The aircraft- and airship-based HAPs are at differing stages of development, and the speed of this HAP development is affected by technical, political, economic, as well as manufacturing constraints. The HAP itself is one of the most costly items in the overall system and it is important that it is fully and widely utilised, enabling volume production to create economies of scale, thereby minimising costs. To aid development it is probable that each HAP will be equipped with a generic multi-module payload bay, designed with ideally a standard set of operating constraints suitable for a wide range of payload modules. This

---

[1] One of the authors (David Grace) of this book participated in this SSA.

Broadband Communications via High Altitude Platforms   David Grace and Mihael Mohorčič
© 2011 John Wiley & Sons, Ltd

approach should make HAPs simpler for nonspecialists to use, boosting the take up of the technology. It will also enable volume production, creating economies of scale, while also providing a political impetus to appropriate adjustment of both the aeronautical and telecommunications regulatory framework.

Currently, the aircraft-based unmanned HAPs, e.g. AV's Global Observer, are at a more advanced state of completion as we discussed in Chapter 1, with airship-based platforms, with the exception of Lockheed Martin's HAA, suffering from a lack of funding and political will to take the necessary risks to develop them further. Thus, it is likely that the first missions will be of short duration, delivered on small-payload HAPs powered by conventional fuel. It may also be, as previously discussed, that platforms are deployed on a fleet basis, by maintaining connectivity through handoff from a retiring HAP to a newly deployed HAP. The development of modular payloads yields further advantages in that they can also be deployed with more advanced HAPs (either aircraft or airship), meaning that the development time, expenditure, and intellectual property created can be best exploited. Thus, the early basic HAP deployments will provide confidence in the technology and business models.

Also possible in the short term is the use of manned HAPs with high payload capabilities. Such craft already satisfy the necessary airworthiness requirements but are costly to run and have missions largely constrained by the flying time of the pilots. Fleet operation is essential for all but the shortest missions. It is likely that such an approach is a technological dead-end but it is likely to advance the political cause for HAPs and may also help make the business case for this means of telecommunications deployment as shown in Chapter 4.

In the medium term aircraft-based HAPs with higher payload capabilities will be developed, extending the flexibility and allowing new missions to be defined. Spurring on this development is likely to be increased funding streams, more HAP friendly regulation, developments in advanced materials, propulsion, energy storage, and power generation. Such HAPs could well be solar powered and capable of operating continuously providing missions without renewal of up to a year.

It is expected in the longer term (>5 years) that aircraft-based HAPs may be replaced by solar powered airship technologies. Such craft can satisfy most of the technical requirements posed by payload developers. Providing the airworthiness of the craft themselves can be guaranteed for periods of up to 5 years, long endurance missions should be possible without platform replacement or maintenance. It is expected that such platforms, just as satellites, will not be designed to land in a conventional sense but will instead be brought down in a controlled descent, with the payload infrastructure (minus envelope) refurbished and installed in a new HAP. It should be noted that such an approach requires a significant advance on the current technology where aircraft require ongoing maintenance after just a few hours.

Anecdotally, HAPs are sometimes described as being similar to satellites in terms of their long endurance capabilities but it is worth noting that while it is true in some aspects, e.g. in terms of payload endurance, satellites have few moving parts and are not subject to the stresses and strains of an aircraft or airship, meaning they can be designed to have a lower mean time between failure.

The four different types of HAPs, manned plane, unmanned plane (fuel), unmanned plane (solar) and unmanned airship (solar), and their operating constraints and payload capabilities (ultimately determining the number of modules they can support) are given in Table 5.1.

Table 5.1 Anticipated HAPs operating parameters and estimated availability (based on [2])

	Manned	Unmanned		
	Plane	Plane (fuel)	Plane (solar)	Airship (solar)
Estimated availability	now	2010	2013	2016
Mission suitability	Short	Short/medium	All	All
Payload – mass (kg)	1000	250	150	1000
Payload – power (kW)	10	1		10
Altitude (km)	20	17		20
Flight duration	12 h	3 days	weeks	years
Number for continuous operation per coverage area	2		1	
Redundancy factor	0.3		0.2	
Fleet/reserve multiplier	0.6		0.2	

Given the likelihood of failure, built in redundancy is required for each HAP type and payload, which will depend on the HAP characteristics and mission length. HAPs with longer flight durations will require higher payload redundancy. HAPs with short flight durations will need to operate in fleets.

5.2 HAP Roadmaps for Different Types of Applications

Before discussing the technical aspects of possible roadmaps for the role out of HAP telecommunications services (business aspects were discussed briefly in Section 4.2). it is helpful first to define three aspects used in describing such roadmaps:

1. Modular payload – this is the physical implementation of the wireless system.
2. Telecommunications application– this is the application that will be used by end users, which will require use of one or more payload modules.

3. Mission – this is used to describe the beginning to end of operations including launch(es) and landing(s) of the HAP(s) and provision of one or more telecommunications applications.

Telecommunications applications invariably involve the delivery of different data types, for instance, speech, video, file downloads, email, and use a variety of data rates and frequency allocations. Conventionally, these have been separated in fixed broadband and mobile applications but this definition is becoming increasing blurred. As well as the pure telecommunications applications, communications delivery acts as an enabler for virtually all of the other HAPs applications, security, remote sensing, etc. Some of the more significant telecommunications applications could form missions in their own right but here we try and separate the application from the mission (missions will be discussed in more detail in the next section). Below are a number of examples of dedicated telecommunications applications, which we have taken from Chapter 4 but here the focus is on the technical aspects.

5.2.1   Application Example 1: WLAN to Trains

With this application it is assumed that the service provider installs conventional WLAN access points and on-board server on trains, which are then connected via mm-wave transceivers on top of the train to the HAP. This enables high speed WLAN access to users located in the carriages of the train. The service provider then sells WLAN capacity in a conventional way on the train. HAPs have the potential to deliver significant capacity and throughput improvements compared with alternative means but the price a customer is willing to pay is very sensitive. Key drivers for the commercial viability of this model are the number of passengers that will use the service on each train, the number of trains that need to be equipped to provide a sensible marketable service, and from a technical viewpoint the availability of usable alternative network connections, e.g. GPRS and satellite to provide a more limited service when out of the HAP coverage area.

5.2.2   Application Example 2: Backhaul for Terrestrial Base Stations/Access Points

The purpose of this application is to provide a backhaul communications link from ground-based terrestrial 3G base stations and WLAN access points on a wholesale basis, with the customer being a retail service provider. It should not be confused with HAPs backhaul that connects the HAPs with the wider network. Conventionally over 60% [3] of links are provided by point-to-point microwave links, but this becomes very costly in remote locations, with the extra factor that there are ever-increasing capacity demands by users. A basic €520 CAPEX [4] cost for each ground terminal is assumed (although this aspect is not too price sensitive for this application). Given that the customer will pay fully for the cost of this equipment, this application is much more lucrative than models where the CPE has to be subsidised.

5.2.3   Application Example 3: Broadband Internet

This application example assumes the delivery of broadband Internet-based service directly to the end user via a dish, with capacity of 10 Mbps being delivered per customer with a contention ratio of 50:1. We assume that the CPE terminal will still cost €520 [4]. Getting a foothold into the already established broadband Internet market is potentially very difficult owing to the ever-increasing speeds offered by xDSL technologies. In order to compete, it would be sensible to exploit the market segment where the user is too remote from the exchange (the more rural user). Alternatively, as we have already discussed in Chapter 4, the mm-wave CPE needs to be heavily subsidised to say €100, which eats significantly into the profits, given that income per user is only likely to be €300 per annum. Use of lower frequency bands may provide a lower cost CPE. Another significant factor is that the overall capacity requirements are likely to be far in excess of that needed for the other applications. Even though a HAPs system has significantly more capacity than say a satellite system, the other applications are much more lucrative in terms of the revenue they can raise per Mbit/s. A HAPs system should not rely on this application alone.

5.2.4   Application Example 4: Broadcast/Multicast

This application example deals with providing the next generation of Broadcast/ Multicast services to the end user. Satellite competitors are obvious but the degree of local control (particularly important in many non-European countries) and the ability to package these services along with others offered on the HAP make it particularly attractive. The application has the potential to be highly profitable. It is assumed that the CPE cost will be €520, which will require a subsidy but this is more than compensated for by the significantly reduced capacity requirements (and costs), due to the same information being supplied to multiple users.

5.2.5   Application Example 5: 3G Mobile Communications

Mobile communications delivery could be a significant application for HAPs, with the ability to substantially reduce ground communications infrastructure, and the potential to allow new service providers to compete with incumbents. 3G mobile has been fully standardised for HAPs and can use conventional handsets. It has a number of advantages compared with the other broadband applications in that users are willing to tolerate a lower quality of service. This makes it ideal for deployment from early stage HAPs.

5.3   Telecommunication Missions

In practice different telecommunications applications will be grouped together to serve a specific mission. Common to all will be the requirement of a quasi-stationary HAP(s) capable of delivering high availability services (typically 95–99.9%) for the duration of a mission. The mission length will range from 1 day through to continuous deployment, depending on the scenario and development timescales. As we have already discussed it is not important that a mission be delivered by a single HAP, instead multiple craft can be used either as replacements for craft to extend the duration of a mission or simultaneously in order to deliver appropriate capacity as shown in Figure 5.1 and discussed further in Chapter 9. This simultaneous deployment over a coverage area can exploit the directional antenna characteristics of the ground-based antenna (not available with 3G) to enable frequency reuse from a number of platforms sharing the same frequency band [5]. Incremental deployment is also possible meaning that costs are kept under control.

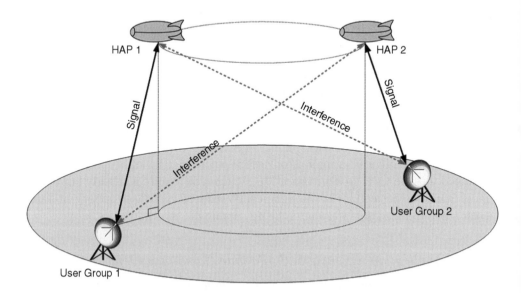

Figure 5.1   Multiple HAP with overlapping coverage area incremental deployment model

Mission examples have been chosen to fit in with the short, medium and long term development of HAPs technological development, discussed earlier based on the advancement of HAPs aeronautical technology. Requirements become increasingly more stringent, and take into account the anticipated progression in HAPs aeronautical technologies discussed above.

- **Short term – Event servicing/disaster relief.** This application is aimed at providing a range of high capacity broadband services for a short period of time. Details are shown in Table 5.2.
- **Medium term – Mobile and broadband communications for fixed users in developing countries.** It is expected that availability in the medium term will not be as high as currently available from conventional infrastructure. Therefore it is important to select markets where no reliable infrastructure exists. Details are shown in Table 5.3.
- **Long term – Mobile and broadband communications for both high-speed mobile and fixed users (developed countries, with established infrastructures).** Details are shown in Table 5.4.

Table 5.2   Example mission possible in the short term and required capabilities [1]

Example mission	Event servicing/disaster relief
Telecom payload	Broadband (WiMAX) either mm-wave (or ideally below 11 GHz) Possible applications: • High speed Internet • Broadcast/Multicast • WLAN on high speed vehicles • 3G
Mission duration	1 day to 2 weeks – availability 95.0% To maintain continuity of service, platforms could be replaced on a regular basis. The replacement interval would be set by platform type
Payload mass	<50 kg (CAPANINA payload is 9 kg)
Payload power	<200 W (CAPANINA payload is 60 W)
Payload volume	0.05 m^3 (CAPANINA payload is 0.01 m^3)
Altitude	Technically no constraints – possible radio regulatory issues (HAPS > 20 km), but may be better to consider it a UAV. Altitude will determine the coverage area, since a minimum elevation angle of 20–30° is desirable especially for mm-wave
Geostationary position	(originally specified by HeliNet project, not mandatory) Position cylinders • Availability 99.0%, height +/−500 m, radius 2.5 km • Availability 99.9%, height +/−1500 m, radius 4.0 km
Platform stability	Currently under discussion (<15°)
Types of platform	Manned plane HAPs, tethered aerostats (not stratospheric)

Table 5.3 Example mission possible in the medium term and required capabilities [1]

Example mission	Mobile and broadband communications for fixed users in developing countries
Telecom payload	Broadband (WiMAX) either mm-wave (or ideally below 11 GHz), limited number of cells (or additional platforms) to increase capacity Possible applications: • High speed Internet • Broadcast/Multicast • WLAN on high speed vehicles • 3G
Mission duration	Continuous – availability 95.0% To maintain continuity of service, platforms could be replaced on a regular basis. The replacement interval would be set by platform type
Payload mass	<250 kg, could be distributed across one or more platforms (50 kg per cell)
Payload power	<500 W, could be distributed across one or more platforms (100 W per cell)
Payload volume	0.25 m³, could be distributed across one or more platforms (0.05 m³ per cell)
Altitude	Technically no constraints – possible radio regulatory issues (HAPS > 20 km), but may be better to consider it a UAV. Altitude will determine the coverage area, since a minimum elevation angle of 20–30° is desirable especially for mm-wave
Geostationary position	(originally specified by HeliNet, project not mandatory) Position cylinders • Availability 99.0%, height +/−500 m, radius 2.5 km • Availability 99.9%, height +/−500 m, radius 4.0 km
Platform stability	Currently under discussion (<15°)
Types of platform	Unmanned planes (solar) operating in fleets, airship HAPs (if available)

It is important to mention that an event servicing/disaster relief mission could be deployed in two ways:

- A dedicated rapid deployment HAP, which is flown in at short notice to provide the broadband services. Such a configuration is technically possible in the short term, using already available HAPs.
- Modules/capacity provided from a HAP already in service, providing other applications. This configuration would make most sense in the longer term, as an event (e.g. Olympics) could well be located within a HAP coverage area.

5.3.1 The Payload for Telecommunications Applications

The payload requirements will depend on the type of mission and applications to be served. It is expected that a range of applications will be served on a mix and

Table 5.4 Example mission possible in the long term and required capabilities [1]

Example mission	Mobile and broadband communications for both high speed mobile and fixed users (developed countries, with established infrastructures)
Telecom payload	High capacity broadband (WiMAX) either mm-wave (or ideally below 11 GHz), high number of cells (and/or additional platforms) to increase capacity Possible applications: • High speed Internet • Broadcast/Multicast • WLAN on high speed vehicles • 3G
Mission duration	Continuous – availability 99.9% To maintain continuity of service, platforms could be replaced on a regular basis. The replacement interval would be set by platform type
Payload mass	<1000 kg, could be distributed across one or more platforms (50 kg per cell)
Payload power	10–20 kW, could be distributed across one or more platforms (100 W/cell)
Payload volume	$5\,m^3$, could be distributed across one or more platforms ($0.05\,m^3$ per cell)
Altitude	Technically no constraints – possible radio regulatory issues (HAPS > 20 km), but may be better to consider it a UAV. Altitude will determine the coverage area, since a minimum elevation angle of 20–30° is desirable especially for mm-wave.
Geostationary position	Ideally position cylinder • Availability 99.9%, height $+/-500$ m, radius 1 km
Platform stability	Currently under discussion (<5°)
Types of platform	Airship HAPs, unmanned planes (solar) operating in fleets

match basis, and to enable this it is important that payload capacity is available in a modular format. One can envisage that one module will be required for each application per cell, with the number of cells being dependent on the capacity to be delivered. Each platform type will be able to carry a variable number of modules depending on its payload capability in terms of power, mass, and volume. It will be the norm for services to be supplied to more than one service provider.

The long term example in particular will require multiple HAP deployments and will enable capacity to be incrementally deployed from an increasing number of HAPs. Early HAP deployments need only carry a small payload and later deployed HAPs can carry larger payloads using the latest technology. Each HAP type also has a varying number of cells depending on its capabilities, providing further frequency reuse, as discussed in Chapter 1, with the key communications parameters reiterated here in Table 5.5.

Currently available licensed spectrum means that mm-wave band communications will be used to supply broadband communications fronthaul and backhaul.

Table 5.5 Anticipated communications operating parameters for different HAP types (based on [2])

	Manned	Unmanned		
	Plane	Plane (fuel)	Plane (solar)	Airship (solar)
Fronthaul				
Max. clear air cell capacity (Mbps)	120			
Number of cells	121	7	19	121
Total fronthaul capacity (Mbps)	14520	840	2280	14520
Diameter of cell (km)	6	15		6
Diameter of coverage area (km)	60	36	59	60
Availability at min. data rate (%)	95–99.9			
Redundancy (%)	0		5	20
Backhaul				
Backhaul capacity per 50 MHz	320			
No. of carriers per gateway station	3			
Gateway capacity (Mbps)	960			
Redundancy (%)	0		5	30

Bands at 48/47 GHz are licensed for worldwide use, with 31/28 GHz permitted for use in over 40 countries worldwide [6]. The use of the bands in the sub-10 GHz range is also being discussed in ITU-R [7]. Optical communications could additionally be used to provide interplatform links and additional backhaul to ground but is not directly included in these models.

The main payload modules are:

- **3G** – provides 3G communications, effectively a Node B, one per cell.
- **Broadband** – provides broadband capacity to users, effectively an access point, one per cell.
- **Broadcast/Multicast** – processing unit capable of delivering Broadcast and Multicast technologies.
- **RF Fronthaul** – the antenna transceiver structure for a fronthaul communications module. This will require some form of motion stabilisation, and pointing, acquisition and tracking technology. It may be possible for more than one module to share the same antenna, depending on the frequency allocations and transmission formats, e.g.:
  - mm-wave broadband – 31/28 GHz (47/48 GHz) frequencies.
  - 3G – around 2 GHz frequency band.
- **RF Backhaul** – the antenna transceiver structure for a backhaul communications module to connect the HAPs into the wider network. This will require some form of motion stabilisation, and pointing, acquisition and tracking technology.

- **Server** – provides on board data processing capabilities, filestore, data cache.
- **Router/switch** – depending on complexity of the payloads, used to ensure data are sent to and from appropriate cells/backhaul links.
- **Backhaul** – provides appropriate data formatting to provide capacity to the HAPs into wider network

References

1. USEHAAS Project Deliverable, USEHAAS Strategic Research Agenda for Development of High Altitude Aircraft and Airships, FP6 USEHAAS Project 2007.
2. D. Grace and P. Likitthanasate, *A Business Modelling Approach for Broadband Services from High Altitude Platforms*, Invited paper at International Conference on Telecommunications (ICT'06), Madeira, Portugal, May 2006.
3. S. Byars, *Using Pseudo-Wires for Mobile Wireless Backhaul over Carrier Ethernet,* Blueprint Metro Ethernet, http://www.convergedigest.com/, 13 February 2006.
4. CAPANINA Project Deliverable, *D12 - Strategy Document: Delivering Broadband for All from Aerial Platforms Including Commercial and Technical Risk Assessments,* www.capanina.org/documents/CAP-D12-WP12-BT-PUB-01.pdf.
5. D. Grace, J. Thornton, G. Chen, G. P. White and T. C. Tozer, *Improving the System Capacity of Broadband Services Using Multiple High Altitude Platforms*, IEEE Trans. Wireless Commun., March 2005, Vol. 4, No. 2, pp. 700–709.
6. ITU Recommendation ITU-R F.1569, *Technical and Operational Characteristics for the Fixed Service Using High Altitude Platform Stations in the Bands 27.5-28.35 GHz and 31-31.3 GHz*, International Telecommunications Union, Geneva, Switzerland, 2002.
7. Resolution 734 (Rev.WRC-07), *Studies for Spectrum Identification for Gateway Links for High Altitude Platform Stations in the Range from 5850 to 7250 MHz*, World Radiocommunication Conference, Geneva, Switzerland, 2007.

Part Two

Broadband Wireless Communications from High Altitude Platforms

6

HAP System Operating Environment

6.1 Operating Environment and Related Limitations

In the design of any wireless communication system an understanding of impairments introduced by a radio propagation channel is of vital importance. Several factors influence the amount and type of radio signal distortions introduced by a propagation channel such as communication scenario, carrier frequency, frequency bandwidth, employed access scheme, antenna types and the operating environment. For example, Doppler spreading and strong multipath are characteristic of terrestrial mobile communications while in satellite mobile communications the blockage and shadowing are the main sources of errors. The strong multipath propagation is typical also for indoor communications. The atmospheric impairments, such as rain attenuation and scintillation, have to be taken into account at Ku frequency band and above, while at lower frequencies they do not have any significant effect on the transmitted signal.

In the past many techniques have been investigated to counteract the multipath. For example, a narrow beam antenna significantly reduces the multipath. Today, however, some communication technologies such as multiple input multiple output (MIMO), rake receivers in CDMA systems and communication systems applying OFDM are taking advantage of multipath. The carrier to interference ratio strongly depends on the employed access schemes. In order to identify channel impairments in HAP communications and design suitable propagation channel models, the system operating environment with respect to platform type, its altitude, carrier

Broadband Communications via High Altitude Platforms   David Grace and Mihael Mohorčič
© 2011 John Wiley & Sons, Ltd

frequency, propagation environment and communication scenarios of a HAP system need to be overviewed first.

HAPs equipped with communications payload are expected to operate in the lower stratosphere, typically at altitudes between 17 km and 22 km, which guarantees high elevation angles. They can take the form of an airship or a plane and can be manned or unmanned. These options importantly influence the main parameters of the mission, including the mission duration, platform altitude, service scenario, and others. From the perspective of propagation channel modelling the basic difference between the airship and plane is in their station keeping capability. While the airship is nearly stationary, and only slight drifts are expected due to wind, the plane is constantly flying along a quasi circular trajectory to keep the altitude nearly constant and stay above the targeted coverage area. This introduces Doppler frequency shift in the received signal and variation of the communication channel in time even if the subscriber terminal is stationary. In addition, small random variations such as pitch, yaw and roll of the platform contribute to propagation channel variations and its randomness. The typical propagation environment for stratospheric communications is shown in Figure 6.1.

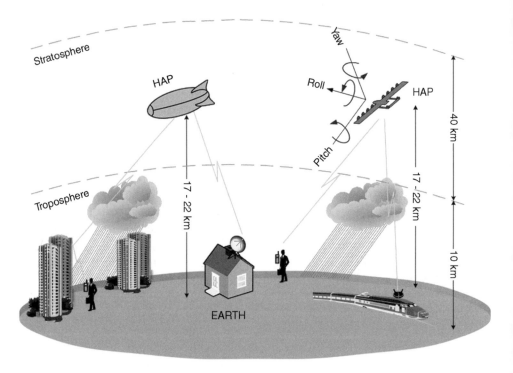

Figure 6.1 HAP operating environment

Due to the operating altitude in the lower stratosphere, the line of sight (LOS) propagation conditions are nearly always achieved in a HAP communication system with low or no multipath and small propagation delay. As a consequence of LOS channel conditions a higher data throughput is achieved at much lower power consumption with less complex terminals, while small propagation delay enables the use of closed loop link adaptation techniques. The differences between stratospheric, satellite and terrestrial communications in terms of LOS/non line of sight (NLOS) conditions, propagation delays and distance between the transmitter and receiver are illustrated in Figure 6.2.

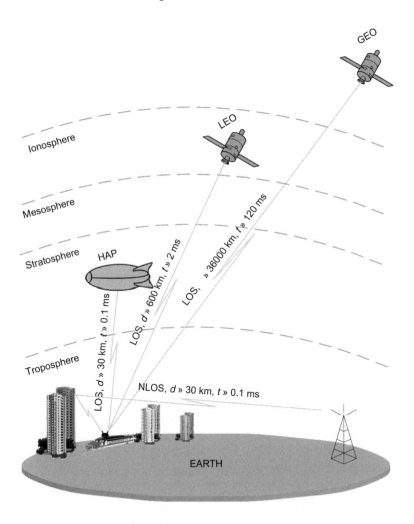

Figure 6.2   Comparison of satellite, terrestrial and HAP propagation environment

The carrier frequency also has an important impact on signal propagation. As discussed in Section 1.3.2 the International Telecommunication Union (ITU) permitted the use of several frequency bands for provision of communication services via HAPS. These include a band at 2 GHz for 3G mobile communication services and two bands for broadband services in millimetre-wave frequencies, namely at 47/48 GHz and 28/31 GHz. At the moment there are no allocations for the use of stratospheric platforms at other frequency bands, but an agenda item for WRC11 is looking for studies of HAP systems at 6 GHz frequency band for gateway links with stratospheric platforms [1].

The signals at allocated frequency bands for stratospheric communications exhibit quite different distortions, including free space loss attenuation and the amount of multipath and atmospheric effects. Higher free space loss at millimetre frequency bands is usually counteracted by applying directional narrow beam antennas with high gains, which also limit the multipath effect. The free space path loss at HAP allocated frequencies is plotted in Figure 6.3 as a function of elevation angle. The significant increase of path loss at lower elevation angles is due to increased distance between the ground station and HAP. At very low elevation angles the LOS communication propagation conditions are replaced by plane earth propagation, characterised by a two-ray propagation model consisting of a direct ray and a ray reflected from the earth's surface. This results in two times higher propagation loss at the same distance. Signal in the Ka frequency band experiences

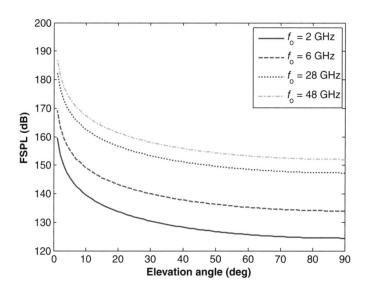

Figure 6.3   The free space path loss (FSPL) as a function of elevation angle

more than 30 dB higher free space attenuation than signal in the L frequency band (i.e. at 2 GHz). Signal at millimetre frequency band is also prone to atmospheric distortions, while in the 2 GHz band signal is not sensitive to atmospheric distortions but exhibits strong multipath propagation.

At the millimetre frequency band an additional attenuation due to particles in the troposphere is superimposed to high free space path loss. This attenuation of radio signal is caused by the conversion of radio frequency energy to thermal energy of the transmission media [2]. The radio frequency energy is absorbed by particles such as molecules of gases or rain drops in the atmosphere. In the 2 GHz frequency band the absorption due to gases and rain drops is negligible. In the millimetre frequency band, on the other hand, the rain drops, whose size are in the millimetre range, are recognised as an important factor that adversely affects the link availability [2] in land and satellite communication systems. The oxygen and water vapour also contribute to the absorption in the millimetre frequency band. The main peaks of oxygen absorption are at 60 GHz and at 118.74 GHz, while the peaks for water vapour are at 22.3 GHz and at 183.3 GHz [2], which is below or above the allocated frequency band for HAPs. Nonetheless, at low elevation angles the radio path through the atmosphere is long, and even if the main peaks of attenuation are out of frequency bands of interest, significant link attenuation due to gases may still occur.

The long term atmospheric effects such as attenuation due to rain and gases are in the millimetre frequency band accompanied by short term effects like scattering and scintillation. Scattering is caused by the refraction and diffraction of radio signal on small particles such as rain drops or gas molecules, whereas scintillation corresponds to rapid fluctuation of the received signal power, phase and angle of arrival, as a consequence of small scale refractive index inhomogeneities [2, 3].

Scattering causes only a fraction of signal power to be transmitted in the direction towards the receiver. Since only signals with wavelength comparable or lower than the size of the particles in the air are scattered, the main scatterers in the HAPs allocated frequency bands are raindrops, fog and clouds [2, 3].

The main source of tropospheric scintillation is small scale refractive index inhomogeneities, which are caused by air turbulence. The scintillation affects radio frequency links above 10 GHz and is very frequency dependent. It is also known from free space optics and is observable as a star twinkling. The intensity of scintillation increases with increasing frequency and with decreasing elevation angle and antenna aperture. Scintillation is significantly increased in cloudy conditions, but it may also appear in rainy weather or clear sky. The intensity of scintillation is the highest at noon. The attenuation due to rain is generally much higher than attenuation due to scintillation; however, in low margin communications, the scintillation could have a major impact. This may especially happen for elevation angles lower than 5° [2, 3].

In LOS propagation conditions, the link capacity can be doubled in a given frequency band if two signals with orthogonal polarisations are applied, either horizontal and vertical polarisations or left and right circular polarisations. However, the isolation between polarisations can be distorted due to imperfect antenna characteristics and various atmospheric effects. The cross polarisation in the Ka band and above is caused by nonspherical particles in the troposphere, such as rain drops, snowflakes and ice crystals, which lead to different attenuation and phase shifts on different main polarisation directions. In the Ka frequency band the main source of cross polarisation effect are ice crystals [2, 3].

Multipath propagation is the next short term effect which has to be considered as a source of distortion of the signal transmitted from HAPs. Narrow beam antennas typically used at the millimetre frequency band reduce the possibility of multipath propagation; however, in the 2 GHz frequency band omnidirectional antennas are usually used, potentially giving rise to significant multipath effect. The amount of multipath effect on the received signal depends also on the propagation environment, i.e. the refractive ability of surrounding obstacles, their geometry and position. The refractive ability is determined by the roughness factor h_c, which is a function of the angle of incidence α and wavelength λ [2]:

$$h_c = \frac{\lambda}{8 \cdot \sin\alpha} \tag{6.1}$$

The material is considered smooth, i.e. the ray optical approach for calculation of the refracted ray can be applied, if the surface height does not exceed the roughness factor. This criterion is also known as the Rayleigh criterion. In the millimetre frequency band nearly all materials of buildings such as concrete, bricks and mortar (except glass or metal parts) are considered rough. This means that signal dispersion occurs instead of signal reflection, which limits the multipath propagation. On the other hand, at the 2 GHz frequency band, the majority of materials that buildings are made of are considered smooth, giving rise to strong multipath.

The HAP allocated frequency band at 2 GHz is foreseen for narrowband mobile communications whereas those in the millimetre frequency band are planned for fixed and mobile broadband communications. Random motion of the platform and user terminals leads to Doppler frequency shift and Doppler frequency spread at both allocated frequency bands. The amount of Doppler effects depends on terminal speed, channel bandwidth, amount of multipath, etc.

The 2 GHz frequency band appears particularly interesting for the provision of a good coverage of wide areas and for mobile 3G services from HAPs, but due to large horizon distances the communications from stratospheric platforms may potentially cause harmful interference over a wide area. The fixed and mobile communications are foreseen in both millimetre frequency bands. However, due to high free space path loss in those bands, highly directional antennas with high gains and narrow beams are required. In the case of mobile terminals, the terminal is moving and rotating, so the mobile station antenna has to be continuously steered towards the base station in order to maintain the connection. Such steerable antennas can be implemented either mechanically or electrically using an antenna array. In both cases this increases the complexity and the size of mobile terminals.

The propagation impairments in the stratospheric propagation channel for allocated frequency bands are summarised in Table 6.1. The effect on transmitted signal is described based on heuristic estimations. Each of these signal distortion phenomena in the HAP propagation channel is described in more detail in Section 6.3 and its influence on the HAP propagation channel is quantified, after the approaches for characterisation and modelling of wireless propagation channels are overviewed in Section 6.2.

Table 6.1 Distortions in HAP propagation channel for allocated frequency bands

Carrier frequency	2 GHz	28/31 GHz and 47/48 GHz	
Mobility	Mobile	Fixed	Mobile
Multipath	Small	Negligible	Negligible
Free space path loss	124–130 dB	148–154 dB	148–154 dB
Rain attenuation	Negligible	Yes	Yes
Gas attenuation	Negligible	Considerable	Considerable
Scintillation	Negligible	Negligible	Negligible
Cross polarisation	Negligible	Negligible	Negligible
Multipath	Yes	Negligible	Negligible
Shadowing/blocking	Yes	Negligible	Yes
Doppler shift	Yes	No	Yes
Doppler spread	Yes	No	Negligible

6.2   Propagation Channel Modelling

A designer of a wireless communication system has to understand the propagation environment for the professional design of a communication system. A test phase of the system follows the initial design to find the system properties and its limits in various radio propagation environments. In this phase a reliable propagation channel model is indispensable, either implemented in software or as a hardware channel simulator, in order to provide the designer with a controlled radio propagation environment. Many reliable propagation channel models exist for communication systems using mature technology that has been available on the market for some time. However, for the emerging communication systems the channel models are not readily available or their reliability is disputable due to lack of communication channel measurements and lack of theoretical knowledge of ray propagation in some specific environment, at specific carrier frequencies or in allocated bandwidth. Taking this into account, propagation channel models for emerging communication systems are typically based on known and proven channel models designed for pre-existing wireless communication systems operating in a similar radio propagation environment with the necessary modifications.

Generally speaking, the wireless propagation channel models may be classified into three main classes, namely [4]:

• empirical wireless propagation channel models;
• deterministic wireless propagation channel models;
• semi-empirical wireless propagation channel models.

The channel model classification is illustrated in Figure 6.4

Empirical channel models are based on a set of measurements in a specific environment. They are represented by a simple path loss equation or a probability density function with a set of parameters, which are obtained by statistical

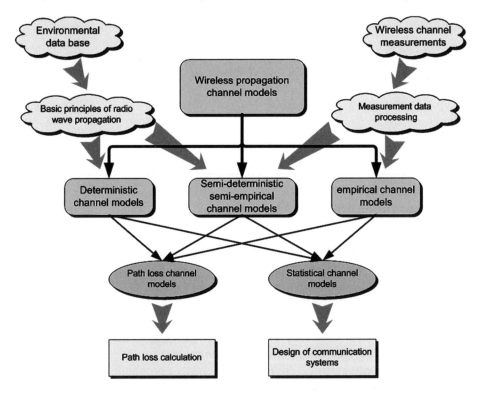

Figure 6.4   Classification of the wireless propagation channel modelling

processing of measured data. These models do not require precise knowledge of the propagation environment. The empirical model accuracy depends on the amount of data obtained by measurement campaign, quality of statistical data processing and the similarity of the propagation environment, which forms the basis for estimating the channel parameters, and the propagation environment, where the communication system of interest operates. Empirical propagation channel models are frequently applied for path loss calculation in mobile terrestrial systems. Typical representatives of the path loss channel model include the well known Okumura–Hata path loss channel model [5], COST 231-Hata path loss channel model [6] and many others [7]. These models are very popular for the calculation of the initial base station coverage of the cellular mobile communication systems due to their simplicity, low computational complexity and easy implementation in software tools.

For the design of communication systems, in particular for the design of fade mitigation techniques, statistical channel models describing statistical properties of the propagation channel are more desirable. In this case the measurement results can be also applied for the statistical description of small and large scale fading in narrowband and broadband wireless propagation channels finding the channel power delay profile, frequency correlation function, time correlation function, Doppler spectral density or also condensed parameters such as the root mean square (rms) delay spread, mean delay, channel coherence bandwidth, channel coherence time, mean Doppler shift or rms Doppler spread [8]. In terrestrial communication systems, the statistical channel models are frequently implemented as tapped delay lines, the length of which depends on propagation environment. The variation of tapped delay line coefficients follows propagation environment specific probability density functions. The tapped delay model can be expressed mathematically [8]:

$$h(t, \tau) = c_0 \cdot \delta(t - \tau_0) + \sum_{n=1}^{N} c_n(t) \cdot \delta(t - \tau_n) \tag{6.2}$$

In Equation (6.2) c_0 is the LOS component and does not vary with time, $c_n(t)$ denotes the nth time variable coefficient which varies with time and usually obeys the zero mean Gaussian probability density function, and τ_n denotes the delay of the nth tap. The temporal variability of channel is modelled as a variation of the coefficients in time and can be represented as a Doppler spectrum. The tapped delay line channel model with one tap and a complex coefficient obeying Gaussian distribution is a well known Rayleigh channel [8]. More complex channel models are used for broadband wireless communication system, such as WiMAX. The Stanford University Interim (SUI) propagation channel model [9], for instance, consists of three taps with tap delay τ_i between $0\,\mu s$ and $20\,\mu s$. The model is applicable in the design of WiMAX communication systems. A general tap delay line wireless propagation channel model is depicted in Figure 6.5.

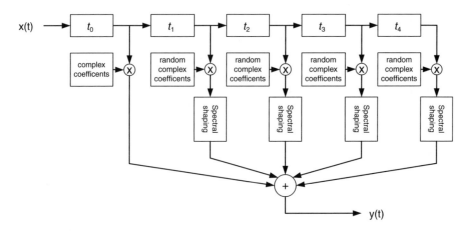

Figure 6.5   General tap delay line wireless propagation channel model

A slightly different approach for statistical channel modelling is known in mobile satellite systems. Due to the motion of either satellite or user terminal the LOS between them is frequently shadowed or even blocked by different obstacles such as buildings, hills or trees. This results in quite different propagation conditions, which have to be modelled with two or more tapped delay lines. The coefficients of the delay lines obey different probability density functions [10–13]. In addition, a function describing the switching between different channel models has to be included in the model. The common name, i.e. switch channel models, is used for these channel models in the scientific literature (Figure 6.6).

In order to determine the propagation channel characteristics such as path loss, scattering, etc., the deterministic channel models use well-known principles of electromagnetic wave propagation, such as Maxwell equations, radio wave reflection, diffraction, refraction and absorption applied on a site-specific environment. The propagation environment has to be described by the geometrical properties of objects such as position, height, width and length, as well as the type of material of the object characterised by its electrical properties, such as electrical permittivity and susceptibility. The accuracy of the environmental description is of high importance for the correctness of the deterministic channel models. While quite accurate geometrical descriptions of the propagation environment can be found in databases of geographical information systems (GIS) and building plans, there are nearly no data available on electrical properties of building surfaces. The deterministic channel models are most often based on various ray tracing methods [4].

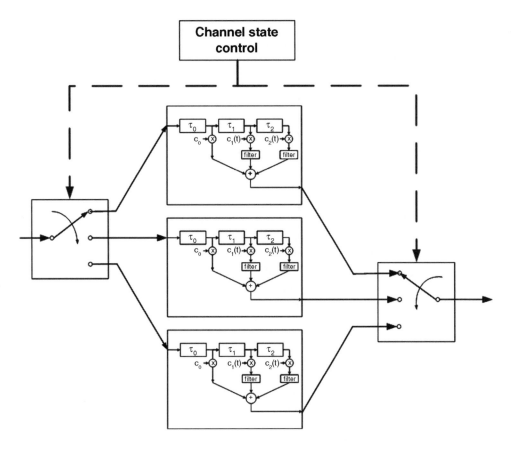

Figure 6.6 Switched wireless propagation channel model

In the case of a propagation environment with a large number of surfaces ray tracing methods become computationally too complex to be applied in software tools for path loss prediction. The computational complexity of the deterministic channel models can be reduced by using simplified ray tracing methods. For instance, instead of using three-dimensional ray tracing only two-dimensional ray tracing can be applied, or alternatively the deterministic channel models are only applied in limited areas such as nano, pico and femto cells for indoor as well as for outdoor environments.

Semi-empirical or semi-deterministic channel models combine statistical description of empirical channel models and one or two most significant site specific propagation effects by applying deterministic channel modelling. By adding the deterministic channel modelling approach to the empirical channel model, the accuracy of the channel model is increased with minimum increase in the propagation channel model computational complexity. A common semi-empirical approach for calculating path loss in cellular communication systems is to use the Okumura–Hata path loss channel model upgraded by the additional attenuation caused by vegetation at the terminal site and applying the digital elevation model to calculate the terminal shadowing due to terrain [14].

The statistical channel models of the propagation environment are required to test designed fade mitigation techniques, modulation and coding schemes, time and carrier synchronisation approaches, access schemes and other elements of communication system in the controlled propagation environment, which can be provided by applying statistical channel models. The majority of statistical propagation channel models are empirical channel models, based on measurement campaigns. For emerging communication systems such measurement campaigns do not exist, hence the design of such statistical channel models is not feasible. The problem can be solved by modifying the existing statistical channel model originally designed for a similar propagation environment. Such an existing channel model is complemented by specifics of the propagation environment of the emerging communication system. In the case of HAP communication systems, which fill the gap between terrestrial and satellite communication systems, either satellite or terrestrial communication channel models can be used as a baseline. However, as mentioned earlier no one of them completely covers the expected characteristics of the stratospheric wireless radio channel.

From the perspective of channel modelling the allocated HAP frequency bands fall into the Ka and L frequency bands. While the Ka band is foreseen for broadband mobile and fixed communications, in the L band only narrow band communications are envisaged as an umbrella cell of 3G communication systems. Radio signal in the Ka frequency band exhibits 30 dB higher free space path loss compared with the L band signal. The high path loss at Ka frequency band is usually compensated by directional antennas at the terminal, while at the L frequency band the terminals are equipped with omnidirectional antennas. Different antennas used as well as different reflection, absorption and dispersion properties of the surrounding objects at the Ka and L frequency bands will result in completely different propagation channel models for the two frequency bands. Since the focus of this book is broadband communication systems, we are limiting our focus to the propagation channel models at the Ka frequency band.

Taking into account predominantly LOS propagation conditions, operation in millimetre frequency band and similar elevation angles at the user terminal side, the HAP propagation channel is most similar to the satellite one [13, 15]. However, there are three important distinctions between HAP and satellite channels that need to be adequately addressed:

- The distance between the receiver and transmitter is significantly shorter in the case of HAP systems resulting in smaller propagation delay and lower free space path loss.
- The signal transmitted from HAP only exhibits distortions due to propagation through the troposphere, while in satellite systems additional distortions of the signal occur when passing the ionosphere.
- In the case of the mobile communication scenario, the rate of and reason for elevation angle variation are different in satellite and HAP systems. In the case of GEO satellites, the elevation angle does not vary significantly as only the mobile terminal is changing its position. In LEO satellite systems the rate of variation of the elevation angle is comparable with that in HAP systems but it is caused mainly by the motion of the satellite and, for that reason, it is easily predictable.

The main differences between LEO and GEO satellite systems, HAP communication systems and terrestrial communications systems with respect to distance, propagation delay and free space losses in the Ka frequency band are illustrated in Figure 6.2. Clearly, the use of either satellite or terrestrial propagation channel models for HAP channel modelling is not possible without adaptations. In this respect, the similarities between satellite and HAP systems led many research groups to the decision to use mobile satellite switched channel models as a baseline for mobile HAP channel modelling [13, 15–17] and the atmospheric distortion models for satellite communications in the Ka frequency band are used as a starting point for modelling meteorological distortions in the broadband HAP propagation channel.

6.3 HAP Radio Frequency Propagation Channel Modelling

In the design of Earth to HAP links for communication systems similar effects to those in Earth to satellite communication links have to be considered, carefully taking into account the difference in altitude between HAPs and satellites. In particular, radio signal coming from a satellite is subject to propagation impairments in the ionosphere and troposphere, whereas radio signal transmitted from a stratospheric platform is distorted only by tropospheric effects. The ITU-R P.618-10 [18, 19] listed the atmospheric effects, which have to be considered in the design of the Earth–space links, some of them being important also for HAP communications. They can be classified in the following three classes:

- Clear air effects
  - Free space path loss; it is elevation angle dependent and may reach high values at low elevation angles.

- – Absorption due to atmospheric gases; it may be important at low elevation angles due to long signal path through the atmosphere.
- – Beam spreading loss due to ray banding; it is caused by the decrease of the radio refractive index of the atmosphere with the increasing altitude.
- – Tropospheric scintillation.

- • Hydrometeor effects
  - – Absorption due to rain, liquid water clouds, fog, ice hydrometeors, ice crystals, snow, hail and melting layer or rain fading.
  - – Scattering by hydrometeors such as water, ice droplets in precipitations, clouds, etc. and noise emitted from absorbing media; the effects are critical for frequencies above 10 GHz. Since the allocated frequency band for broadband communications via HAPs is in the Ka frequency band, the absorption due to atmospheric gases significantly contributes to HAP link attenuation.
  - – The influence of hydrometeor effects on cross polarisation.

- • The effects of the surrounding environment
  - – Attenuation and shadowing by local environment such as trees, buildings, hills and mountains.
  - – Multipath propagation due to reflections of radio signal from obstacles in the local environment near receiving antenna on the ground and on the platform.
  - – The effects due to motion of either platform or terminal; such motion causes Doppler frequency shift and the variable elevation angle and variable distance between the terminal and platform.

The clear air effects, hydrometeor effects and effects of the surrounding environment have been deeply studied for satellite and terrestrial systems. Adjustment of these effects to HAP communication systems has to take into account the geometry of the HAP operating environment, which is depicted in Figure 6.7. The most important parameter is the path of the radio wave through the atmosphere, which depends on the elevation angle of the platform and effective height of water vapour and gases in the atmosphere.

The free space path loss dependency on elevation angles is plotted in Figure 6.3 for different carrier frequencies. It is mainly compensated by the transmitted power and antenna gains at transmitter and receiver.

6.3.1   Absorption Due to Water Vapour and Atmospheric Gases

Absorption due to water vapour and atmospheric gases is classified among the clear air effects. The total attenuation produced by atmospheric gases is calculated using

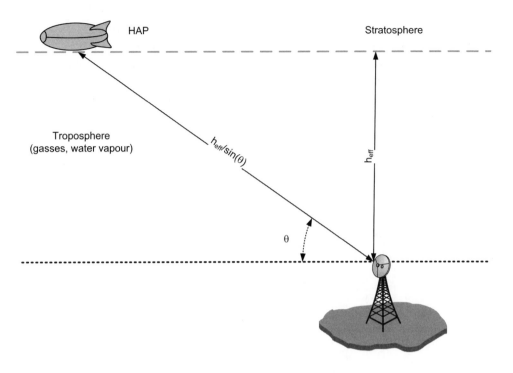

Figure 6.7   Geometry of the HAP operating environment for attenuation calculation due to gases and water vapour

the specific attenuation (in dB/km) for each gas, and equivalent path length through the atmosphere, which is calculated using equivalent height. The equivalent height is applied because gas is not distributed uniformly with height but its density decreases with increasing height. For example, in the Recommendation ITU-R P.676 'Attenuation by atmospheric gases' the exponential decrease in density of water vapour and oxygen is assumed to calculate the equivalent heights for dry air and water vapour [17, 19]. A detailed procedure for the calculation of losses caused by dry air and water vapour can be found in the Recommendation ITU-R P.676, where the Earth to space path loss due to atmospheric gases absorption is given as:

$$A = \frac{A_o + A_w}{\sin \theta}; \quad A_o = h_o \cdot \gamma_o \quad \text{and} \quad A_w = h_w \cdot \gamma_w \qquad (6.3)$$

where h_o and h_w are equivalent heights for oxygen and for water vapour respectively, while γ_o and γ_w are specific attenuations for oxygen and water vapour. The elevation angle is denoted by θ. For example, the water vapour equivalent height is 1.6 km for clear weather and 2.1 km for rain [19]. Assuming data for specific attenuation from the Recommendation ITU-R P.676, the cumulative gas attenuation can be calculated as a function of elevation angle (see Figure 6.8). The path loss due to absorption of gases in the atmosphere can be negligible at high elevation angles and carrier frequencies below 10 GHz, but at frequencies higher than 10 GHz and elevation angles lower than 10° the water vapour and atmospheric gases may cause significant link degradations, and as such they have to be considered in link dimensioning.

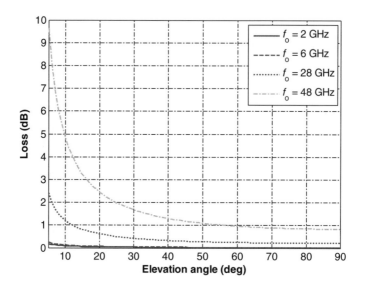

Figure 6.8   Attenuation due to absorption of the atmospheric gases

The decrease of refractive index with the increasing height causes banding of radio waves and results in defocusing effect at low elevation angles. The loss due to beam spreading is very low for elevation angles higher than 3° and can be neglected [18, 19]. In HAP communication systems the elevation angle is always above 3°, so the defocusing effect does not need to be taken into account.

6.3.2   Scintillation

Scintillation is a fast variation of the received signal due to turbulent eddies in the atmosphere, which mix the air masses whose temperature, pressure and humidity slightly vary, causing random variation in the refractive index. The magnitude and structure of the refractive index variations have high impact on the magnitude of the tropospheric scintillation. The longer the path length through the propagation media, which occurs at low elevation angles, the higher the scintillation. The antenna with higher aperture, on the other hand, decreases the scintillation effect due to averaging the signal over space. At very low elevation angles, i.e. below a couple of degrees, the scintillation may be transformed into atmospheric multipath, which is characterised by deep slow fading (> dB).

The scintillation amplitude depends on the wet term of the refractivity N_{wet}, which is sensitive to partial vapour pressure e of water vapour in the air, while water vapour is a function of relative humidity H (%):

$$e = e_{sat} \cdot H/100 \tag{6.4}$$

where e_{sat} denotes the saturated vapour pressure, i.e. the maximum amount of vapour that air at the mentioned temperature can contain. Typical values for mean temperature and relative humidity, taken from [20], and calculated values for the standard deviation of the signal amplitude variations σ, calculated following the procedure in [18], are summarised in Table 6.2. As shown in Table 6.2, the scintillation is higher during summer months.

Table 6.2 Standard deviations of signal amplitude variations as a function of meteorological parameters for some more critical locations in Europe

City	Month	Time	Temperature (°C)	Humidity (%)	N_{wet}	σ (dB)
Rome	August	13:00	30	43	74.2	0.357
Rome	December	13:00	13	70	47.8	0.272
Rome	August	7:00	20	73	74.2	0.357
Rome	December	7:00	6	85	38.1	0.240
Gibraltar	August	14:30	29	60	98.4	0.434
Nicosia	August	14:00	37	35	85.4	0.398

The scintillation may affect also the angle of arrival. The procedure to estimate the angle of arrival is given in the Recommendation ITU-R P.834 'Effects of tropospheric refraction on radiowave propagation' for frequencies up to 20 GHz and in [21]. The estimated value is 0.1° for the link elevation of 1°, which is much less than the antenna beamwidth specified for HAP broadband communications.

The Recommendation ITU-R P.618 describes a general method for predicting the cumulative distortion of the tropospheric scintillation at elevation angles higher than 4°. The method is derived for frequencies up to 14 GHz and is based on the ambient temperature and relative humidity of the site. The procedure in ITU-R P.618 is recommended up to a carrier frequency of 20 GHz. For higher frequencies the method proposed in [18] may underestimate the influence of scintillation. For higher frequencies better results can be obtained using the Tatarskii theory on propagation through turbulent media [22, 23], which also provides information about the spectral properties of the scintillation.

According to the theory, the power spectrum density of scintillation in clear sky is flat for frequencies lower than the corner frequency f_c and it has a slope of -80 dB/ decade at higher frequencies, all in log-log scale of the graph. The corner frequency can be deduced either from the theoretical model [24] or estimated from measurement campaigns [22]. Under some assumptions the theoretical value of the corner frequency can be calculated using:

$$f_c = 1.425 \cdot f_0 \left[\frac{1 - \frac{z_1}{z_2}}{1 - \left(\frac{z_1}{z_2}\right)^{11/6}} \right]^{3/8} \tag{6.5}$$

$$f_0 = \frac{1}{2\pi} \cdot V_t \cdot \sqrt{\frac{k}{z}} \tag{6.6}$$

where V_t is the transverse velocity of wind and k is the wave number. Furthermore, the relationship between the thickness L of the turbulent layer extending from z_1 to z_2 along the path and the path elevation angle θ can be expressed by:

$$L = z_2 - z_1 = H/\sin\theta \tag{6.7}$$

$$z = \frac{1}{2} \cdot (z_1 + z_2) \tag{6.8}$$

where H is the mean turbulent layer altitude. Typical values for the corner frequency are between 0.1 Hz and 0.6 Hz. The geometry of the model is shown in Figure 6.9.

Similar to satellite links, the signal fading due to scintillation depends on the length of the path through the turbulent layer also in HAP links. As shown in Figure 6.9, the length of the path through the turbulent layer strongly depends on the elevation angle. Consequently, the results obtained for scintillation from the satellite model are not directly applicable to HAP communications for the same reason as in the case of rain attenuation. Again a similar procedure can be used to introduce the elevation angle dependency in the expression of the corner frequency as for the rain attenuation. According to Figure 6.9, we express the path lengths z_1 and z_2 as functions of the mean turbulent layer altitude H, elevation angle θ and the altitude of the turbulent layer h:

$$z_1 = \frac{h}{\sin\theta} \quad z_2 = \frac{H+h}{\sin\theta} \tag{6.9}$$

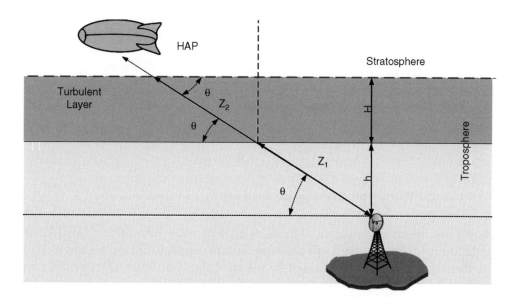

Figure 6.9 Geometry of the scintillation model for HAP operating environment

By inserting (6.9) in (6.5) the corner frequency can be expressed as:

$$f_c(\theta) = \frac{1.425}{2\pi} \cdot V_t \cdot \sqrt{k \cdot \frac{2 \cdot \sin\theta}{2h + H}} \cdot \left[\frac{1 - \frac{h}{H+h}}{1 - \left(\frac{h}{H+h}\right)^{11/6}} \right]^{3/8} \qquad (6.10)$$

Assuming the corner frequency $f_c(\theta_0)$ is known for an elevation angle θ_0, the calculation of the corner frequency for an arbitrary elevation angle is straightforward:

$$f_c(\theta) = \sqrt{\frac{\sin\theta}{\sin\theta_0}} \cdot f_c(\theta_0) \qquad (6.11)$$

The channel model block diagram for scintillation modelling is shown in Figure 6.10. It consists of a complex Gaussian noise random generator with adjustable standard deviation and a low-pass filter, which aim to guarantee the spectral density of the scintillation distortions.

6.3.3 Rain Fading

The water appears in the atmosphere in different forms either as clouds or precipitations such as rain drops, snowflakes, ice crystals, hail and graupel. At

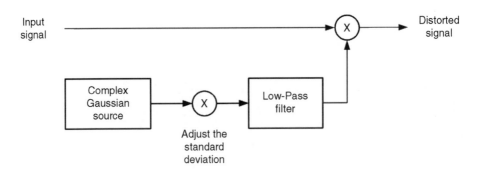

Figure 6.10   Block diagram of the scintillation channel simulator

any given instance of time more than one type of water form affects the HAP or satellite radio link, so their effects on radio link cannot be measured and treated separately. The attenuation caused by different forms of water is known as rain fading. More precisely, rain fading is a form of signal attenuation caused by precipitation, clouds and other meteorological phenomena, where precipitations have the most significant influence. Rain fading causes sporadic signal attenuation in satellite, terrestrial and HAP communication systems operating on all frequency bands, but the effect is becoming more detrimental at higher frequency bands in particular in millimetre and higher frequency bands. The impact of rain fading on communication system performance depends mainly on the rain rate, type of the rain (showers, heavy rain, drizzle, etc.) and the thickness of clouds. The meteorological data are available for all regions of the world, but mapping those data to radio channel models is not straightforward. Numerous methods are published in the scientific literature and a good overview is provided in [[19], Ch. 6], but the detailed study of proposed methods is beyond the scope of this book. However, ITU-R recommendations give methods to estimate the effects of rain fading on earth to space telecommunication systems, which may be applied also for HAP communication systems taking into consideration the constrains of HAP links.

Rain fading is typically characterised by the corresponding climate zone as defined by ITU-R. The ITU-R provides recommendations concerning the effects of rain on radio wave propagation. The Recommendation ITU-R P.618 [18] gives the method which may be applied to investigate the dependence of attenuation statistics on elevation angle, polarisation and frequency. This method uses rainfall maps given in Recommendation ITU-R P.837 'Characteristics of precipitation for propagation modelling' and the frequency dependent coefficient from Recommendation ITU-R P.838 'Specific attenuation model for rain for use in prediction methods'.

In addition, the Recommendation ITU-R P.618 [18] gives also an empirical formula for frequency scaling, for the case if reliable measured data on attenuation is available for one frequency and the channel model is required for another frequency:

$$A_2 = A_1 \cdot (\varphi_2/\varphi_1)^{1-H(\varphi_1,\varphi_2,A_1)} \tag{6.12}$$

where

$$\varphi(f) = \frac{f^2}{1 + 10^{-4} \cdot f^2} \tag{6.13}$$

and

$$H(\varphi_1,\varphi_2,A_1) = 1.12 \cdot 10^{-3} \cdot (\varphi_2/\varphi_1)^{0.5} \cdot (\varphi_1 \cdot A_1)^{0.55} \tag{6.14}$$

A_1 and A_2 represent rain attenuation with equal probability at frequencies f_1 and f_2 (in GHz), respectively. The method can be used in the frequency range from 7 to 55 GHz.

The methods proposed in ITU-R recommendations refer to average conditions and long term statistics, which are appropriate for calculating year outage probability. However, they are not sufficient for the design and analysis of communication systems, in particular the physical and MAC layers, where the time series generators are required for modelling long and short term propagation effects. Typical approaches for modelling rain attenuation are based either on converting meteorological data into channel attenuation time series or on setting the parameters of a time series attenuation generator according to the measurement results obtained by long term measurement campaigns [25].

In satellite communication systems, two basic approaches that constitute a good compromise between channel model accuracy and complexity have been used to model rain fading with the time series of attenuation:

- The Markov chain approach used in the ONERA simulator [25, 26];
- The DLR segment channel approach [25, 27].

The ONERA Markov channel model [26] is a switched channel model which consists of macroscopic and microscopic sub-models, with a component that combines the outputs of each. The microscopic sub-model gives the short term dynamic behaviour of rain attenuation. It is modelled as an N-state Markov chain, which generates the time series with high resolution [26]. The macroscopic sub-model, on the other hand, is a 2-state Markov chain representing the rain state and clear sky state. It follows the long term behaviour of rain fading, taking into account the mean duration of rain events. The parameters of macroscopic models can be derived from the radio meteorological data banks such as ITU-R or the European Centre for Medium-Range Weather Forecasts (ECMWF) [28]. The ITU-R Recommendation P.837 provides the probability of rain averaged over 1 year, from which all necessary macroscopic parameters of the model can be calculated.

The switched channel model proposed by DLR, known as the DLR segment channel approach, is also based on a Markov chain and a Gaussian random variables generator [25, 27]. It consists of a generic part and a specific set of parameters that allow adjustment of the channel model to different elevation angles, carrier frequencies and climatic zones. The model specifies three types of channel attenuation segment:

- The channel attenuation segment with almost constant received power, referred to as the C-segment.
- The channel attenuation segment with mainly monotonously decreasing channel attenuation, referred to as the D-segment.
- The channel attenuation segment with mainly monotonously increasing channel attenuation, referred to as the U-segment.

Each segment is characterised by the conditional probability distribution $P(y|x)$, which is estimated by statistical processing of data obtained from a measurement campaign. The conditional probability distribution $P(y|x)$ denotes the likelihood that the current channel attenuation (in dB) is y conditioned that $\Delta\tau$ seconds before the attenuation has been x. Satellite channel measurements show that $\Delta\tau$ is around 1 s and that $P(y|x)$ obeys a Gaussian distribution. The standard deviation and mean value of the Gaussian distribution depend on the segment (C/D/U), the current attenuation and on location (latitude, longitude) of measurements and satellite elevation angle. Switching between attenuation segments of the channel is determined by calculating the difference between channel attenuation in successive time intervals:

$$\Delta a(iT) = a(iT) - a[(i-1) \cdot T] \tag{6.15}$$

where $a(iT)$ denotes the attenuation at ith time (interval):

$$\Delta a(iT) = \begin{cases} |a(iT) - a[(i-1) \cdot T]| < 1 \text{ dB}, & \text{C segment} \\ a(iT) - a[(i-1) \cdot T] \geq 1 \text{ dB}, & \text{D segment} \\ a(iT) - a[(i-1) \cdot T] \leq 1 \text{ dB}, & \text{U segment} \end{cases} \tag{6.16}$$

The algorithm of the DLR segment model can be summarised in the following three steps [25]:

1. Calculate the attenuation segment type from $a(iT)$, $a[(i-1)T]$.
2. Obtain the mean value and standard deviation from the lookup table, taking into account attenuation segment type and current attenuation $a(iT)$.
3. Generate the attenuation for the next time interval $a[(i+1)^.]$.

A rain fading event with the C, D and U segments is illustrated in Figure 6.11. The lookup tables [16, 25, 27] for mean value and standard deviation of the Gaussian process are calculated by processing measurements obtained from the measurement campaign at DLR Obrepfaffenhofen (11.3° East, 48.1° North) with the satellite elevation angle equal to 34.8°, the ITU-R climate zone K and the carrier frequency 40 GHz. The analysis shows that the standard deviation and mean value depend significantly on the season of the year.

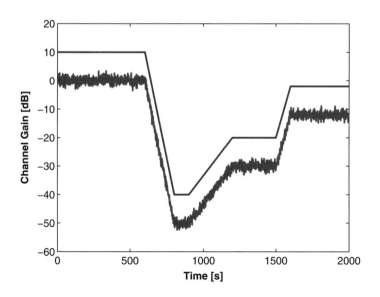

Figure 6.11   DLR segment approach for rain fading model

The DLR segment model based on the measurement campaign described above can be adapted for rain fading modelling in HAP communication systems. In particular, the carrier frequency and the elevation angles for HAP communication systems are significantly different from those used in the DLR measurement campaign [25, 27], resulting in different signal path length affected by rain. Thus, the DLR segment model should be adjusted by introducing the frequency correction term and normalising the rain attenuation on a per kilometre basis.

The model proposed by the CCIR Report 721-1 [29] and accepted by the ITU-R P-series [30] can be applied to correct the attenuation due to different carrier frequencies:

$$\frac{A_1}{A_2} = \frac{g(f_1)}{g(f_2)} \tag{6.17}$$

where A_1 and A_2 are attenuations at frequencies f_1 and f_2, respectively, and $g(f)$ is defined by:

$$g(f) = \frac{f^{1.72}}{1 + 3 \cdot 10^{-7} \cdot (f^{1.72})^2} \tag{6.18}$$

where f is the carrier frequency (in GHz).

The results published in [25], were obtained for fixed elevation angle and consequently for fixed length of the signal path affected by the rain. On the other hand, HAP can be seen at different elevation angles due to its movement around the nominal position, thus the length of the signal path affected by rain is changing. In order to capture this effect, the results obtained from the DLR sequence model should be normalised to 1 km, using Recommendation ITU-R P. 618 [18] for slant path length calculation according to the following procedure.

1. Calculate the effective rain height h_R (in km):

$$h_R = \begin{cases} 3.0 + 0.028 \cdot \phi & 0° \le \phi < 36° \\ 4.0 - 0.075 \cdot (\phi - 36) & 36° \le \phi \end{cases} \tag{6.19}$$

where ϕ is the latitude of the Earth station.

2. Calculate the slant path length L_S (in km) below the rain height:

$$L_S = \begin{cases} \dfrac{h_R - h_S}{\sin\theta} & \theta \le 5° \\ \dfrac{2 \cdot (h_R - h_S)}{\left(\sin^2\theta + \dfrac{2 \cdot (h_R - h_S)}{R_e}\right)^{1/2} + \sin\theta} & \theta > 5° \end{cases} \tag{6.20}$$

where θ is the elevation angle, h_S is the ground station's altitude above sea level (in km), and R_e is the equivalent radius of the Earth.

3. Multiply the distance from HAP (in km) by the normalised attenuation to obtain the rain attenuation (in dB).

An example of a time series representing channel gain is illustrated in Figure 6.12, calculated for 15 days in spring and for a fixed link in a HAP communication system. Sporadic rain events can be observed causing attenuation from a few dB to almost 15 dB. Such rain attenuation typically results in severe degradation of system performance, so it needs to be compensated by suitable fade mitigation techniques. The dynamic range of channel variation shown in this example, i.e. 15 dB (Figure 6.12), can easily be treated with the adaptive coding and modulation approach proposed for instance in DVB-S2 standard [31], assuming slow channel variation.

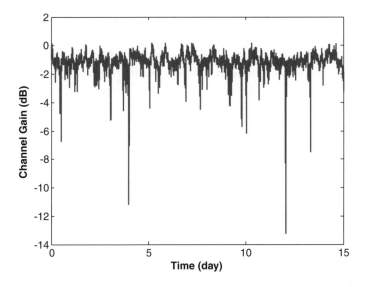

Figure 6.12   Rain attenuation time series generated for HAP operating environment

6.3.4   Rain Fading and Scintillation

The scintillation may become a relevant source of errors only for low elevation angles. However, in combination with rain fading the scintillation may significantly contribute to signal distortions also at higher elevation angles. The conditional average standard deviation of tropospheric scintillation seems to be linked to attenuation by the power law [22]. It can be predicted by a very simple physical model, such as a turbulent thin layer aloft during the rain event that indicates an increase of tropospheric scintillation during rain. The correlation between the rain attenuation and the scintillation is summarised in the following equation [22]:

$$\sigma = \begin{cases} \sigma_0 & \text{if } A < 1 \text{ dB} \\ \sigma_0 \cdot A^{5/12} & \text{if } A > 1 \text{ dB} \end{cases} \tag{6.21}$$

where σ_0 (in dB) is the standard deviation of scintillation before the rain event, σ is the standard deviation of the scintillation and A is the attenuation due to the rain event. The correlation between the rain attenuation and scintillation does not provide any knowledge about the spectral properties of the scintillation. This is given in [22, 32] where Tatarskii's theory [23] on propagation through a turbulent media is applied to obtain the spectrum of scintillation in clear sky. In the example illustrating scintillation effect in [17] median values were used of the standard deviation of scintillation σ_0, elevation angle θ_0 and corner frequency $f_c(\theta_0)$. These values, i.e. $\theta_0 = 37.8°$, $f_c(\theta_0) = 0.45$ Hz and $\sigma_0 = 0.099$, were obtained by the measurements during the Intelsat experiment in 1994 [22].

Three examples of fading due to scintillation are illustrated in Figure 6.13 [17] for the elevation angle of 38°. The lower trace represents the scintillation in the channel without rain (rain attenuation equal to 0 dB), exhibiting the range of amplitude variation equal to approximately 0.01 dB. The upper trace shows shifted

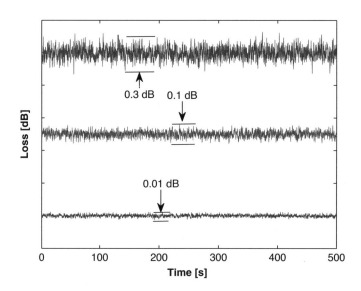

Figure 6.13 Attenuation time series due to the scintillation effect generated for HAP operating environment

scintillation loss for rain attenuation of 20 dB. In the latter case the amplitude variation is in the range of approximately 0.3 dB. The middle trace (shifted) is generated for rain attenuation of 17 dB which results in amplitude variation of 0.1 dB.

According to the proposed model the standard deviation of the scintillation is higher for higher rain attenuation, which is clearly seen from the generated time series of scintillation fading. Due to the same elevation angles in both cases the corner frequency is the same for both examples. At low elevation angles the path through the turbulent layer is longer, resulting in higher attenuation due to rain and lower corner frequency, with the consequence of slower variation of fading due to scintillation.

6.3.5 Influence of Hydrometeor Effects on Cross Polarisation

The space communication link capacity can be increased by applying orthogonal polarisation and thus reusing frequencies. However certain propagation conditions, such as the presence of ice crystals and nonspherical raindrops, cause the unwanted interference appearing from the signal transmitted over one polarisation in the orthogonal one. In addition, the nonideal antennas will also contribute to cross polarisation effect.

A common measure of the amount of cross polarisation is the cross polarisation discriminator (XPD) defined as a logarithmic ratio between the desired received power P_{cpol} and unwanted received power due to cross polarisation P_{xpol}:

$$\text{XPD} = 10 \cdot \log \frac{P_{\text{cpol}}}{P_{\text{xpol}}} \qquad (6.22)$$

The antenna contribution to cross polarisation discrimination is 30 dB in the boresight direction and down to 20 dB at cell edge for antennas used for broadband HAP communications.

Recommendation ITU-R P.618 [18] gives a method for prediction of the long term statistics of cross polarisation considering the long term statistics of rain attenuation. The cumulative XPD at the millimetre frequency band depends on:

1. A frequency dependent term $C_f = 26 \cdot \log f + 4.1$ for frequencies $9\,\text{GHz} \leq f \leq 36\,\text{GHz}$.
2. A term C_A which depends on rain attenuation exceeded for the required percentage of time A_p: $C_A = 22.6 \cdot \log A_p$ for frequencies $20\,\text{GHz} \leq f \leq 40\,\text{GHz}$.
3. An elevation angle dependent term $C_\theta = -40 \log (\cos\theta)$ for $\theta \leq 60°$.
4. A term that gives polarisation improvement $C_\tau = 10 \cdot \log \{1 - 0.484 \cdot [1 + \cos (4 \cdot \tau)]\}$.
5. A rain drop canting term $C_\sigma = 0.0053 \cdot \sigma$, where σ is the effective standard deviation of the raindrop canting angle distribution.

The XPD due to rain can be expressed as:

$$\text{XPD}_{\text{rain}} = C_f - C_A + C_\theta + C_\tau + C_\sigma \tag{6.23}$$

The ice dependent XPD term is:

$$C_{\text{ice}} = \text{XPD}_{\text{rain}} \cdot (0.3 + 0.1 \cdot \log p)/2 \tag{6.24}$$

where p denotes the probability the XPD does not exceed the value. The XPD not exceeded for $p\%$ of time is calculated by:

$$\text{XPD}_p = \text{XPD}_{\text{rain}} - C_{\text{ice}} \tag{6.25}$$

The influence of the XPD on HAP communications is important only for fixed communications with LOS propagation conditions from the transmitter to the receiver. The preliminary calculation for broadband communications between HAP and fast train shows that most of the time the XPD will exceed 25 dB, which is comparable with distortions introduced by nonideal antennas. Additional measures to improve the XPD can be added in the system with additional signal processing based on the latest discovery in multiple input multiple output communication scenarios. In addition, reflected radio signal may cause more significant distortion than the cross polarisation and hydrometeor effects.

6.3.6   The Effects of Surrounding Environment

In the millimetre frequency band a directional antenna is applied to compensate high free space loss by high antennas gain. The directional antennas can be seen as a form of spatial filter, as they filter out reflected radio waves, which are approaching the receiver out of the main lobe. Consequently, the amount of multipath distortions in the millimetre frequency band is notably reduced. However, any obstacle, such as a tree, building, hill or mountain, causes shadowing or even blocking of the radio signal.

In fixed HAP communication systems, similar as in space communications, the highly directional antennas are applied and the transmitted signals typically exhibit LOS propagation conditions. Consequently, the distortions due to reflected radio waves are negligible or very small. The additive white Gaussian noise channel is a common approach in channel modelling in the former case, while the Ricean channel is used for the latter case [2, 12, 33].

In mobile HAP communication systems, mobile terminals are equipped with antennas that have wider beamwidth in order to be able to track the satellite or aerial platform. This results in receiving a higher amount of reflected rays and in the need for a more advanced probability density function to model the space and HAP channels [12, 34]. In addition to the movement of the terminal, Doppler frequency shift is introduced into the signal and the channel is alternating between the LOS, shadowed and completely blocked states. The phenomenon is well known from land mobile satellite communications, but the rate and reason why the channel is switching from one state to the other is different.

The channel model for mobile communication systems can be derived from the satellite mobile channel [13, 15]. In land mobile satellite communications, the visibility of satellites is often blocked by different obstacles such as buildings, hills, mountains, etc. As a result the channel model can be in two extreme propagation conditions: LOS conditions with strong direct component, and shadowed conditions, where the received signal is a superposition of reflected rays. So, the propagation channel cannot be characterised by only one set of parameters or by a single statistical distribution. The communication channel behaves similarly in HAP communication systems, where environmental properties change when the mobile moves from one location to another. Channel measurements obtained for the land mobile satellite systems show that statistical channel properties do not change over an area with similar environmental attributes, for example when the signal is shadowed. Considering this, the HAP and Land Mobile Satellite (LMS) channel can be modelled in two stages [12, 35]:

- The first stage is concerned with modelling a very slow channel variation by dividing the area of interest into sub-areas with constant environmental attributes and finding the transition probabilities from one sub-area to another.
- The second stage is focused on modelling slow and fast channel variation by determining the statistical channel properties for each sub-area.

The first-order Markov chain is widely applied [10, 12, 34–36] to model the first stage and different probability density functions such as Rice, Rayleigh and lognormal to model the second stage. A general channel model considering the above approach is depicted in Figure 6.14. The model consists of two main parts: a finite state machine and a set of statistical distributions with associated parameters for different operating conditions. In Figure 6.14 we assume LOS, shadowed and blocked channel conditions.

The first-order Markov chain is a stochastic process with a number of discrete states, in which the probability of being in a given state depends only on the previous state. The Markov chain is characterised by:

- state probability vector $\mathbf{W} = [w_1, w_2, \ldots, w_M]$
- state transition probability matrix

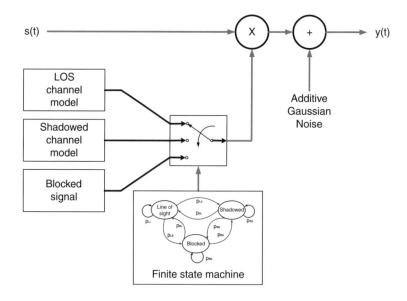

Figure 6.14   A general propagation channel model for HAP communication systems

$$\mathbf{P} = \begin{bmatrix} p_{11} & p_{12} & \cdots & p_{1M} \\ p_{21} & p_{22} & \cdots & p_{2M} \\ \vdots & \vdots & \ddots & \vdots \\ p_{M1} & p_{M2} & \cdots & p_{MM} \end{bmatrix} \qquad (6.26)$$

The elements of the matrix **P**, p_{ij}, denote the probability that the state is changed from ith to jth and can be calculated as $p_{ij} = \frac{N_{ij}}{N_i}$, where N_i is the number of frames (i.e. time intervals) in state i, while N_{ij} is the number of transitions from state i to state j. The elements w_j of the vector **W** are calculated as $w_i = \frac{N_i}{N_f}$, where N_f is the total number of frames. The vector **W** and matrix **M** describe the following matrix equations: **W·P** = **W** and **W·E** = **I**, where **E** is a column vector whose entries are 1s and **I** is the identity matrix.

In a mobile operating scenario at millimetre frequency bands, the user terminal is usually equipped with narrow beamwidth highly directional antenna. The narrow beamwidth antenna limits the multipath effect, making reflected signal components negligible with respect to directly received signal. Thus we can assume that the received signal is mainly distorted by the additive Gaussian noise. Due to the motion of the user terminal, the vegetation, such as trees, can block the LOS towards the HAP. When this happens the direct path does not exist and the received radio signal is highly attenuated and scattered on leaves. The signal can also be completely blocked by obstacles such as buildings, hills or mountains. In such a scenario one can use a 3-state flat fading model to describe a stratospheric channel. The first state represents the LOS propagation conditions, the second state denotes the situation when the signal is partially shadowed by trees, while the third state represents the channel that is completely blocked.

Direct application of parameters measured for GEO satellite systems and published in [34] is not appropriate for HAP communication systems since the elevation angle of a GEO satellite is constant. For that reason the use of a digital relief model was proposed to estimate the parameters of a HAP channel model using the ray-tracing method [13]. A channel state variation for such systems is depicted in Figure 6.15.

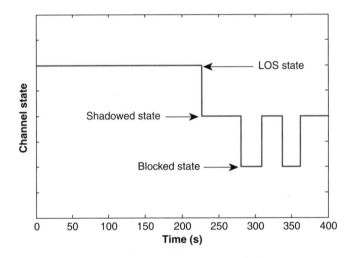

Figure 6.15 Channel state variation for HAP communication systems

Parameters of a statistical channel model can be derived from the parameters of the empirical channel model by calculating the state transition probabilities from the channel state to time dependency. If we assume that the channel property changes every 10 ms, i.e. the channel stays in a given state at least 10 ms, the elements of the state transition probability matrix **P** can be obtained by counting the state transitions and applying normalisation. The state probability vector **W** is obtained by simulating the finite state machine for 100 000 time intervals. As an example the vector **W** and matrix **P** obtained for a railway line in Slovenia [13] are given below:

$$\mathbf{P} = \begin{bmatrix} 0.958469945 & 0.03442623 & 0.007103825 \\ 0.011483067 & 0.966329311 & 0.022187622 \\ 0.002448862 & 0.015989628 & 0.98156151 \end{bmatrix} \quad (6.27)$$

$$\mathbf{W} = \begin{bmatrix} 0.1319 & 0.3722 & 0.4959 \end{bmatrix} \quad (6.28)$$

A similar matrix $\mathbf{P}$ and vector $\mathbf{W}$ can be obtained for an arbitrary propagation environment such as mountainous, hilly, flat and urban.

The Markov chain can model very slow changes of the channel characteristics caused by large obstacles such as hills and mountains. The slow channel variation caused by nonuniform receive antenna patterns, changes of mobile orientation with respect to the stratospheric platform in LOS conditions, or shadowing variation behind a group of trees due to different orientation of leaves or branches, are usually modelled by lognormal distribution. The typical correlation distance, describing how fast the channel is changing, is for slow fading in satellite channels between 1 m and 3 m [34]. The phenomenon is similar for HAP communication systems and can be directly applied to the HAP channel model.

The different probability density functions are proposed for modelling slow channel variations in blocked, shadowed and LOS state. A straightforward approach is to use Rice distribution with k factor around 20 dB for LOS state and Rayleigh distribution with additional attenuation for shadow state of approximately 20 dB, as

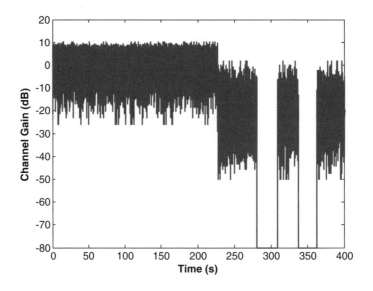

Figure 6.16   Channel attenuation of the HAP channel

used in [36]. The Loo distribution proposed in [34], which is the combination of lognormal and Rayleigh distribution, opens opportunities to use one distribution for all channel states applying only state specific parameters. The channel attenuation for a time interval of 400 s is plotted in Figure 6.16. The LOS parameters and deep shadow parameters for leaf trees from Figure 6.16 are inserted in a channel simulator. The channel attenuation is assumed constant for 10 ms, which corresponds for instance to 10 frames in the IEEE 802.16 standard, suggested as one of candidate standards for HAP communications. The switching between different channel states is visible. The attenuation of the channel in the blocked state is set to −80 dB. Note that the free space loss is not considered in results shown in Figure 6.16 but that is only an additive constant dependent on the distance between the platform and the mobile terminal.

6.4 Conclusion

The understanding and modelling of channel impairments are of vital importance for the design of any communication system. This chapter provides an overview of some activities in HAP channel modelling, taking as the main assumption the situation where no data are available from measurement campaigns. This means that the HAP channel design is based on results from satellite and terrestrial mobile channel measurements and the corresponding channel models.

Although no data from long term measurement campaigns are available at the moment, the development of a HAP channel model remains one of the most important research topics. For instance, based on some successful trials [37] the short data sequences are available, which may be applied to modelling the effects of surrounding objects on the radio signals [38, 39]. In addition to previously mentioned activities in channel modelling, the theoretical approach using basic principles of physics is always attractive. Up to date information about ongoing research in the field of channel modelling in stratospheric communications can be found in [40].

References

1 Resolution 734 (Rev.WRC-07), *Studies for Spectrum Identification for Gateway Links for High Altitude Platform Stations in the Range from 5850 to 7250 MHz*, World Radiocommunication Conference, Geneva, Switzerland, 2007.
2. S. R. Saunders, *Antennas and Propagation for Wireless Communication Systems*, 1st Edn, John Wiley & Sons, Ltd, London, 2005.
3. G. E. Corazza, *Digital Satellite Communications*, Springer, New York, 2007.
4. M. F. Catedra and J. Perez, *Cell Planning for Wireless Communications*, Artech House, Inc., Norwood, MA, 1999.

5. M. Hata and T. Nagatsu, *Mobile Location Using Signal Strength Measurements in a Cellular System*, IEEE Trans. Veh. Technol., May 1980, Vol. 29, No. 2, pp. 245–255.

6. P. E. Mogensen, P. Eggers, C. Jensen and J. B. Andersen, *Urban Area Radio Propagation Measurements at 955 and 1845 MHz for Small and Micro Cells*, Proc. IEEE Global Telecommunications Conference, GLOBECOM '91, Phoenix, AZ, USA, 1991, Vol. 2, pp. 1297–1302.

7. G. L. Stuber, *Principles of Mobile Communications*, Kluwer Academic, Boston/Dordrecht/London, 2001.

8. A. F. Molisch, *Wireless Communications*, John Wiley & Sons, Ltd, Chichester, 2005.

9. V. Erceg, K. V. S. Hari, M. S. Smith, D. S. Baum, P. Soma, L. J. Greenstein, D. G. Michelson, S. Ghassemzadeh, A. J. Rustako, R. S. Roman, K. P. Sheikh, C. Tappenden, J. M. Costa, C. Bushue, A. Sarajedini, R. Schwartz, D. Branlund, T. Kaitz and D. Trinkwon, *Channel Models for Fixed Wireless Applications*, IEEE 802.16.3c-01/29r1, 23 February, 2001.

10. E. Lutz, D. Cygan, M. Dippold, F. Dolainsky and W. Papke, *The Land Mobile Satellite Communication Channel-Recording, Statistics and Channel Model*, IEEE Trans. Veh. Technol., May 1991, Vol. 40, No. 2, pp. 375–386.

11. M. A. Vazquez-Castro and F. Perez-Fontan, LMS Markov model and its use for power control error impact analysis on system capacity, IEEE J. Selected Areas Commun., 2002, Vol. 20, No. 6, pp. 1258–1265.

12. E. Lutz, M. Werner and A. Jahn, *Satellite Systems for Personal and Broadband Communications*, Springer, Berlin, 2000.

13. S. Plevel, T. Javornik, M. Mohorcic and G. Kandus, *Empirical Propagation Channel Model for High Altitude Platform Communication Systems*, Proc. 11th Ka and Broadband Communications Conference, Rome, Italy, 25–28 September 2005, pp. 187–194.

14. Ericsson Radio Systems AB, *TEMSTM CellPlanner 3.4 User Guide*, 2001.

15. T. Javornik, M. Mohorčič, A. Švigelj, I. Ozimek and G. Kandus, *Adaptive Coding and Modulation for Mobile Wireless Access via High Altitude Platforms*, Wireless Personal Communications, 2005, Vol. 32, No. 3–4, pp. 301–317.

16. A. Aragon-Zavala, J. L. Cuevas-Ruiz and J. A. Delgado-Penin, *High-Altitude Platforms for Wireless Communications*, John Wiley & Sons, Ltd, Chichester, 2008.

17. G. Kandus, M. Mohorčič, M. Smolnikar, E. Leitgeb and T. Javornik, *A Channel Model of Atmospheric Impairment for the Design of Adaptive Coding and Modulation in Stratospheric Communication*, WSEAS Trans. Commun., April 2008, Vol. 7, No. 4, pp. 311–326.

18. ITU-P618, 97. *ITU-R Propagation Recommendation P.618-10: Propagation Data and Prediction Methods Required for the Design of Earth–Space Telecommunication Systems,* International Telecommunication Union, Geneva, Switzerland, 2009.

19. L. Barclay, *Propagation of Radio Waves*, The Institution of Electrical Engineers, London, 2003.

20. E. A. Pearce and C. G. Smith, *The World Weather Guide*, Hutchinson, London, 1990.

21. E. Vilar, G. Weaver, P. Lo and H. Smith, *Angle of Arrival Fluctuations in High and Low Elevation Earth Space Paths*, Proc. International Conference on Antennas and Propagation (ICAP 85), Coventry, UK, 1985.

22. E. Matricciani, M. Mauri and C. Riva, *Relationship between Scintillation and Rain Attenuation at 19.77 GHz*, Radio Science, 1996, Vol. 31, No. 2, pp. 273–279.

23. V. I. Tatarskii, *Wave Propagation in a Turbulent Media*, McGraw-Hill, New York, 1961.

24. D. Vanhoenacker-Janvier and H. Vasseur, *Prediction of Scintillation Effects on Satellite Communications above 10 GHz*, IEE Proc. Microwaves, Antennas and Propagation, 1995, Vol. 142, No. 2, pp. 102–108.

25. U.-C. Fiebig, L. Castanet, J. Lemorton, E. Matricciani, F. Pérez-Fontán, C. Riva and R. Watson, *Review of Propagation Channel Modelling*, Proc. 2nd Workshop of the COST 272-280 Action 'Propagation Impairment Mitigation for Millimetre Wave Radio Systems', Noordwijk, The Netherlands, May 2003.

26. L. Castanet, T. Deloues and J. Lemorton, *Methodology to Simulate Long-Term Propagation Time-Series from the Identification of Attenuation Periods Filled with Synthesized Events*, Proc. 2nd Workshop of the COST 272-280 Action 'Propagation Impairment Mitigation for Millimetre Wave Radio Systems', Noordwijk, The Netherlands, May 2003.

27. U.-C. Fiebig, *A Time-Series Generator Modelling Rain Fading*, Proc. Open Symposium on Propagation and Remote Sensing, URSI Commission F, Espoo, Finland, October 2002.

28. European Centre for Medium-Range Weather Forecasts, http://www.ecmwf.int.
29. CCIR Report 721-1, Attenuation by Hydrometers, in Precipitation, and Other Atmospheric Particles, Propagation in Non-Ionized Media, CCIR, Geneva, Switzerland, 1980.
30. ITU-R P Radiowave propagation (P-series) of recommendations, 2002.
31. ETSI EN 302 307 (V1.1.2), *Digital Video Broadcasting (DVB): Second Generation Framing Structure, Channel Coding and Modulation Systems for Broadcasting, Interactive Services, News Gathering and Other Broadband Satellite Applications*, European Telecommunications Standards Institute (ETSI), June 2006.
32. F. Lacoste, J. P. Millerioux, L. Castanet and C. Riva, *Generation of Time Series of Scintillation Combined with Rain Attenuation*, Proc. ECPS 2005 Conference, Brest, France, March 2005.
33. G. Maral, *Satellite Communications Systems: Systems, Techniques and Technologies*, John Wiley & Sons, Ltd, Chichester, 2002.
34. F. P. Fontan, M. Vazquez-Castro, C. E. Cabado, J. P. Garcia and E. Kubista, *Statistical Modeling of the LMS Channel*, IEEE Trans. Veh. Technol., November 2001, Vol. 50, No. 6, pp. 1549–1567.
35. B. Vučetić and J. Du, *Channel Modeling and Simulation in Satellite Mobile Communication Systems*, IEEE J. Selected Areas Commun., October 1992, Vol. 10, No. 8, pp. 1209–1218.
36. G. Sciascia, S. Scalise, H. Ernst and R. Mura, *Statistical Characterization of the Railroad Satellite Channel at Ku-Band*, International Workshop of Cost Actions 272 and 280, Noordwijk, The Netherlands, May 2003.
37. CAPANINA Project Website, http://www.capanina.org/.
38. T. Celcer, T. Javornik, M. H. Capstick, M. Mohorcic and G. Kandus, *Analysis of HAP Propagation Channel Measurement Data*, Proc. 15th IST Mobile and Wireless Summit, Myconos, Athens, Institute of Communications and Computer Systems, National Technical University of Athens, 4–8 June 2006, p. 5.
39. J. Holis and P. Pechac, *Elevation Dependent Shadowing Model for Mobile Communications via High Altitude Platforms in Built-Up Areas*, IEEE Trans. Ant. Prop., 2008, Vol. 56, No. 4, pp. 1078–1084.
40. COST 297 - High Altitude Platforms for Communications and Other Services Website, http://www.hapcos.org.

7

FSO in HAP-Based Communication Systems

7.1 Applicability of FSO Technology to HAP Networks

Optical communication is now accepted as the primary form of high-speed high-capacity broadband communication, with optical fibre deployed on the ocean bed connecting continents across the globe. The wireless implementation of optical communication cannot achieve the same performance parameters as fibre but is nevertheless a promising candidate for broadband wireless connectivity, under certain conditions.

In optical wireless communication, commonly termed free space optics (FSO), the transmitted data are modulated onto a light beam, usually in the form of a laser. The light beam propagates unguided through the transmission channel, which is the atmosphere for terrestrial applications such as metropolitan building-to-building links. These links are often dubbed 'last mile solutions' as they connect end-users to the fibre backbone when all-fibre solutions are not possible or not practical and they are typically of ranges of more or less a mile. In order to benefit from the multitude of hardware developed for optical fibre communication systems a common wavelength for urban FSO is 1.55 μm, which is the preferred wavelength for fibre communication due to the material characteristics of the fibre. However, other carrier wavelengths have also been used in FSO systems. In all-space applications, namely satellite-to-satellite communications, the propagation channel is, indeed, free space and thousands of kilometres can be spanned by a single FSO link as illustrated in Figure 7.1. At the other end of the scale of transmission ranges, indoor FSO using infrared radiation has become a familiar feature of domestic life. For

Broadband Communications via High Altitude Platforms   David Grace and Mihael Mohorčič
© 2011 John Wiley & Sons, Ltd

instance, wireless remote control units for televisions and air conditioners operate using both line of sight (LOS) (also termed 'point and shoot) and non line of sight (NLOS) diffuse links. In contrast, LOS links are imperative for outdoor FSO. This is one of the major differences between outdoor FSO and classic radio frequency (RF) systems.

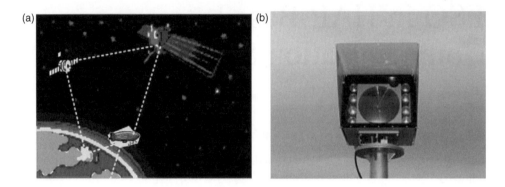

Figure 7.1 (a) Space and (b) terrestrial FSO (reproduced from [1] and [2], respectively)

Over the last half century, FSO has gained recognition as a feasible high-bandwidth wireless communication alternative with distinct advantages over the more familiar RF systems. The major meritorious features of FSO, which will be discussed in the context of HAP applications, are:

- very high bandwidths;
- very small and compact equipment;
- low power consumption;
- high security with low probability of interception;
- exemption from licensing restrictions and tariffs.

The high transmission bandwidth afforded by optical frequencies facilitates data rates of up to several Gbps in a point-to-point link under the most favourable circumstances, although 155 Mbps is often quoted as a typical high-speed FSO link capacity. This capacity can be significantly enhanced using wavelength division multiplexing (WDM) techniques, whereby data are transmitted simultaneously over the same channel on different carriers. High capacity is a desirable feature for HAP-based communication and expands the potential utility of this technology.

The lightweight and compact hardware associated with FSO has won it a place of respect in satellite communication systems, where bulk and mass are at a premium.

In HAP applications these features are equally important and render FSO systems worthy of careful consideration.

The highly directional and narrow laser beam that is the carrier for FSO is the source of high energy efficiency in the transmission budget and leads to low power consumption. However, it is also the cause of one of the drawbacks of FSO, which is the stringent requirement for aligning the transmitter and receiver in LOS links. In the HAP context the low power consumption leads to significant savings in on-board energy sources but the need for alignment presents a major challenge, as discussed below.

An additional benefit of the directionality of the transmitter beam is the robustness of FSO to interference, jamming and eavesdropping. The high security offered by FSO has made it the modality of choice for applications where these factors are critical. The further facility of quantum cryptography, a unique optical technology for making data transmissions secure, is an emerging technology with great promise for guaranteeing privacy and confidentiality. This feature may not be central to HAP-based systems per se, but is of increasingly recognised importance within wireless communication altogether.

The last-mentioned of the major meritorious features of FSO is the exemption of FSO transmissions from licensing fees and tariffs. This is primarily an economic rather than a technical aspect of FSO. However, awaiting regulatory approval for the deployment of a wireless communication system can delay operation substantially and this can be very significant when rapid deployment is important such as covering a disaster area. Rapid deployment is known to be one of the advantageous characteristics of HAPs, so that exemption from regulatory restrictions is, in fact, a significant factor promoting the implementation of HAP broadband links using FSO.

One drawback of FSO has been mentioned – the strict alignment requirements – and this is nontrivial in the case of HAP communication where station-keeping is a major challenge. Even in the case of relatively stable HAP positioning each vessel may vibrate due to mechanical vibrations and tracking noise. Pointing, acquisition and tracking (PAT) subsystems would be an essential part of an FSO communication system on-board the HAP and this would unfortunately increase the payload. However, it may well transpire that PAT subsystems are also necessary for RF communications and hence would be an unavoidable element in all HAP implementations, whether RF or FSO. Furthermore, in optical communication, where the beam is highly directional, the PAT system could provide accurate real-time position data of the HAP that may be useful for time-stamping remote sensed imagery and for other location-sensitive applications.

Two more features challenge the operability of FSO. The first is the potential health hazard of laser optic power. This can be checked by maintaining power density levels below the maximum allowable for eye safety in the vicinity of a human population. However, eye-safety risks in HAP FSO require careful consideration of human presence in the proximity of the transmitter and along the LOS trajectory, such as pilots in manned high altitude aircraft. Luckily, extremely sensitive optic receivers are available and these reduce the power requirement for transmissions, which is in the interest both of safety and of energy savings.

The second, and major, issue restricting the widespread implementation of FSO is the detrimental effect of the atmosphere on optical wireless communication. This is manifested in several ways and can be divided into absorption, scattering and turbulence effects. Atmospheric degradation of optic signals will be expanded upon below.

7.1.1   Atmospheric Effects

The atmosphere enfolding the planet Earth is a protective layer of gases containing a suspension of aerosols and is the propagation channel for HAP-to-ground and HAP-to-HAP wireless links. The molecules and particles in the atmosphere affect the optic power in a propagating light beam by means of absorption and scattering. The atmospheric composition changes over latitude and is subject to seasonal changes as well as random variations due to natural and anthropogenic causes. As a consequence, any comment on the impact of the atmosphere on light propagation must be recognised as a general guideline and experimental studies are necessary to verify the precise characteristics of any given channel. Predictions of atmospheric transmissions on a vertical path, derived using simulation software, can be seen in Figure 7.2.

Absorption is a quantum process whereby photons of light are absorbed by molecules in the propagation path. It is highly wavelength sensitive and can severely challenge the operability of an FSO link. The absorption characteristics of the atmosphere have been studied for many decades and the preferred wavelengths for FSO are well known. Notably, light absorption by water molecules, oxygen, carbon dioxide and ozone has precluded the use of much of the visible and infrared spectrum. The major windows are found between 1.5 μm and 1.8 μm and in the mid and far infrared as well as at selected bands in the visible range. The net effect of absorption in the propagation path is attenuation. Increasing the transmission power, within the limitations imposed by eye safety, can overcome moderate signal degradation due to absorption, once a suitable radiation wavelength has been selected.

It can be seen from Figure 7.2 that transmission varies with radiation wavelength from near zero to near unity for a given propagation path. In Figure 7.2(a) we see the transmission in the visible and near infrared spectral range. The afore-mentioned

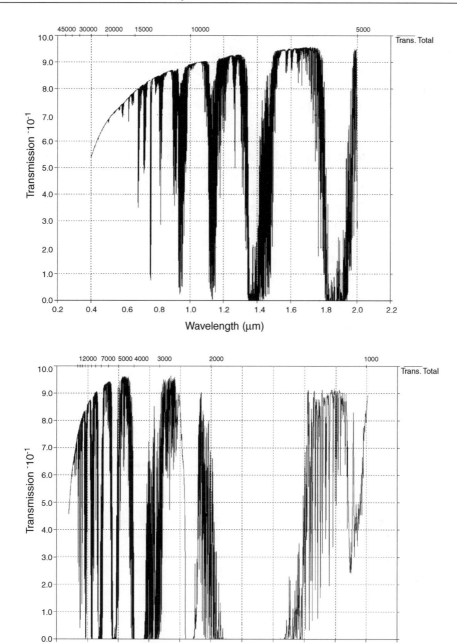

Figure 7.2 Transmission of a 20 km vertical path to Earth for radiation ranging from 400 nm to 10 μm; derived using MODTRAN software from the Ontar Corporation and assuming a rural aerosol content with visibility 23 km. (a) Transmission for the range 400 nm–2.0 μm. (b) Transmission for the range 0.4–10 μm

preferred wavelength of 1.55 μm can be seen to be well situated within an absorption window. Other wavelengths with high transmission are around 850 nm and in the region of 1 μm. In Figure 7.2(b) we see a broader spectral range with somewhat lower resolution. The absorption window between 8 μm and 13 μm is clearly evident - juxtaposed to a spectral range of 5–8 μm where the radiation is almost totally absorbed.

Scattering of a propagating light beam is also a process that is a result of the atmospheric composition and, likewise, causes signal attenuation. However, in a highly scattering medium multiply scattered light may reach the receiver by different optic paths and result in pulse stretching in the time domain – in a similar way to multipath distortion in RF propagation. This is illustrated in Figure 7.3. Note that the delay in signal reception is given by the range divided by the speed of propagation of electromagnetic radiation, as is the case in RF propagation. When the particles in the atmosphere are smaller than or of the same order of magnitude as the wavelength of the light isotropic and nonisotropic scatter (Rayleigh and Mie scattering, respectively) can be dominant sources of signal degradation. This is shown schematically in Figure 7.4. Common examples of multi-scattering media are fog, haze, smoke and dust from sandstorms or volcanoes. Clouds in the propagation

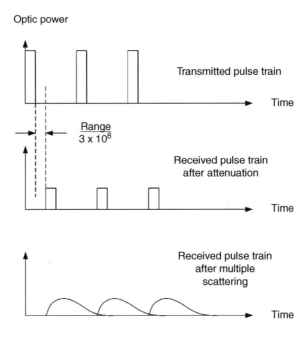

Figure 7.3   Schematic illustration of pulse attenuation and stretching due to multiple scattering

path are so harmful to FSO that they threaten the availability of a link altogether. For this reason satellite-to-ground links are primarily RF or microwave.

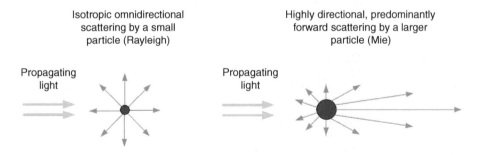

Figure 7.4   Schematic illustration of Rayleigh and Mie scattering regimes

Turbulence is a third major source of signal degradation in outdoor FSO but is not due to the particulate composition of the atmosphere. Turbulence, in contrast, is the result of random fluctuations in the temperature, pressure and humidity of the atmosphere, induced by temperature gradients and air currents. At sunrise and sunset the ground and surrounding air temperatures and humidities are most at equilibrium, while at midday these gradients are usually maximal. At higher altitudes temperature and humidity gradients are generally less pronounced but wind speed can affect the degree of air mixing and consequent air homogeneity. The optical property of the atmosphere affected by these parameters is the refractive index. Even if the fluctuations in the refractive index are small, the atmospheric propagation path becomes analogous to a succession of many tiny prisms that bend the light beam to and fro. A multiply deformed wave front reaches the receiver and the resultant distortions are both spatial and temporal. Time-varying light intensity causes scintillation in the received signal and, in the spatial dimension, the beam-spot 'wanders' around over the receiver telescope aperture and beyond.

Extensive research efforts have been conducted over the past years to gain a deeper understanding of the impact of atmospheric effects on FSO performance and to evaluate performance limitations in various scenarios [3–9].

To gain a better understanding of HAP communication scenarios in the presence of turbulence a number of extensive field tests have been performed [10]. In particular, The STROPEX (Stratospheric Optical Payload Experiment), conducted within the CAPANINA Project, yielded valuable data on FSO HAP-to-ground link feasibility. The experiment was performed in Kiruna, Sweden, where the HAP reached 23 km altitude and the coverage spanned a distance of 64 km to the ground station, where the elevation angle was around 21°. An optical downlink of 1.25 Gbps was demonstrated with impressive eye diagram results (bit-error-rates could not be measured at this data rate due to equipment limitations). At a more modest data rate of 622Mbps bit-error-rates of 10^{-9} and less were measured. A sample of the results of the experiments is shown in Figure 7.5.

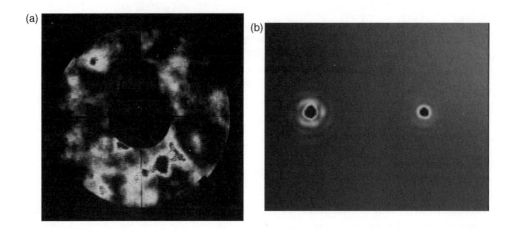

Figure 7.5 (a) Received intensity image (central occluded circle is due to telescope design). Reproduced from [10] by permission © German Aerospace Centre; and (b) Differential Image Motion Monitor (DIMM) image used for coherence length computation. Reproduced from [10] by permission © German Aerospace Centre

The turbulence features of the propagation path were characterised from data in terms of structure constant. Derivations from measurements of received intensity can be seen in Figure 7.6. The mean size of the intensity speckles was also measured. The coherence length was computed from the differential image motion of two spots and is shown in Figure 7.5(b). As is common, the turbulence is minimal at sunrise and sunset [in Figure 7.5(b) only the morning data are displayed including results for sunrise, which was around 08.00] manifested in the high coherence length.

7.1.2   HAP FSO-Link Configurations

We now distinguish between four types of HAP links where FSO could contribute:

- HAP-to-HAP
- HAP-to-ground
- HAP-to-satellite
- Hybrid FSO-RF links

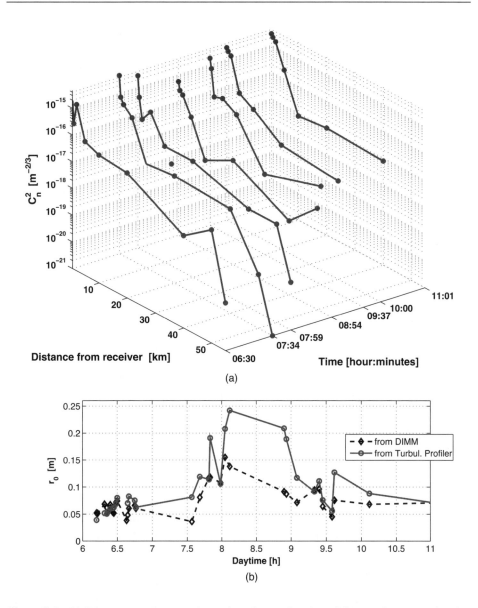

Figure 7.6 (a) Structure constant experimental results as a function of distance from ground station and time during the morning. Reproduced from [10] by permission © German Aerospace Centre; and (b) computed coherence length. Reproduced from [10] by permission © German Aerospace Centre

HAP-to-HAP links operate at a fixed altitude in the region of 17–22 km above the Earth's surface. At this altitude cloud cover is uncommon but, due to the temperature inversion and accompanying minimal winds, there tends to be considerable aerosol loading, including volcanic ash, that may well cause enhanced scatter.

The behaviour of laser beams propagating along horizontal paths at this altitude has not been extensively studied and this lack of understanding should be redressed before definite statements are made regarding performance of FSO in HAP-to-HAP links. Turbulence may also be encountered since the presence of winds often generates homogeneity in the temperature and humidity of the prevailing atmosphere. The homogeneity reduces refractive index variations and turbulence phenomena. In the absence of winds the turbulence may be considerable, but this conjecture requires experimental verification. The range of the HAP-to-HAP link is also a major factor in determining the feasibility of an FSO solution and the consequent transmitter power required. This is due to the inevitable reduction in power density as a result of beam divergence. The operability of a HAP-to-HAP optical link would necessitate the altitudes of the two HAPs and the range between them to comply with the requirement for the LOS between them to be above the 'cloud ceiling'. This derives from the curvature of the Earth and is not usually a concern for satellite-to-satellite links due to the considerably higher altitudes. Notwithstanding, the considerable research and practical experience with implementation of optical satellite-to-satellite links is a rich source of data and intuition for the emerging technology of HAP-to-HAP communication.

An additional challenge for HAP-to-HAP FSO links lies in the aggravation of the station-keeping difficulty. Both HAPs in a HAP-to-HAP link are at a nondeterministic location at any moment in time so that continuous PAT efforts would be required to maintain a viable FSO link. This is a familiar issue for satellite-to-satellite links. However, in contrast to satellite-to-satellite links, the HAPs are nominally stationary with a fixed distance separating them. The impact of vessel vibration is recognised as a major obstacle to reliable satellite-to-satellite communication and much research has been conducted to evaluate the phenomenon and explore mitigation techniques. The findings will undoubtedly be helpful in HAP-to-HAP FSO link analysis.

Acquisition of the target HAP prior to transmission will involve both coarse and fine tracking and some form of handshake and authentication protocol, so that some latency will need to be tolerated. However, once transmission has commenced tracking requires agile beam-steering to maintain continuous connectivity during motion events. Numerous methods for agile beam-steering using compact and lightweight hardware are available or under research and include acousto-optic devices, liquid crystal technology and fast steering mirrors.

Despite the challenges to be met, it appears to be feasible to achieve extremely high data rate and high capacity links with extremely lightweight and compact equipment. This would be particularly advantageous within a HAP network where multiplexed signals from the coverage area of one HAP could be transmitted to a neighbouring HAP and further afield as necessary.

HAP-to-ground scenarios would suffer less than HAP-to-HAP links in terms of transceiver alignment, but the atmospheric degradation of the propagating signal would be problematic in most locations and climates due to cloud cover and turbulence. However, if deployments are selected in areas of benign climatic conditions an FSO link may well prove successful and vertical ranges of 17–22 km have been shown to be quite manageable (see Section 7.1.1 describing the results of the CAPANINA STROPEX tests).

HAP-to-satellite links are probably the 'favourite' for FSO in HAP scenarios. Unimpaired by atmospheric effects, the HAP-to-satellite link can harness the tremendous advantages of satellite communication, such as very long transmission 'hops', massive geographical coverage and immense remote sensing capabilities. Local data transmissions uploaded to the HAP can traverse the globe rapidly and reliably with the help of HAP-to-satellite and inter-satellite links and then be downloaded at their destination. Conversely, satellite-sensed data can be downloaded locally to a ground station as needed using HAP infrastructure as a relay station. Hence, the very high capacity applications such as broadband backhaul network transmissions could be facilitated with the aid of HAP-to-satellite links.

The hybrid FSO-RF link is the fourth scenario addressed. The notion of hybrid wireless communication link has been discussed in the scientific literature extensively [6–8]. The underlying concept is to harness the benefits of two complementary technologies so that, at the price of some degree of redundancy, availability and throughput are maximised. For instance, if millimetre-wave radiation is coupled with FSO then in the case of fog or cloud in the propagation channel the millimetre-wave link is operated, while, when the clouds clear, the link will switch to FSO for increased data-rate transmission. A rainstorm, threatening the viability of the millimetre-wave transmission would also trigger a partial or total switch to FSO in accordance with prevailing channel conditions accompanied by appropriate adaptation of the data rates on each link. An intelligent controller with access to channel-monitoring data, current communication link performance, traffic data and system requirements would reduce/increase data rates so as to maximise throughput at all time.

7.2 Physical Layer Aspects for FSO Links in HAP Networks

FSO is implemented by means of a modulated laser beam in the transmitter and an opto-electronic device at the receiver. A schematic representation of an FSO transceiver is shown in Figure 7.7.

The light source in the transmitter is modulated to carry information. Simple on-off keying (OOK) is a robust alternative, and is the most common form of modulation in FSO. Another modulation method that is quite easy to implement is

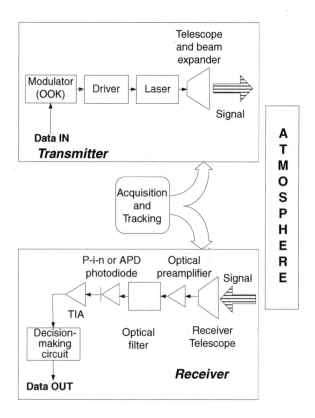

Figure 7.7   Block diagram illustrating basic optical wireless communication system

pulse position modulation (PPM). This method requires stringent synchronisation, which may be hard to implement.

The driver provides the energy for the light source and stabilises its performance.

The light source may be a laser or a light emitting diode (LED). The latter is used for low-power transmissions where the attenuation is minimal and is not really suitable for HAP scenarios. The laser has coherence properties, which make phase modulation possible in theory but commercial lasers do not usually have adequate coherence lengths so this is not a realistic alternative for practical HAP applications. Again, advances in satellite communications may open new opportunities for HAP communication. For instance, an optical BPSK modulation method has been field tested in satellite links so that this may prove a viable alternative on condition that turbulence in the propagation channel does not perturb the wavefront and high coherence lasers are readily available. The more commonly used lasers are 'edge-emitting' semiconductor laser diodes, which produce light from an active region sandwiched between two similar layers with lower refractive indices forming a waveguide. The resultant beam is somewhat oval in cross section. Circular cross-section beams can be produced by vertical cavity lasers (VCSELs). An array of VCSELs can be integrated onto a single chip facilitating multiple transmitter operation and enhanced power generation as well as improved robustness in the presence of turbulence in the channel. The wavelengths of light sources for communication purposes range from blue light at around 400 nm to the far infrared at 10 μm or more. In addition to continuous wave (cw) lasers, some lasers operate in 'pulsed mode' emitting streams of short-duration pulses of light. Much progress has been made in this field for ultra-low power-reception regimes. A very high peak power pulse can penetrate a harsh propagation medium and close a communication link where a cw laser of the same average power would not succeed.

A telescope collimates the emitted laser beam and determines the beamwidth and beam divergence. Some optical losses and distortions are incurred due to the optical elements. The divergence of the beam determines the free space loss. A narrow beam is more energy efficient and will therefore be suitable for long range and/or low power links. However, as mentioned, the narrow beam will pose a problem for PAT system performance.

The transmitted light is collected at the receiver by the receiver telescope. The widespread reception method is direct detection (DD), where incident photons are converted to electric current in accordance with the detector's quantum efficiency. Coherent modulation is usually impossible, as mentioned, as a result of the limited coherence length of the laser. The maximum angle of reception is determined by the telescope field of view (FOV). The FOV and the telescope aperture size determine the amount of light received, both from the signal and due to undesired background illumination. Multiply scattered light may reach the receiver when the FOV is large but this is at the cost of additional background light reception.

An optical amplifier may follow the receiver telescope to increase the light power entering the next stage of the receiver, which is the optical filter (the optical filter may also be placed prior to the amplifier). The optical amplifier is particularly efficatious when thermal noise in the receiver is a significant source of performance degradation. However the light amplification stimulates amplified spontaneous emission (ASE), which is a further source of signal degradation and mixes with other light to produce beat noises.

The optical filter allows light from a predetermined spectral band to pass, and thus reduces the unwanted background light.

The optical power is converted to electrical power in the opto-electronic device, which is usually a semiconductor photodetector in the form of a reverse-biased semiconductor diode. The two commonly used photodiodes are the simpler semiconductor p-i-n diode, where a layer of intrinsic material separates the p and n layers of a p-n junction and the higher-gain but noisier avalanche photodiode (APD). The APD is advantageous in the presence of high thermal noise but has a lower bandwidth due to the time duration of the internal amplification mechanism. Extremely sensitive 'photon counters' are becoming available for very low power reception systems. In conjunction with a high peak power pulsed laser and excellent synchronisation this can increase the link availability and/or transmission range.

A transimpedance amplifier (TIA) is used to convert the photocurrent output of the photodiode to a voltage signal, which is then fed into the decision-making device.

The decision-making device decodes the received signal and determines the nature of the data. In digital communication, each received bit is interpreted as a binary '1' or '0', and each erroneous decision leads to increased bit-error-rate (BER). The errors are a result of noise sources obscuring the signal as well as signal attenuation and beam spread, which combine to reduce the signal-to-noise ratio (SNR).

The major sources of noise in an FSO receiver are thermal noise due to the receiver circuitry, background illumination-induced shot noise, thermal current noise and amplifier noise. In the case of HAP links the background illumination will be highly variable depending on the time of day, location, season, etc., and narrow linewidth lasers in the transmitter together with narrow bandpass optical filters in the receiver are advisable to minimise the deleterious effects of background illumination. In HAP-to-HAP scenarios the background illumination includes the special feature of sunlight reflected from the target HAP itself, which is, ironically, particularly severe when the sun is *behind* the receiving HAP. The reflectivity of the HAP's outer coating is a critical factor in determining the amount of reflected solar radiation, and should be considered in the HAP design. In HAP-to-ground scenarios sunlight reflected from a lake or metallic rooftop could also severely obscure a signal from the ground station and this should be taken into consideration when selecting ground station locations.

The SNR and BER are, clearly, dependent on the amount of received light as well as the noise levels. The received optic power is calculated from the link budget. The link budget in an FSO link is given by the following expression:

$$P_R = P_T \eta_T \eta_R \exp(-c(\lambda)R) \frac{A_{\text{Rec}}}{\pi[R\tan(\theta_0)]^2} \qquad (7.1)$$

where P_T is the average transmitter optical power, η_T and η_R are the optical efficiencies of the transmitter and the receiver, respectively, R is the transmission range, A_{Rec} is the receiver aperture area and θ_0 is the laser beam divergence angle. This expression assumes that the receiver aperture is smaller than the spot size of the divergent beam. If a large aperture receiver is placed a short distance from the transmitter and the transmitter beam is narrow in the interest of power efficiency (7.1) becomes:

$$P_R = P_T \eta_T \eta_R \exp[-c(\lambda)R] \tag{7.2}$$

since all the propagating optic power is collected by the receiver. However, in the case of HAP scenarios the latter is not appropriate. $[c(\lambda)]$ is the extinction coefficient of the propagation medium and is a function of the radiation wavelength, λ. In general, the reduction in radiant flux, Φ, with distance r from a source is proportional to the initial radiant flux, the distance traversed and the coefficients governing extinction by absorption and scattering. The attenuation of light propagating is the result of the cumulative effects of absorption and scattering, governed by the absorption and scattering coefficients $a(\lambda)$ and $b(\lambda)$, respectively. The total attenuation is described by the extinction coefficient $c(\lambda)$, which is related to $a(\lambda)$ and $b(\lambda)$ by the simple relation:

$$c(\lambda) = a(\lambda) + b(\lambda) \tag{7.3}$$

This can be summarised as:

$$c(\lambda) = -\left\{ \frac{[\partial\Phi(r,\lambda)]_{\text{abs}} + [\partial\Phi(r,\lambda)]_{\text{scatt}}}{\Phi(r,\lambda)\partial r} \right\} \tag{7.4}$$

where the loss of radiant flux has been separated into loss due to absorption $[\partial\Phi(r,\lambda)]_{\text{abs}}$ and loss due to scattering $[\partial\Phi(r,\lambda)]_{\text{scatt}}$.

The critical importance of LOS between transmitter and receiver in an outdoor FSO link has been mentioned several times. The subsystem that is responsible for maintaining LOS is the PAT system. Initial acquisition is an issue in itself but in the case of HAP links, where the HAP is nominally stationary, this is a once-off operation that may be achieved with the help of GPS. Typically a beacon signal is emitted by one transceiver, while the other transceiver is equipped with a quadrant or matrix detector. Simple image processing locates the centre of the beacon spot and computes the angular position of the other transceiver. Coarse pointing is actuated by mechanical means (a gimbal), while fine pointing can be implemented in a number of ways. Fast steering mirrors (FSD) are an excellent solution for agile beam steering minimising BER due to link loss. The mirror could be a galvanometer mirror driven electrically in two dimensions or a piezzo mirror that responds to the application of a voltage. Other methods exist and are the subject of ongoing research and include optical phased arrays, liquid crystal devices, diffraction gratings and acousto-optic devices.

It should be noted that inadequate pointing leads to pointing loss, that degrades the signal in addition to the noise sources discussed. The pointing error may be composed of both transmitter and receiver pointing errors when both transceivers track their partner and are generally described by:

$$L_{T/R} = e^{(-G_{T/R}\theta_{T/R}^2)} \qquad (7.5)$$

where $\theta_{T/R}$ is the transmitter/receiver pointing error and $G_{T/R}$ is the transmitter/receiver diffraction limited gain that is proportional to the aperture diameter and inversely proportional to the square of the radiation wavelength. If both the elevation and azimuthal pointing angles can be assumed to be distributed with a Gaussian probability function and to be independent and identical, then the overall pointing error probability distribution is Rayleigh and given by:

$$f(\vartheta) = \left(\frac{\vartheta}{\sigma^2}\right) e^{\left(-\frac{\vartheta^2}{2\sigma^2}\right)} \qquad (7.6)$$

where σ is the standard deviation of each of the independent and identically distributed pointing angles.

An extended introduction to FSO implementation can be found in [9].

During the STROPEX experiment described earlier, PAT performance was studied for the special HAP application, where high rotation is encountered in the yaw angle (up to 9 rpm), while the vessel tends to oscillate in the roll and pitch directions. The severe demands of the HAP-to-ground application necessitated proprietary design, despite the policy for performing the experiment with COTS (components off the shelf) that are described in [10]. Initial acquisition was obtained within 30 s in different trials. During 75% of the experiment duration the tracking error was under 142 μrad and continuous connectivity was achieved.

Lastly, it is important to mention that all optical equipment will be required to withstand all the environmental conditions to which they will be subjected and to function satisfactorily. In addition to the temperature, pressure, vibrations, ionising radiation, etc., in the stratosphere, the optical terminal will have to endure the various stresses encountered during launch. Notably, high humidity during launch could precipitate a problem of icing of optical components at the sub-zero temperatures encountered, so all components should be adequately protected.

7.3 Free Space Optics for Optical Transport Networks

It will be clear from the preceding sections that the primary niche for FSO in HAP scenarios will be to supply the ultra-high capacity needed for backhaul links, particularly within HAP networks. This implies the existence and smooth functioning of adequate transport layer solutions linking HAPs and interfacing with gateways to other networks. Despite the drawbacks of HAP-to-ground optical links in adverse weather, they nevertheless appear to be plausible in sufficiently benign conditions and are a major asset where backhaul links are required to transport the aggregate traffic from multiple users. The next step in implementing HAP broadband communication is to explore solutions for optical transport networks, including all-optical networks that can seamlessly provide the high capacities anticipated without opto-electrical and electro-optical conversion. A typical all-optical transport network architecture with optical interplatform links and HAP-to-ground backhaul links operating alongside HAP-to-user RF links is shown in Figure 7.8 [11].

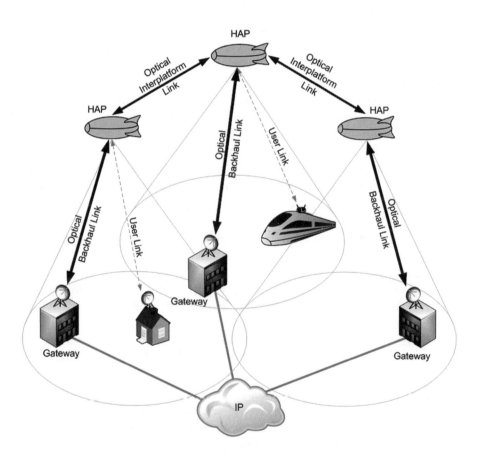

Figure 7.8 Schematic illustration of possible all-optical HAP network operating alongside HAP-to-user RF links (reproduced from [11])

The transport layer can be facilitated exploiting unique optical features such as WDM and all-optical routers and switches. Additionally, wavelength assignment algorithms enable adaptive routing in response to changing traffic conditions and demands.

WDM is a form of frequency division multiplexing in the optical domain. By transmitting signals simultaneously through a common propagation channel over optical carriers of different wavelengths the overall capacity is enhanced in proportion to the number of carriers. The major applications of WDM are found in fibre optic communications but significant inroads in the wireless sphere have been achieved as well. As many as 64 different wavelengths have been studied for HAP scenarios. The different transmissions are multiplexed at the transmitter and demultiplexed at the receiver. Combining the two functions in a single device yields and add-drop multiplexer that is an essential component in multi-hop transmissions where routing is necessary at each node.

Numerous network topologies are possible between HAPs and in accordance with the spatial distribution of the ground stations. It should be noted that a HAP may have more than one ground station within its coverage area. In an extreme case of an isolated HAP there may even be no ground station within the coverage area, in which case a HAP-to-HAP link is essential to render the isolated HAP operable for communication with terrestrial users. The number of HAPs and the number of ground stations per HAP comprise the total number of network nodes. In [11] expressions for the number of optical carriers in a WDM scheme needed to serve N HAPs connected in a mesh topology when k ground stations are located in the coverage area of each HAP is developed analytically. For a robust optical network that can circumvent an inoperable HAP-to-ground link (due to adverse weather, etc.) it is desirable that HAPs are interconnected with single hops if possible to avoid the shortcomings of multiple hop end-to-end connectivity. It is assumed that WDM is not required at the interplatform link level since each HAP-to-HAP link is exclusive and dedicated. In contrast, each ground station requires N wavelengths in order to establish a communication link with each HAP. $(Nk - 1)$ wavelengths are needed at each ground station in addition to enable unequivocal communication lines with each ground station. Table 7.1 shows the minimum number of WDM carriers required in a full mesh topology between N HAPs and k ground stations per HAP.

At the opposite extreme of interconnectivity, with little robustness in the face of a failed link, an alternative topology would be a bus formation, where N HAPs are stringed together, each communicating with k ground stations in their coverage area. On condition that it is considered important to construct a homogeneous network, all HAPs on the bus backbone will need to support the maximum required number of wavelengths; this is the number required by the HAP(s) in the middle of

Table 7.1   The number of optical carriers needed in a WDM scheme for a mesh topology between N HAPs and k ground stations per HAP

Number of HAPs in network, N	Number of ground stations per HAP, k		
	1	2	3
2	4	9	16
3	5	9	16
4	7	11	16
5	9	14	19
6	11	17	23
7	13	20	27
8	15	23	31
9	17	26	35
10	19	29	39

Table 7.2   The number of optical carriers needed in a WDM scheme for a bus topology between N HAPs and k ground stations per HAP

Number of HAPs in network, N	Number of ground stations per HAP, k		
	1	2	3
2	4	9	16
3	8	18	32
4	16	36	64
5	24	54	96
6	36	81	144
7	48	108	192
8	64	144	256
9	80	180	320
10	100	225	400

the string. Table 7.2 shows that the minimum number of WDM carriers required indicates a significant increase in the number of wavelengths.

An extensive treatment of network issues can be found in [11], including simulation results for different loading conditions, network topologies and network dimensions. Further reading on network issues can be found in [12–15].

References

1. HAPCOS Website, http://www.hapcos.org/DOCS/wg2/wg2_home.php, accessed 20 January 2010.
2. Wikipedia Website, http://en.wikipedia.org/wiki/Free_Space_Optics, accessed 20 January 2010.

3. D. Kedar and S. Arnon, *Urban Optical Wireless Communication Network: the Main Challenges and Possible Solutions*, IEEE Commun. Mag., May 2004, pp. S1–S7.
4. D. Bushuev and S. Arnon, *Analysis of the Performance of a Wireless Optical MIMO Communication System*, J. Opt. Soc. Am. A, July 2006, Vol. 23, No. 7, pp. 1722–1729.
5. J. Farserotu, G. Kotrotsios, I. Kjelberg and A. Prasad, *Scalable, Hybrid Optical-RF Wireless Communication System for Broadband and Multimedia Service to Fixed and Mobile Users*, J. Wireless Personal Communications, 2003, Vol. 24, No. 2, pp. 327–323.
6. H. Izadpanah, T. ElBatt, V. Kukshya, F. Dolezal and B.K. Ryu, *High-Availability Free Space Optical and RF Hybrid Wireless Networks*, IEEE Wireless Commun., 2003, Vol. 10, No. 2, pp. 45–53.
7. H. Wu and M. Kavehrad, *Availability Evaluation of Ground-to-Air Hybrid FSO/RF Links*, Int. J. Wireless Inform. Networks, 2007, Vol. 14, No. 1, pp. 33–45.
8. I.L. Kim and E. Korevaar, *Availability of Free Space Optics (FSO) and Hybrid FSO/RF Systems*, Proc. SPIE, 2001, Vol. 4530, pp. 84–95.
9. S. Arnon, *Optical Wireless Communication*, in *The Encyclopedia of Optical Engineering (EOE)*, R. G. Driggers ed., Marcel Dekker, 2003, pp. 1866–1886.
10. FP6-IST-2003-506745 CAPANINA, Deliverable Number D2, *Final Optical Terminal Design Report.*
11. FP6-IST-2003-506745 CAPANINA, Deliverable Number D27, *Network Architecture and Protocols.*
12. Z. Jia, J. Yu, G. Ellinas and G.-K. Chang, *Key Enabling Technologies for Optical-Wireless Networks: Optical Millimeter-Wave Generation, Wavelength Reuse, and Architecture*, J. Lightwave Technol., 2007, Vol. 25, No. 11, pp. 3452–3471.
13. Z. Jia, S.D. Milner, A. Desai, T.-H. Ho, J. Llorca, S. Trisno and C.C. Davis, *Self-Organizing Broadband Hybrid Wireless Networks*, J. Opt. Netw., 2005, Vol. 4, No. 7, pp. 446–459.
14. D. Kedar, D. Grace and S. Arnon, *Laser Nonlinearity Effects on Optical Broadband Backhaul Communication Links for High Altitude Platform Systems*, IEEE Trans. Aerospace Electron. Systems submitted.
15. M Mohorcic, C. Fortuna, A. Vilhar, J. Horwath, *Evaluation of wavelength requirements for stratospheric optical transport networks*, Journal of Communications, 2009, vol. 4, no. 8, pp. 588–596.

8

Advanced Communication Techniques as Enablers for HAP-Based Communication Systems

8.1 Modern Wireless System Design Concepts

Some operating scenarios described in Section 3.1 consider relatively independent operation of HAPs, especially in the case of disaster relief and short term event servicing in remote areas. However, as discussed in Section 4.2, in the longer term HAPs are expected to operate also in developed countries with existing communications infrastructure. Thus, in order to represent an acceptable complementary solution to currently existing and developing wireless technologies, HAP-based communication systems will have to efficiently integrate in the future heterogeneous wireless access infrastructure using one of the widespread communication standards. This inevitably calls for the application of modern wireless system concepts when designing HAP-based systems to increase the link and system capacity, to improve spectrum efficiency and optimise utilisation of resources, to guarantee harmless operation alongside and with other wireless networks, and to ensure robust operation in hostile radio propagation environment. Some of these concepts include the use of different diversity techniques, adaptive coding and modulation schemes, advanced radio resource management techniques, smart

Broadband Communications via High Altitude Platforms   David Grace and Mihael Mohorčič
© 2011 John Wiley & Sons, Ltd

antennas techniques, multiple input multiple output (MIMO) concept, spatial multiplexing, space-time processing and coding, cross-layer design and optimisation, the cognitive radio concept and dynamic spectrum management.

Due to their elevated position and consequently large coverage area and predominantly line of sight (LOS) propagation conditions, not all of these concepts are directly applicable to HAP-based communication systems. Some require adaptation to the specific HAP operating environment, in which special attention needs to be paid to the interference received from other systems as well as interference caused to other systems. Namely, as discussed in Section 1.3, the millimetre wave frequency bands are allocated to HAPs on a secondary basis provided they shall not cause harmful interference to primary systems operating in the same band, nor shall they claim protection from those systems. In this respect smart antennas have a potential to ensure harmless coexistence of HAP and terrestrial systems on the short term, whereas the cognitive radio concept with dynamic spectrum management can facilitate harmless side-by-side operation of heterogeneous wireless access systems in potential white space frequency bands. In order to do so the cognitive radio and dynamic spectrum management depend on cross-layer design and optimisation and make use of techniques from the areas of artificial intelligence, machine learning, etc.

8.1.1   Smart Antennas

Smart antennas are in fact cooperative antenna arrays capable of advanced signal processing to support spatial beam forming and spatial coding with the aim to avoid causing/cancel interference in selected directions. Smart antennas are typically based on using simple finite impulse response (FIR) tapped delay line filters with adaptively changing weights so as to steer the radiation pattern in the direction of a desired signal and at the same time minimise, or even better notch out, the radiation pattern in the direction of interfering sources or targets. The alternative implementation assumes several predefined fixed radiation patterns available in the antenna, which is switching between them upon the requirements of the system. Clearly, this second approach gives much less flexibility and is only suitable for the environment with fixed users, and even in this case it only minimises and not cancels the interference. Implementation of smart antennas can also be difficult in the mm-wave bands due to the maturity of the technology, and the small wavelength, which can present problems given that array elements are often designed to be comparable with the size of the wavelength. More discussion on antenna aspects in general can be found in Section 3.2.

8.1.2 Cognitive Radio and Dynamic Spectrum Management

The cognitive radio concept assumes that user terminals and related access networks comprehend sufficient computational intelligence about radio and network resources, as well as on wireless services, to be able to infer the user needs based on the context, and to adapt transmission or reception parameters (e.g. frequency, power, waveform) in an opportunistic manner to meet those needs in most efficient way, taking into account also the environment in which it operates [1]. From the perspective of harmless coexistence of heterogeneous wireless access systems in the same frequency band with other primary and nonprimary systems, the cognitive radio concept facilitates spectrum sharing through the dynamic spectrum management techniques for cooperative optimisation. These techniques tend to avoid causing and receiving interference by searching for a portion of the spectrum that is not being used, instead of shaping the radiation pattern. They are choosing the most appropriate channel to meet user's needs, negotiate the use of this channel with other potential users and vacate the channel once not needed or upon detection of a primary user [2]. This approach is based on the presumption that frequency bands are only scarce resources because they are allocated on a permanent basis, but in fact most of the time in most locations heavily underutilised. So, by releasing some frequency bands for the use of cognitive radio supported wireless systems, even if only on secondary basis, user terminals could share the available spectrum with neighbouring primary and secondary user terminals. Clearly, in the process of searching for available spectrum, user terminals will contend and interact, some will be selfish, others fair, some even altruistic. Such behaviours can efficiently be analysed by game theory, which can also be used to assign the spectrum and set the operating parameters [3].

The use of cognitive radio for harmless coexistence of HAP-based and terrestrial WiMAX systems has been investigated in [4]. In particular, two spectrum etiquettes have been proposed for the downlink to enhance the coexistence performance and to balance the usage priority of shared spectrum between completely overlapped HAP and terrestrial systems through the exploitation of directional user antennas. The parameter settings for coexistence performance are determined by antenna beamwidths and multiple modulation scheme levels. The additional interference from HAP-based to terrestrial WiMAX system can be accommodated by the transmitter power margin of the terrestrial system, allowing incumbent users to maintain or even improve their modulation scheme levels.

In [5] two CINR-based cognitive radio schemes are investigated for spectrum allocation to users in a multiple HAPs system serving fixed users with directional antennas in a common coverage area to increase the available capacity. One scheme aims at improving channel quality and the other at improving trunking efficiency.

Both schemes can simultaneously operate within a shared spectrum environment without detrimentally affecting the performance of the other scheme.

HAPs with their inherently large coverage area with respect to terrestrial systems can also play an assistive role in enabling dynamic configuration and policy enforcement on cognitive radio capable wireless nodes in heterogeneous terrestrial wireless access infrastructure. Depending on the operating scenario, policies for wireless nodes may have to change, so the radio device should be capable of getting the up-to-date policies. In [6] cognitive radio devices in terrestrial networks are dynamically configured and policies for dynamic spectrum access are beamed with the assistance of HAP, thus helping the cognitive radio system to be optimised in a global domain. At the same time HAP can receive the feedback information from the wireless nodes in the footprint about the cognitive radio parameters such as interference level, throughput and bandwidth requirements of applications, etc. Using this information HAP can either synthesise an image of the frequency usage and use it to optimise transmission parameters of terrestrial wireless nodes, or alternatively send this information for processing and optimisation to the ground control centre. When updating software defined wireless nodes or distributing optimised transmission parameters HAP makes use of its inherent broadcast capabilities.

8.1.3   Cross-Layer Design and Optimisation

Advanced techniques in wireless communication systems are being developed to improve spectrum efficiency and optimise utilisation of resources. However, these techniques found the layering paradigm of ISO/OSI reference model or TCP/IP protocol stack, both originally developed for wired networks, overly restrictive. These paradigms are based on layered protocols that provide services with pre-defined functionality and interfaces only to the layer above, and request services only from the layer below. This approach historically proved advantageous for wired networks as it provided connectivity of devices from different manufacturers. In wireless networks, however, it prevents consideration of and optimisation to specific operating conditions stemming from the fact that different connections share the same frequency spectrum. Along with dependency on propagation conditions this causes wireless networks to suffer link impairments resulting in high BER, large delays in case of required retransmissions and especially in satellite links, and difficulties in meeting QoS requirements. This is the reason for worst case link dimensioning and inefficient use of resources, since the layering paradigm does not give the possibility to optimise the functioning of interdependent processes (e.g. error correction schemes, QoS mechanisms, etc.) residing at different layers. In this respect, cautious cross-layer design and optimisation of interdependent processes has a potential to significantly improve the wireless system performance and utilisation of resources.

According to [7] two basic approaches need to be distinguished with respect to cross-layer operation of protocols, also having an impact on the protocol design. In the case of joint protocol optimisation taking into account different layers, processes and interactions during the design phase only, without any additional signalling exchange during the operation and thus not supporting dynamic adaptation, we talk about implicit cross-layer design. Conversely, if additional signalling beyond adjacent layers is employed in order to facilitate run-time dynamic adaptation at different layers, we talk about explicit cross-layer design.

In theory cross-layer optimisation can be implemented between any pair or even set of nonadjacent layers, or in some cases also between adjacent layers when acquiring or setting internal, otherwise unavailable states. In practice, however, most common cross-layer interactions in wireless networks are implemented between:

- Physical and MAC layers, e.g. supporting interactions between the resource management and adaptive coding and modulation taking into account also propagation conditions.
- Physical and network layers, e.g. physical layer supporting mobility management protocols on the networking layer to minimise delays in rerouting the IP datagrams.
- MAC and network layers, e.g. implementing appropriate access protocols and scheduling policies to support QoS mechanisms and mobility management on the network layer.
- MAC and transport layers, e.g. helping the transport layer to distinguish between end-to-end packet losses due to congestion in the network, requiring reduction of traffic load, and due to link impairments caused by deteriorating propagation conditions and shared medium, not needing any traffic reduction.
- Physical, MAC and higher layers, e.g. trying to control the service priority and application source generation bit rate so as to match requested QoS with resource management capabilities and propagation conditions.

Given that HAP-based communication systems are in most parameters but propagation conditions very similar to terrestrial wireless systems, and that they are expected to use existing or developing terrestrial or satellite wireless communications standards, the same cross-layer approaches can be applied as in terrestrial or satellite systems, just optimised to HAP specifics.

In the rest of this Chapter some more mature modern wireless system design concepts and how they relate to HAPs are discussed in more detail. These include diversity techniques, MIMO systems, adaptive coding modulation schemes, and linking in to the already discussed cognitive radio and dynamic spectrum management, an example of some elements of their use in advanced radio resource management techniques.

8.2   Diversity Techniques

Diversity is a technique which improves the transmission reliability by making use of two or more statistically independent radio paths. The probability that all radio paths exhibit deep fade is very low, so if all received signals are properly combined the system performance can be significantly improved. The statistically independent paths can be achieved by [8]:

- separation of receive or transmit antennas, i.e. *spatial diversity*;
- transmitting the same information at two carrier frequencies, i.e. *frequency diversity*;
- retransmitting signal in different time instance, i.e. *temporal diversity*;
- transmitting the radio signal in different directions, i.e. *angle diversity*;
- applying different polarization at transmit antennas, i.e. *polarisation diversity*.

Spatial diversity is the oldest and most often used form of diversity technique. Receive or transmit antennas have to be sufficiently separated to make the radio channels statistically independent. In order to counteract the small scale fading, which is caused by the interference effect of numerous reflected rays arriving at the receiver with different phases and amplitudes, the distance between antennas should be at least one half of the signal carrier's wavelength, and this is known as micro diversity. In the case of macro diversity for counteracting large scale fading, which is caused mainly by signal shadowing or even blocking of the transmitted signal, the distance between antennas should be significantly larger. The improvement of the system performance in the case of micro spatial diversity depends on correlation between the channels, which on one side depends on distance between antennas and also on the amount of multipath propagation and antenna types.

In the case of frequency diversity the same information is transmitted at different frequencies. If the channel frequencies are separated more than the coherence bandwidth of the channel, the multiple transmitted versions of the same signal suffer independent fading. Apart from transmitting on two separate carrier frequencies the common way to apply frequency diversity is to spread the signal over a large frequency band by: (i) applying code division multiple access (CDMA); (ii) compressing the information in time; (iii) frequency hopping; or (iv) using multicarrier CDMA. Using the above mentioned methods, the information is transmitted without wasting additional bandwidth, which is the case with retransmitting the same information on different frequencies.

The temporal diversity takes advantage of the mobile radio channel time-variability. Multiple versions of the same signal received at different time instances exhibit independent fading, if the time difference is at least equal to the inverse of twice the maximum Doppler frequency. The simplest form of temporal diversity to counter small scale fading is the repetition code. A more sophisticated approach is a combination of coding and interleaving. The automatic repeat request (ARQ) method is more appropriate for large scale fading mainly caused by shadowing and blocking of the transmitted signal.

Deep fades are usually caused by multipath propagation, when radio signals arrive at the receiver from all directions. If directional antennas are used we can direct them in different directions so as to receive different reflections of radio signals. As a consequence the fading radio channels become uncorrelated. This method is known as angle diversity and is usually combined with space diversity to achieve better performance.

The reflection and diffraction of the radio signal depend on its polarisation. The radio signal transmitted with vertical polarisation experiences different fading than the radio signal transmitted with horizontal polarisation. A certain amount of diversity can be achieved if the same radio signal is transmitted using both polarisations. Such signal is received by dual polarised antenna and each polarisation is processed separately. After basic signal processing both instances of received signal are combined in the base band.

A common aspect of all diversity techniques is that on one hand they increase the transmission reliability whereas on the other hand they reduce the capacity of the communication system by sacrificing radio resources such as radio spectrum, time or polarisation. Clearly, in spatial and angle diversity they also increase the complexity of the system by doubling radio frequency (RF) chains.

In the case of diversity reception two or more replicas of the transmitted signal are available at the receiver. In order to improve the system performance, i.e. reduce the BER at given SNR, one of the signals should be selected for estimation of the transmitted data, or all received signals should be combined into one signal for data estimation. The first method is known as selection diversity, while we refer to the second method as combining diversity.

In the selection diversity all received signals are monitored simultaneously, typically measuring a parameter which represents the channel quality such as received signal strength indicator (RSSI) or BER. Based on this, the signal from the receiver chain with the best channel is selected to estimate the data. In order to implement selection diversity with RSSI monitoring two antennas and two complete RF chains are required, while in the case of BER monitoring these need to be complemented also by two signal processing chains. The selection diversity principle is illustrated in Figure 8.1.

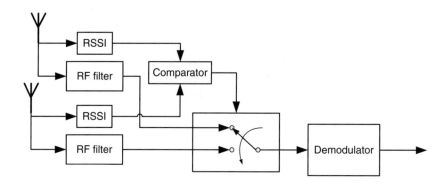

Figure 8.1 Selection diversity principle

An alternative solution to selection diversity is switched diversity, where only one RF and signal processing chain is required. The signal is received using this chain connected to the antenna until the transmission quality is satisfactory. In the case the transmission quality is poorer than expected the receiver chain is switched to another antenna. In this approach some hysteresis should be introduced to prevent continuous switching between antennas. The switched diversity principle is shown in Figure 8.2.

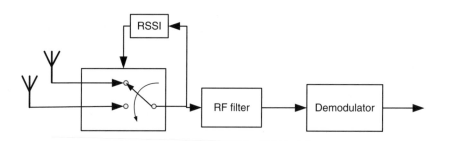

Figure 8.2 Switched diversity principle

In the case of selection diversity part of the received signal energy is thrown away by discarding all except one received signal. This drawback is improved in combining diversity, where a sum of all received signals weighted by complex coefficient gives a signal with better signal to interference ratio than individual received signals. If the coefficients in all branches have the same gains the combining is known as equal gain combining. The equal gain combining performance is not satisfactory, if radio signals in different branches have significantly different SNR. If additive white Gaussian noise (AWGN) is the only source of disturbance to signals in all branches the optimum combining technique is maximum ratio combining, where the branch weights are equal to the complex conjugate of the channel coefficients. This means that the branches with better SNR are multiplied by higher weight than branches with lower SNR. The combining diversity principle is depicted in Figure 8.3.

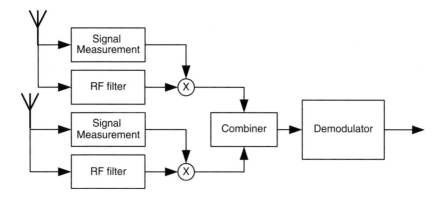

Figure 8.3 Combining diversity principle

The first approach to implement the diversity principle was to place two antennas at the receiver, resulting in receive diversity. However, in mobile communications, the mobile terminal equipped with two antennas would be unpractical. In order to implement the diversity in the downlink, i.e. on the link from the base station to the mobile terminal, multiple antennas are placed on the transmitter side at the base station. This approach is known as transmit diversity. If the channel state information is completely known at the transmitter, there is no basic difference between transmit and receive diversity. In the case of transmit diversity, before transmission the signals are multiplied by a complex coefficient obtained from channel state information. Thus, the received signal is the sum of the optimally weighted transmitted signals which guarantees maximum diversity gain. However, in many cases the channel state information is not available on the transmitter site, so alternative methods are applied to take advantage of transmit diversity such as delay diversity, where the second antenna transmits a delayed replica of the signal transmitted by the first antenna.

8.2.1   Diversity Techniques in Broadband HAP Communications

The broadband HAP communications are characterised by predominantly LOS propagation conditions, occasionally blocked by large objects such as hills, mountains, and man-made objects, shadowed by trees, or attenuated due to rain. The small scale fading is caused by radio signal reflected only by objects in the vicinity of transmit and receive antennas. Due to directional antennas used at the user terminals and at the platforms, the small scale fading is nearly negligible, thus no micro diversity is necessary in broadband HAP communications. On the other hand, blocking and shadowing of the radio signal may cause outage of the system. In order to minimise the probability of the system outage the spatial diversity promises the best improvement of the system performance. However, the distance between antennas should be comparable with the size of the obstacle to achieve statistically independent fading channels.

Other forms of diversity such as temporal, frequency, angular and polarisation are not effective in countering the large scale fading because it is impossible to find uncorrelated channels in the case of signal shadowing. For example if the signal is blocked by obstacles it is blocked for all frequencies of interest and both polarisations. Time diversity may give some improvement of system performance for user terminals in motion. However, the time delay introduced to mitigate shadowing and blocking will reduce the use of the system only to noninteractive applications.

8.2.1.1 Spatial Diversity in Broadband HAP Communications

A HAP communication system can have different system architecture configurations [9,10]. In the simplest configuration HAPs are used as standalone platforms, providing broadband wireless access only to terminals located in their coverage area. In this case the spatial diversity can be implemented either at the platform or user terminal. Mounting two antennas on the HAP does not lead to two statistically independent radio channels due to comparatively small HAP size with respect to the path length and due to LOS propagation condition. Thus, the spatial diversity can be provided only if two antennas are mounted on the user terminal, which should have the size of obstacles causing signal shadowing or blocking. Clearly, for macro diversity this condition can only be fulfilled for fixed and vehicle mounted terminals. For instance, in the case of the CAPANINA project [11] a collective terminal was assumed to provide wireless access to passengers on fast trains. Trains allow the terminal antennas to be mounted on the roof, where they can be sufficiently separated to promise statistically independent channels at least in an urban environment.

In the case of multi-platform constellations HAPs can be interconnected via ground stations or by interplatform links forming a network of HAPs, thus arbitrarily extending the system coverage and enabling to apply a spatial diversity using multiple HAPs. In order to investigate the potential gain offered by exploiting space and platform diversity, a configuration with two HAPs and a train with two antennas are considered in [10] and with four HAPs in [12]. The corresponding system architecture alternatives are graphically depicted in Figure 8.4.

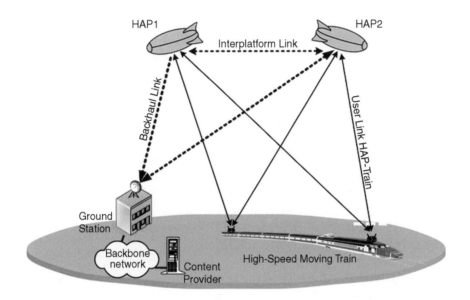

Figure 8.4 System architecture for exploiting space and platform diversity

From the perspective of space and platform diversity the above system architecture enables four operating scenarios [10]:

1. Single HAP – single train antenna
   This scenario does not provide any diversity so the transmitted signal is very sensitive to all kinds of fading, including fading due to electrical bridges, trellises and posts along the railway [13].
2. Single HAP – multiple train antennas
   Train antennas are mounted on the roof of the train at an appropriate distance from each other to exploit space diversity and thus counteract fading due to electrical trellises, bridges, posts, i.e. obstacles with dimensions smaller than

the train length. Such a system may exploit the receive diversity in downlink and the transmit diversity in uplink. However, in spite of using space diversity on the train, the radio signal may be shadowed or even completely blocked if the propagation channels are correlated, i.e. the obstacle is larger than the distance between train antennas (deep canyons, localised heavy rain, etc.).

3. Multiple HAPs – single train antenna

   In a multiple HAP system there may exist another platform providing LOS link to the train. In such a case the link reliability can be significantly improved by using platform diversity, where the same signal is transmitted from multiple platforms and received by a single antenna on the train. Such a system may exploit the transmit diversity in downlink and the receive diversity in uplink. Platform diversity is particularly attractive in the case where the network of platforms exists. These can be interconnected via interplatform links, via a ground station suitably located in the overlapping part of HAP coverage areas, or even via two separate ground stations in their respective coverage areas, which are connected via a terrestrial network.

4. Multiple HAPs – multiple train antennas

   In this scenario the same signal is transmitted from multiple platforms and received by multiple antennas on the train. The system can exploit full space diversity in transmit and receive directions on uplink and downlink. In this scenario the HAP communication system can be considered a MIMO system [14]. The network of HAPs as a MIMO system requires close cooperation between physical, MAC and network layers.

The impact of spatial diversity on the spectral efficiency of the HAP communication system is illustrated in Figure 8.5. As a case study for investigating the trade-off between blocking probability and spectral efficiency a system consisting of four HAPs was assumed with HAPs placed in the vertices of the parallelogram above the two perpendicular railway lines, which intersect in the middle [12]. The three state switched HAP propagation channel model is assumed [15] with 50% of the time in the LOS state (using Rician channel model), 37% of the time in the shadowed state (using Rayleigh channel model) and 13% of the time in the blocked state. The blocking probability in the case without platform diversity is very high. It is represented by the CDF value at spectral efficiency 0 (bits/s/Hz) in Figure 8.5, i.e. 13%. A significant improvement is achieved by using platform diversity from two HAPs. The blocking probability is decreased nearly to zero. In the case of the diversity from 4 HAPs, no significant further improvement is observed in blocking probability compared with the diversity from 2 HAPs. However, the difference between the diversity from 2 and 4 HAPs is significant at higher spectral efficiency.

For example the cumulative density function (CDF) in Figure 8.5 shows that the probability of achieving lower spectral efficiency than 4 bit/s/Hz is 60% when the same information is transmitted simultaneously from 2 HAPs, and reduces below 50% for the diversity from 4 HAPs. In other words, in the case of diversity from 4 platforms, the probability to achieve spectral efficiency higher than 4 bits/s/Hz is 50% and only 40% for the diversity from 2 platforms. In the case of no diversity, this percentage is approximately 30%.

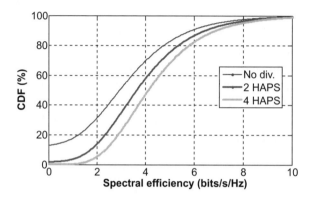

Figure 8.5   Cumulative density function of spectral efficiency for no diversity, diversity from 2 HAPs and diversity from 4 HAPs

8.3   MIMO Systems

A MIMO concept refers to a wireless communication system where at both sides of the communication link more than one antenna is applied [16–18]. The MIMO approach promises a significant increase in the system capacity and also in the reliability of communication links in the multipath propagation environment. The approach is an extension of diversity principles often applied in microwave radio links to improve link reliability. Combining the transmit and receive diversity results in a new concept, which not only increases the link reliability but also offers a potential to increase the radio link capacity. The MIMO concept can be used for three different aims, namely: (i) to increase the communication link capacity by using spatial multiplexing; (ii) to improve the communication link reliability by applying spatial diversity; and (iii) to cancel the interference using beam forming. While the concepts of beam forming and spatial diversity have been well known for decades, the concept of spatial multiplexing was a complete novelty at the end of the 1990s.

8.3.1   Spatial Multiplexing

An increase in the link capacity is achieved by spatial multiplexing. The principle of spatial multiplexing is shown in Figure 8.6. The input data sequence is converted to M parallel sub-sequences. Each sub-sequence is modulated and transmitted via single antenna. In a rich scattering propagation environment each signal from the ith transmit antenna to the jth receive antenna exhibits its own fading channel. In a system with M transmit and N receive antennas, there exist $M \times N$ sub-channels between the transmitter and receiver. In general, each sub-channel exhibits a selective fading which can be modelled as a linear discrete time FIR filter with complex coefficients. In the case of flat fading, the signal in each sub-channel is only attenuated and phase shifted due to different propagation times between each receive and transmit antenna. Thus, a sub-channel can be represented by one complex coefficient. In the majority of analyses the MIMO channel is assumed to be time invariant within one time slot but it varies from time slot to time slot. Such a MIMO channel is said to be quasi-static. In this section the MIMO channel is assumed quasi-static for capacity analysis.

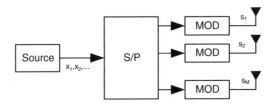

Figure 8.6   Spatial multiplexing

In a MIMO system with M transmit and N receive antennas, the received signal on the jth receive antenna y_j can be expressed as:

$$y_j = \sum_{i=1}^{M} h_{ij} \cdot x_i + n_j \tag{8.1}$$

where x_i is the transmitted signal from the ith antenna, and variable n_j denotes samples of circularly symmetric complex Gaussian noise with variance σ_n^2 at the jth receiver. The fading channel is described as a sum of complex paths h_{ij} between receive and transmit antennas. The complex gain coefficient h_{ij} obeys Gaussian distribution. This relationship can also be expressed in a matrix form:

$$\mathbf{y} = \mathbf{Hx} + \mathbf{n} \tag{8.2}$$

where **y** is the column vector of the received signals, **H** is the channel matrix, **x** is the column vector of the transmitted signal and **n** is a column vector of the AWGN.

The capacity of the MIMO link, assuming an ideal knowledge of channel state information only at the receiver side and the same power allocated across the transmit antennas, can be expressed as [19]:

$$C = \log_2 \left[\det \left(\mathbf{I} + \frac{\rho}{M} \mathbf{H} \mathbf{H}^* \right) \right] \tag{8.3}$$

where $\rho = P/\sigma_n^2$ denotes the SNR at one receive antenna. P is the cumulative power transmitted by all antennas and σ_n^2 is the noise power at each receive antenna. Inserting nonzero eigenvalues of the matrix $\mathbf{H}\mathbf{H}^*$, $\lambda_1, \lambda_2, \ldots, \lambda_K$, where K is the rank of matrix $\mathbf{H}\mathbf{H}^*$ into Equation (8.2) yields:

$$C = \sum_{i=1}^{K} \log_2 \left(1 + \frac{\lambda_i}{M} \rho \right) \tag{8.4}$$

It is obvious from Equation (8.4) that the MIMO channel can be explained as K orthogonal sub-channels with channel gains λ_i. If the sub-channel gains are known at the transmitter, the power can be allocated to each channel optimally, applying the water filling algorithm. The water filling algorithm allocates more power to channels with higher gains. Thus, slightly higher channel capacity can be achieved by channel knowledge at the transmitter. According to Equations (8.3) and (8.4) the channel capacity increases linearly with the rank of the matrix $\mathbf{H}\mathbf{H}^*$, or in the case of uncorrelated channels linearly with min(N, M), which gives great potential for applying the MIMO approach to increase the channel capacity.

The spatial multiplexing has been implemented at Bell Laboratories [17] and it is often called Vertical Bell Laboratory LAyered Space Time architecture (V-BLAST). While the implementation of the V-BLAST transmitter is straightforward, the decoding of the MIMO signal is computationally very complex. The optimum decoding of spatial multiplex data is maximum likelihood (ML) decoding, denoted by:

$$\mathbf{x}_{ML} = \arg \min_{\mathbf{x}_j \in \{\mathbf{x}_1, \ldots, \mathbf{x}_K\}} \left\| \mathbf{y} - \mathbf{H} s_j \right\|^2 \tag{8.5}$$

The ML decoding requires an extensive search through all possible transmit symbols and therefore its computation time is exponentially proportional to the number of transmit antennas and number of bits encoded in each spatially multiplied symbol. A more straightforward method is the zero forcing (ZF) method. In the ZF detector the vector of the received signal **y** is multiplied by the pseudo inverse of the channel matrix **H**. Due to the noise enhancement the performance of the (ZF) detector is poor, especially in the case of the highly correlated MIMO sub-channel. A good compromise between detector performance and algorithm complexity is successive interference cancellation (SIC) decoding and sphere decoding. SIC decoding is an iterative detection method which consists of three steps: zeroing usually by ZF algorithm, quantisation, i.e. estimation of the transmitted signal, and interference cancellation [20]. These steps are applied per individual transmit antenna and they are iteratively repeated until all transmitted symbols are detected. The idea of sphere decoding [21] is to search the solution of ML decoding just for those points **Hx** that are inside the defined M dimensional hyper-sphere with defined radius and centre in **y**. The size of the radius defines the compromise between computational time and efficiency of decoding.

8.3.2 Space–Time Coding

While the multiplexing gain of the MIMO system is exploited by spatial multiplexing, the spatial diversity can be simply utilised by transmitting the same data via all transmit antennas. However, in a fading wireless channel the joint exploitation of spatial and temporal diversity yields high performance gains without increasing frequency bandwidth requirements. The joint explanation of spatial and temporal diversity can be achieved by space–time codes. The space–time codes can be classified into space–time block codes (STBCs) and space–time trellis codes (STTCs).

The block diagram of the STBC is illustrated in Figure 8.7. Space–time block encoder maps a block of input symbols into spatial and temporal sequences. It is usually represented by a matrix **S** in which a row represents a time slot and a column represents an antenna's transmission over time:

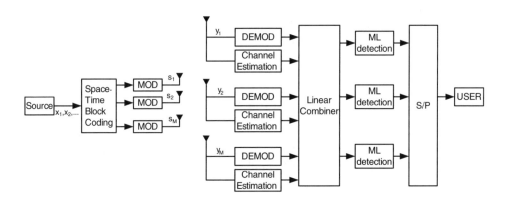

Figure 8.7 Space–time block code transmitter

$$
\mathbf{S} =
\overset{\text{transmit antennas}}{
\begin{bmatrix}
s_{11} & s_{12} & \cdots & s_{1M} \\
s_{21} & s_{22} & \cdots & s_{2M} \\
\vdots & \vdots & \ddots & \vdots \\
s_{T1} & s_{T2} & \cdots & s_{TM}
\end{bmatrix}}
\text{ time slots}
\tag{8.6}
$$

where s_{ij} is the transmitted symbol in time slot i and on antenna j. The number of time slots are T and number of transmit antennas are M. The matrix $\mathbf{S}$ usually follows the orthogonal design. This means that vectors represented by matrix columns are orthogonal to each other, which leads to optimal decoding with linear complexity. The disadvantage of described orthogonal design is that all except one of the codes sacrifice some fraction of coding rate [22]. The nonorthogonal design of STBC also exists but in this case some inter-symbol interference is introduced. The first STBC with two transmit antennas was discovered by Alamouti [23], and is now widely known as the Alamouti code. Subsequently this code was, with some limitations, generalised to other different numbers of transmit antennas [24].

Alamouti introduced a simple orthogonal STBC, which is composed of two transmit and an arbitrary number of receive antennas [23]. The Alamouti scheme with one receive antenna has a coding rate one and can achieve a diversity order $d = 2$. The diversity order denotes the slope of BER curve. The Alamouti transmission scheme is illustrated in Figure 8.8. The source data are encoded in two symbol intervals. In the first interval, symbol s_1 is transmitted over the first antenna and symbol s_2 over the second. In the second interval, complex conjugated symbols are transmitted, $-s_2{}^*$ over the first antenna and $s_1{}^*$ over the second one. Thus, the entire symbol block has the form:

$$
\mathbf{S} =
\begin{bmatrix}
s_1 & s_2 \\
s_2{}^* & -s_1{}^*
\end{bmatrix}
\tag{8.7}
$$

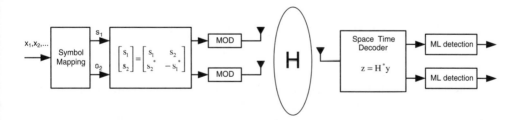

Figure 8.8 Alamouti transmission scheme

where the element $[S]_{ij}$ is transmitted over the jth antenna in the ith symbol period. The matrix equation of the received signal is:

$$\mathbf{y} = \begin{bmatrix} y_1 \\ y_2^* \end{bmatrix} = \begin{bmatrix} h_1 & h_2 \\ h_2^* & -h_1^* \end{bmatrix} \begin{bmatrix} s_1 \\ s_2 \end{bmatrix} + \begin{bmatrix} n_1 \\ n_2^* \end{bmatrix} = \mathbf{Hs} + \mathbf{n} \qquad (8.8)$$

where h_i is the attenuation between the ith transmit antenna and the receive antenna and $\mathbf{n}$ is the noise vector.

The received signal is decoded by multiplying the received vector $\mathbf{y}$ with transposed and complex conjugated channel matrix $\mathbf{H}^*$.

$$\mathbf{H}^*\mathbf{H} = h^2\mathbf{I}_2 \qquad (8.9)$$

Due to the orthogonality of the matrix, the noise remains white and decoding is done independently for s_1 and s_2. The channel gain achieved is $h^2 = |h_1|^2 + |h_2|^2$ as long as the channel remains static during two symbol periods. Multiplying $\mathbf{y}$ with $\mathbf{H}^*$ yields:

$$\mathbf{z} = \mathbf{H}^*\mathbf{y} = h^2 \begin{bmatrix} s_1 \\ s_2 \end{bmatrix} + \mathbf{H}^*\mathbf{n} \qquad (8.10)$$

The orthogonal Alamouti scheme with code rate 1 exists only in the case of two transmit antennas. Schemes with a higher number of transmit antennas have to sacrifice orthogonality in order to achieve the same code rate [22]. An extended Alamouti scheme implementing four transmit antennas at the transmitter, and hence achieving a diversity order $d = 4$, was proposed in [25]. The transmission scheme extends over four symbol periods and two Alamouti codes are used to build the symbol block:

$$\mathbf{S} = \begin{bmatrix} \mathbf{S}_1 & \mathbf{S}_2 \\ \mathbf{S}_2^* & -\mathbf{S}_1^* \end{bmatrix} = \begin{bmatrix} s_1 & s_2 & s_3 & s_4 \\ s_2^* & -s_1^* & s_4^* & -s_3^* \\ s_3^* & s_4^* & -s_1^* & -s_2^* \\ s_4 & -s_3 & -s_2 & s_1 \end{bmatrix} \qquad (8.11)$$

The same approach is used to decode the signal as for the Alamouti space–time code scheme with two transmit antennas. However, after the multiplication of $\mathbf{y}$ with $\mathbf{H}^*$ the channel is not decoupled perfectly due to lack of orthogonality of the original matrix $\mathbf{H}$ [25]:

$$\mathbf{z} = \mathbf{H}^*\mathbf{y} = h^2 \begin{bmatrix} s_1 + Xs_4 \\ s_2 - Xs_3 \\ s_3 - Xs_2 \\ s_4 + Xs_1 \end{bmatrix} + \mathbf{H}^*\mathbf{n} \tag{8.12}$$

where the channel gain is $h^2 = |h_1|^2 + |h_2|^2 + |h_3|^2 + |h_4|^2$ and

$$X = 2Re(h_1 h_4^* - h_2 h_3^*)/h^2 \tag{8.13}$$

As shown in Equations (8.12) and (8.13) X is a channel-dependent real-valued random variable which leads to partial interference between different symbols.

The STBC can achieve maximum diversity gain but no coding gain. This disadvantage is improved by the STTC), which is an extension of trellis coded modulation (TCM) to multiple transmit antennas. It combines transmit diversity and TCM to achieve reliable, high data rate transmission in wireless channels. STTC was introduced by Tarokh *et al.* in 1998 [26]. The STTC can be illustrated in a trellis diagram, in which vertices are defined with diagram of state transitions. Besides the maximum diversity gain, coding gain can also be achieved. The design criteria are discussed in [26] for slow flat fading, fast flat fading, and spatially correlated channels assuming high SNRs. The Viterbi algorithm is applied for STTC decoding. The complexity of this decoding is considerable which is why STBCs are more attractive for the implementation.

The MIMO system aims to achieve multiplexing gain, beam forming gain or diversity gain. All gains cannot be reached simultaneously. A trade-off between diversity and beam forming gain depends on the environment the communication system operates in. In a pure LOS propagation environment the achievable beam forming gain is given by the product $M \times N$. The gain is attained by forming beams at the transmitter with gain M, and at the receiver with gain N and pointing transmitter and receiver beam at each other. No diversity gain can be achieved in LOS propagation conditions. However, in a highly scattered environment, the diversity gain is $N \times M$ [27] and maximum beam forming gain is upper limited to $(\sqrt{M} + \sqrt{N})^2 (\sqrt{M} + \sqrt{N})^2$. The fundamental trade-off between diversity gain and multiplexing gain has been discovered by Zheng and Tse [18]. They have shown that the optimum trade-off curve between diversity order d_{div} and rate r is piecewise linear connecting points:

$$d_{\text{div}} = (M-r) \cdot (N-r)$$
$$\text{where } r = 0, 1, 2, \ldots, \min(M, N). \tag{8.14}$$

The maximum diversity d_{max} and the maximum rate r_{max} are given as:

$$d_{max} = M \cdot N \tag{8.15}$$

and

$$r_{max} = \min(M, N) \tag{8.16}$$

8.3.3 MIMO Systems in HAP Broadband Communications

The classical MIMO approach in HAP broadband communication systems has limited applicability due to predominately LOS channel conditions, poorly scattered communication channel, directional antennas, allocated frequency bands and limited size of the platform. However, two or more platforms can be placed in a formation and interconnected by interplatform links can communicate to a multi-antenna system on the ground, thus forming a virtual MIMO system.

The study in [12] investigates a virtual MIMO (V-MIMO) approach, based on a constellation of multiple HAPs, providing broadband wireless access to a collective terminal with multiple antennas mounted on high-speed trains under predominantly LOS propagation conditions. Due to the effect of random blocking or shadowing of the platforms caused by train operating environment, the channels toward individual HAP are statistically independent, yielding some diversity or multiplexing gain. The study analyses the performance of transmit diversity based on STBC). In particular, Alamouti and extended Alamouti schemes are investigated, using fixed wide-lobe receive antennas, and compared with the reference receive diversity scheme based on best HAP selection that requires highly directional and steerable antennas. The conceptual system architecture making use of the V-MIMO approach in the HAP system is depicted in Figure 8.9.

The system architecture in Figure 8.9 comprises (i) a network of HAPs, placed above the main railway lines and connected by error free interplatform links, (ii) one or more ground stations connected with HAPs by fixed backhaul links and hosting gateways to external networks and/or application servers, and (iii) collective terminals mounted on the train and acting as intermediate nodes to the local wired or wireless system and HAP communication system. The reference HAP constellation with railway lines is plotted in Figure 8.10. HAPs forming a multiple HAP constellation are arranged equidistantly at an altitude of 17 km along two linear lines. The distance between two neighbouring HAPs is 30 km. Each HAP has a coverage area with a diameter of 90 km. All HAP antennas are pointing at the sub-platform point. HAP antennas are based on a $\cos^n(\theta)$ model, setting n to 2.18, which guarantees the $-10\,$dB antenna gain at cell edge.

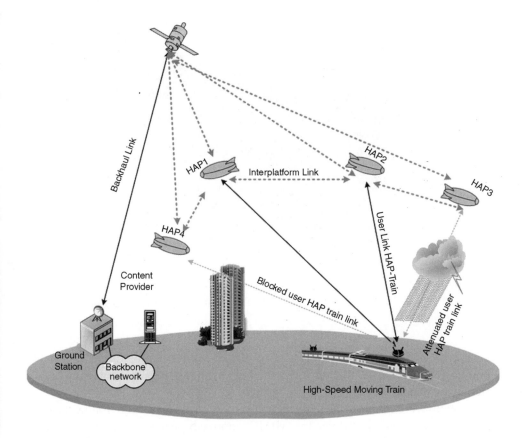

Figure 8.9 The system architecture for the MIMO approach in HAP systems

The reference constellation can be used to analyse the possibility of using V-MIMO with extended Alamouti code for mitigation of shadowing effects in a predominantly LOS mobile environment [12], however considering the switched 3-state channel model. The performance of three diversity schemes has been evaluated and compared, namely (i) the selection diversity approach, where the user terminal chooses the best available HAP, (ii) the V-MIMO approach based on the Alamouti code transmitted from two HAPs and (iii) the V-MIMO approach based on the extended Alamouti code transmitted from four operating HAPs.

The performance evaluation shows that the use of STBC diversity schemes in HAP communication systems can be applied to mitigate shadowing and blocking in a LOS propagation environment [12]. Applying two and four transmit antennas (HAPs) considerably outperforms a scheme based on the selection of a single HAP. Nevertheless, even though link outage probability is reduced considerably, it

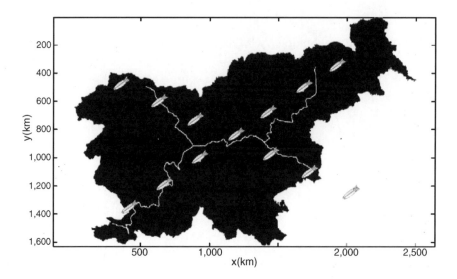

Figure 8.10   Reference constellation of 13 HAPs above two railway lines

cannot be eliminated completely due to tunnels, relief configuration and/or the potential presence of a large rain cell with heavy rainfall in the case of using higher frequency bands.

In satellite communication systems with similar propagation environment as in a HAP communication system, the MIMO approach seems to be applicable by applying two MIMO concepts. These are the concept of cooperative diversity and the concept of polarisation space–time coded diversity (PCT) [28]. The concept of polarisation coded diversity is making use of the differences of orthogonally polarised radio waves at reflection, refraction and diffraction. It promises fourfold diversity gain and twofold increase in the capacity applying double polarized transmission and reception, and sixfold diversity gain and twofold increase in the capacity applying 3D polarized receive antennas [28]. The results published in [28] assume the omnidirectional antenna is applied at the user terminal and directional patch antenna at the satellite. Moreover, the environment at the user terminal is highly scattering with poor LOS component. Such a scenario may occur if the user terminal is located indoors or shadowed by trees and moving at high speed. However, the allocated frequency bands for broadband HAP communications are in the millimetre frequency band, which forces the use of mostly outdoors mounted directional antennas, resulting in a low scattering propagation environment. Thus, cooperative diversity seems the only possible solution to use the MIMO approach in the broadband HAP communication system.

8.4   Adaptive Coding and Modulation Schemes

Adaptive coding and modulation (ACM) is a powerful technique used for mitigating slow variation of the wireless communication channel. Channel variation requiring link adaptation is typically caused by meteorological effects, shadowing and blocking, and higher path loss due to increased distance between transmitter and the receiver, or due to additional path loss caused by moving from an outdoor to indoor environment [29–36]. The ACM schemes can be classified into two classes:

- **Pre-estimated ACM schemes** – in the pre-estimated ACM scheme the modulation type, burst structure and data transmission rate are assigned at the time of connection set-up and the transmission parameters do not adapt to the variation of the channel characteristics during the connection.
- **Dynamic ACM schemes** – in the dynamic ACM scheme the modulation and coding parameters are controlled slot-by-slot and can be changed adaptively during the connection.

The pre-estimated ACM schemes are applied in wired links and sometimes in fixed wireless links, where the link characteristics are not changing during the connection. The dynamic ACM schemes, on the other hand, are used in wireless communication systems where the wireless channel is time varying and in general requires link adaptation.

A typical ACM scheme is a closed loop control system. It logically consists of the adaptive transmitter, the receiver, the forward propagation channel and the return channel. A logical block diagram of a communication system with ACM is depicted in Figure 8.11.

The transmitted burst is composed of a frame which consists of the payload part carrying the information and the pilot–synchronisation part. The pilot–synchronisation symbols are added to the data part of the frame for fast channel estimation and synchronisation. In order to guarantee system synchronisation and the channel estimation even in harmful propagation conditions, the most robust modulation scheme is applied for the pilot–synchronisation part of the frame. The payload part of the frame is encoded and modulated, i.e. transformed to the signal waveforms, using the most appropriate coding and modulation (CM) mode according to the information about the channel quality. The signal waveforms are subsequently upconverted to the carrier frequency and transmitted via the communication channel towards the receiver. During the transmission the transmitted waveforms are distorted by the propagation channel.

At the receiver, the propagation conditions are estimated in the channel estimator and, in the CM mode selector block, the most appropriate CM mode for the next

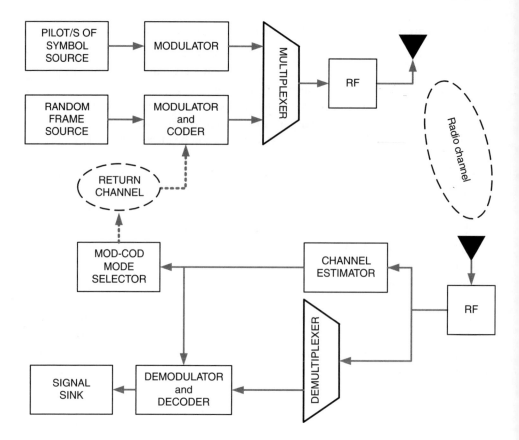

Figure 8.11   Block diagram of a communication system with ACM

transmission frame is determined applying channel prediction. The channel estimation methods can be either data aided or data nonaided. The data nonaided
channel estimation methods do not require the knowledge of transmitted data and
can be based on the RSSI, measuring eye closure, estimating residual BER from the
channel encoder, estimating Euclidean distance at the detection, etc. In the data
aided methods the channel quality is estimated from the pilot symbols added to the
transmitted data sequence and known at the receiver. The data aided methods are
more precise, fast, and can be applied to estimate not only channel attenuation
but also other channel parameters such as delay spread, Doppler shift, etc. Due to
these reasons the data aided channel estimation methods are more often applied in
contemporary wireless communication systems.

The advanced detection techniques, implemented in the receiver, may take
advantage of the estimated channel conditions to improve the reliability of the

communication link. The data concerning the selected CM mode are transmitted from the receiver via the return channel back to the transmitter. Typically the return channel is assumed to be ideal, i.e. error free, in the majority of ACM scheme analyses. This assumption is usually assured by applying a high power efficient coding and modulation scheme to transmit important information in the forward and return radio channel.

When designing the ACM scheme it is necessary to take into account:

- the ratio between round trip delay and rate of channel variation;
- the dynamic range of channel variation;
- the amount of additional signalling introduced in the return and forward link.

In order to make the closed control loop required for ACM to work properly the variation of channel characteristics requiring link adaptation should be slower compared with the round trip delay. The round trip delay is the sum of uplink and downlink propagation delays and the signal processing delays at the transmitter and at the receiver. The propagation delays are critical in satellite communication systems, where one way propagation delay to the geostationary satellites is in the range of 120 ms. Thus, the ACM in a satellite communication system is applied to mitigate very slow channel variations only, such as hydro meteorological effects (rain fading) or shadowing and blocking caused by a large object. In terrestrial and HAP communication systems, on the other hand, one way propagation delay is below 1 ms and is not a critical parameter in the design of the ACM scheme. So, in terrestrial and HAP communication systems the ACM can also mitigate faster channel variations such as transmitter shadowing by man-made objects.

The dynamic range of channel attenuation determines the set of CM modes. The analysis of channel losses due to meteorological effects in HAP communication systems reveals that in time intervals during which the communication link is not affected by heavy rain, the channel loss varies between 0 dB and 2 dB, and that the heavy rain occasionally causes deep fades, in some cases larger than 10 dB. Complete mitigation of such deep fades can be achieved by the ACM scheme consisting of CM modes with operating SNR range covering the expected maximum variation of the channel loss. However, due to implementation limitations and available transmit power, the designers are typically limited to a finite set of CM modes covering only part of the required SNR range. Thus, complete mitigation of rain fading is typically not possible. Instead, systems are designed for a given maximum probability of system outage or expected number of hours per year with the received signal below receiver sensitivity.

Considering an ACM scheme that cannot cover the entire range of channel attenuation with its limited number of CM modes, a trade-off needs to be made between the probability of system outage and system throughput. In extreme cases the link budget can be designed either to maximise the system reliability or to maximise the system throughput. In the approach for maximising the system reliability, the system is designed so as to guarantee communications link with the required BER level even under the worst expected channel conditions. This means that the most robust (i.e. the most power efficient) CM mode is applied in the worst channel state and more bandwidth efficient CM modes are used in better conditions. The application of such an approach results in a certain loss of system throughput. In the alternative approach for maximising system throughput, the link budget is designed to provide the highest possible throughput by applying the most bandwidth efficient CM mode under the best channel conditions. Less bandwidth efficient modes are applied for worse channel conditions to partially compensate for fading. When the received SNR is lower than the SNR threshold of the most power efficient CM mode, the system becomes unavailable. This approach may result in increased system outage.

Upon the decision of the CM mode to be applied for the next frame, the transmitter has to inform the receiver about this decision. This can be done by explicit transmission, implicit transmission or blind detection. In the case of explicit transmission the transmitter can send in the header of the frame the information about the CM modes to be used in the rest of the frame. The implicit method is applied, when the receiver takes the decision about the CM mode to be used in the next frame and sends it to the transmitter, so the necessary information about CM is already available at the transmitter. The blind method estimates the various statistical properties of the signal, such as peak to average power ratio, autocorrelation, etc., to find out the CM mode. The blind method is slow and thus inappropriate to be applied even in moderate time varying radio channels. From the discussion above, it is obvious that in order to have a fully functional ACM scheme, some additional information has to be exchanged between the user terminal and the platform. If the rate of CM mode switching is too high, a notable portion of the link capacity can be wasted. This degradation of link capacity, often neglected in ACM analysis, should be considered in the comparison of the ACM schemes with fixed coding modulation schemes.

8.4.1 ACM in HAP Broadband Communications

As discussed in Section 1.4, there are no definite standards that need to be used with HAPs architecture, or that need to be developed taking into account specific characteristics of HAPs. This fact enables the use of most suitable existing or developing terrestrial or satellite wireless communications standards. The concept of ACM is included in nearly all contemporary standards for broadband communication systems such as IEEE 802.16 [37] or its subset WiMAX [38], standards for digital video broadcasting over satellites DVB-S2 [39] and DVB-RCS [40] and High-Speed Downlink Packet Access (HSDPA) [41,42]. With respect to the allocated frequency bands for broadband communications via HAPs and propagation conditions, which are predominately LOS, the HAP communication channel is not expected to be frequency selective. Thus, the implementation of the OFDM, or even OFDMA, would increase the system complexity and consequently the energy consumption, which may cause problems on HAPs. Consequently, a set of investigations [43–47] limit their selection to standards specifying single carrier transmission with ACM scheme:

- the IEEE 802.16 single carrier part of standard [37,38];
- the digital video broadcasting standard over satellites DVB-S2/RCS [39,40].

Both standards support ACM to cope with large scale fading and channel attenuation due to meteorological distortions. The CM schemes specified are different but they cover nearly equal SNR range of approximately 20 dB.

IEEE standard 802.16 was designed to evolve as a set of air interfaces based on a common MAC protocol. The IEEE standard 802.16 [37] specifies different physical layers for the 2–11 GHz and 10–66 GHz frequency ranges. The 10–66 GHz physical layer assumes LOS propagation, which is very close to propagation conditions in the HAP channel. The standard supports ACM profiles in which modulation (QPSK, 16-QAM, or 64-QAM) and various coding types may be assigned dynamically on a burst-by-burst basis. The most widely used coding types are Type 1, with Reed–Solomon (RS) codes over GF(256), and Type 2, with concatenated RS codes over GF(256), and the block convolutional code (24,16) as an inner code. A subselection of CM modes of IEEE 802.16 standard is listed in Table 8.1 in the increasing order of SNR switching threshold at a target BER equal to 10^{-4} and calculated spectral efficiency [43]. The SNR range covered by CM modes is over 20 dB, offering spectral efficiency from 1.25 up to 6.00 bit/s/Hz, which is sufficient to counter impairments caused by heavy rain events.

The Digital Video Broadcasting-Satellite (DVB-S2) standard, among others, specifies the physical layer for video broadcasting over satellites [39]. It is characterised by variable CM modes, which allow different data rates and error protection levels to be used and changed on a frame-by-frame basis. Combined with the use of a DVB-RCS standard [40] for the return link this may be used for closed-loop ACM. The powerful low density parity check codes (LDPCs) concatenated with BCH (Bose–Chaudhuri–Hocquenghem) codes are used to provide quasi error free transmission for upper protocol layers. The modulation schemes with low peak to average power ratio, namely QPSK, PSK, 16APSK and 32APSK, give the

Table 8.1 SNR thresholds and spectral efficiencies for IEEE 802.16 CM modes at a target BER of 10^{-4}

Modulation	RS code	BCC code	SNR (dB)	Spectral efficiency (bits/s/Hz)
QPSK	(255,239)	2/3	4.77	1.25
QPSK	(255,247)	2/3	5.31	1.29
QPSK	(255,251)	2/3	5.98	1.31
QPSK	(255,253)	2/3	6.42	1.32
QPSK	(255,239)	1	9.23	1.87
QPSK	(255,247)	1	9.87	1.94
16QAM	(255,239)	2/3	10.48	2.50
QPSK	(255,251)	1	10.54	1.97
16QAM	(255,247)	2/3	11.12	2.58
QPSK	(255,253)	1	11.18	1.98
QPSK	(255,255)	1	11.81	2.00
16QAM	(255,251)	2/3	11.99	2.62
16QAM	(255,253)	2/3	12.43	2.65
64QAM	(255,239)	2/3	14.94	3.75
64QAM	(255,247)	2/3	15.88	3.87
16QAM	(255,239)	1	15.94	3.75
16QAM	(255,247)	1	16.68	3.87
64QAM	(255,251)	2/3	16.75	3.94
64QAM	(255,253)	2/3	17.39	3.97
16QAM	(255,251)	1	17.45	3.94
16QAM	(255,253)	1	17.99	3.97
16QAM	(255,255)	1	18.92	4.00
64QAM	(255,239)	1	22.00	5.62
64QAM	(255,247)	1	22.64	5.81
64QAM	(255,251)	1	23.31	5.91
64QAM	(255,253)	1	24.25	5.95
64QAM	(255,255)	1	24.78	6.00

possibility of using an efficient high power amplifier at the transmitter. The set of CM modes specified in the DVB-S2 standard covers nearly 20 dB of SNR range and spectral efficiency from 0.490 to 4.453 bits/s/Hz. The set of CM modes in increasing order of SNR thresholds for a target BER equal to 10^{-4} is given in Table 8.2 [43].

In order to test if the range and the number of CM modes, specified in contemporary broadband wireless access standards can counter rain fading, the usage of the DVB-S2 CM modes are tested over a period of 1 month in the channel which models the rain attenuation. Four subsets of the CM modes specified in DVB-S2 standard are selected for analysis. The main parameters of four ACM schemes are listed in Table 8.3.

Table 8.2 SNR thresholds and spectral efficiencies for DVB-S2 CM modes at a target BER of 10^{-4}

Modulation	LDPC code	SNR (dB)	Spectral efficiency (bits/s/Hz)
QPSK	1/4	−2.70	0.490
QPSK	1/3	−1.45	0.656
QPSK	2/5	−0.55	0.789
QPSK	1/2	0.85	0.989
QPSK	3/5	2.05	1.188
QPSK	2/3	2.95	1.322
QPSK	3/4	3.90	1.487
QPSK	4/5	4.53	1.587
QPSK	5/6	5.05	1.655
8PSK	3/5	5.65	1.780
QPSK	8/9	6.06	1.766
QPSK	9/10	6.26	1.789
8PSK	2/3	6.59	1.980
8PSK	3/4	7.86	2.228
16APSK	2/3	8.81	2.637
8PSK	5/6	9.26	2.479
8PSK	8/9	10.05	2.646
8PSK	9/10	10.48	2.679
16APSK	3/4	10.82	2.967
16APSK	4/5	10.92	3.166
16APSK	5/6	11.47	3.300
32APSK	3/4	12.56	3.703
16APSK	8/9	12.72	3.523
16APSK	9/10	12.93	3.567
32APSK	4/5	13.49	3.952
32APSK	5/6	14.07	4.120
32APSK	8/9	15.46	4.398
32APSK	9/10	15.69	4.453

Table 8.3 Parameters of the ACM schemes

ACM scheme	SNR range (dB)	Spectral efficiency (bits/s/Hz)	SNR range similar to
ACM 1	18.0	0.5–4.5	DVB-S2 (all CM modes)
ACM 2	13.5	0.5–3.5	DVB-S2 (QPSK-8PSK CM modes)
ACM 3	9.0	0.5–2.5	DVB-S2 (QPSK CM modes only)
ACM 4	4.5	0.5–1.5	DVB-S

The SNR ranges and spectral efficiencies are chosen so as to match the SNR range and spectral efficiencies of DVB-S [48] and DVB-S2 [39] standards, selecting either the entire range of CM modes or only a subset, e.g. only CM modes with QPSK modulation in DVB-S2 standard. The number of CM modes in a given ACM scheme varies between 2 and 30. Other CM modes are spread equidistantly on the line connecting the most bandwidth efficient and the most power efficient CM modes on the spectral efficiency/SNR plane.

Time variations of the spectral efficiency for the reference ACM schemes ACM 1 and ACM 4 with different SNR range are given in Figures 8.12 and 8.13, respectively. In the case of a SNR range of 18 dB (Figure 8.12), the reliability of the communication link is guaranteed for deep fades, while at the same time using the most bandwidth efficient CM modes for good channel conditions. The two most robust CM modes are not used at all. In the case where the channel loss represents the heavy rain part of the year, the SNR range of the ACM may be overestimated and a smaller range might yield similar results with much lower system complexity. While the ACM 3 scheme with an SNR range of 9 dB still provides acceptable system performance with a negligible system outage time of 0.018% (see Table 8.4), the system outage probability for ACM 4 scheme with an SNR range of 4.5 dB is not acceptable. As shown in Figure 8.13, deep fades cause system outage time of approximately 0.29%, represented by spectral efficiency equal to zero. The chosen SNR range is underestimated and an increase is required to provide reliable communication in the heavy rain periods. The system outage probability of a 1 month period of rain fading is calculated for representative ACM schemes and listed in Table 8.4.

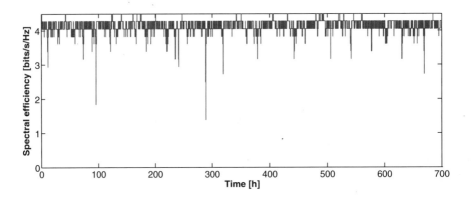

Figure 8.12 Time variation of spectral efficiency of the scheme ACM 1 with SNR range 18 dB and 18 CM modes

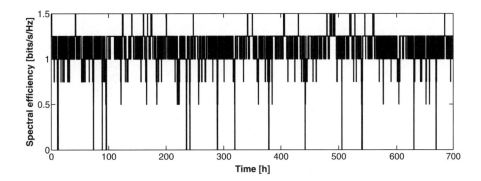

Figure 8.13 Time variation of spectral efficiency of the scheme ACM 4 with SNR range 4.5 dB and 4 CM modes

Table 8.4 System outage probability due to rain fading for different ACM schemes [43]

ACM scheme	SNR range	System outage probability
ACM 1	18 dB	0
ACM 2	13.5 dB	0
ACM 3	9.0 dB	1.8×10^{-4}
ACM 4	4.5 dB	2.9×10^{-3}

In theory, the infinite number of CM modes in a given SNR range would give the highest spectral efficiency. However, in practice an upper limit exists, above which an increase in the number of CM modes does not contribute significantly to the average spectral efficiency. A graph of average spectral efficiency versus the number of CM modes and SNR range of the ACM scheme is depicted in Figure 8.14. As expected, the ACM schemes with larger SNR range provide higher average spectral efficiency. An increase of the SNR range beyond the expected channel loss variation does not bring any benefit in terms of average spectral efficiency. At low numbers of CM modes in the ACM scheme, an addition of a new CM mode significantly increases the average system spectral efficiency, while at higher numbers of CM modes no significant increase of the average spectral efficiency is observed.

The ACM concept combined with platform diversity technique or virtual MIMO approach can efficiently mitigate the shadowing due to natural or man-made obstacles. An example of combined usage of the V-MIMO approach and ACM using the CM modes specified in mobile WiMAX standard is investigated in [12]. The CDF of the achieved spectral efficiency is depicted in Figure 8.15 for the

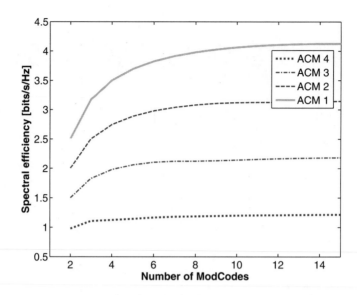

Figure 8.14 The average spectral efficiency versus the number of CM modes for ACM schemes with different SNR ranges

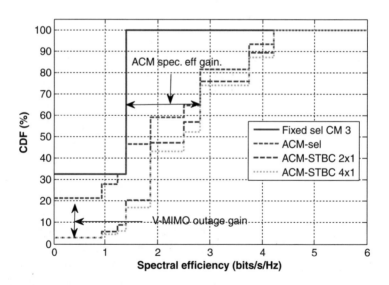

Figure 8.15 Cumulative density function (CDF) of the achieved spectral efficiency for WiMAX combining the V-MIMO approach and ACM

selection diversity with ACM (ACM-sel), V-MIMO system with ACM, more precisely ACM combined with the standard Alamouti code (ACM-STBC 2×1) and the extended Alamouti code (ACM-STBC 4×1), and selection diversity with a single 3rd CM mode (Fixed sel 3 CM). The CDF value at the spectral efficiency 0 bit/s/Hz denotes the outage probability of the communication system, so the lower the value of CDF the better the performance. The results in Figure 8.15 show that by applying the V-MIMO approach the outage probability decreases significantly. The usage of the selection combining with fixed CM scheme either results in high probability of outage, if the system is designed to maximise the link throughput, or in very low link throughput, if the communication system is designed to maximise its reliability. The gain achieved by the V-MIMO concept is represented in Figure 8.15 by the vertical shift of the CDF curve towards the bottom of the graph, while the gain due to the usage of the ACM concept corresponds to the horizontal shift of the CDF curve towards the right-hand side of the graph.

The ACM concept opens an opportunity to increase the system capacity for the subscribers with good channel conditions and, in addition, it enables the subscribers with poor channel conditions to access the base station. The ACM concept combined with other fade mitigation techniques, such as space diversity or V-MIMO, brings additional improvement of the link capacity. However, the ACM concept cannot be combined with all fade mitigation techniques. For example, in OFDMA systems such as standardised in mobile WiMAX, the sub-channel permutation schemes prevent the usage of ACM on the group of sub-carriers.

8.5 Advanced Radio Resource Management Techniques

8.5.1 Introduction

We have already discussed earlier that one of the ways to increase the capacity of a HAP architecture is to use multiple spot-beam antennas, effectively creating a cellular structure on the ground [49]. Base stations are co-located on the HAP with interference in the spot beams arising from antenna side lobes.

The most straightforward way of using resources on the HAP would be to operate each cell in the same way a cell (with its own base station) would be operated on the ground. In this case the base station would be given a dedicated channel or set of channels to operate on. A good example of this is the operation of IEEE 802.16 (WiMAX) [50], where each cell would have a dedicated WiMAX base station, operating largely independently of other cells. Here we look at advanced ways of delivering more advanced radio resource management techniques, where the centralised nature of co-located base stations on the HAP can be more readily exploited.

It has been shown that there is considerable useable overlap between cells generated by HAP spot-beams, due to the way the antenna main beam rolls-off, and the potentially low side lobe level [49]. Previous work [51] has illustrated the benefits of exploiting cell overlap using a resource allocation scheme based on Personal Access Communications – Unlicensed B (PACS-UB) [52], developed originally for indoor terrestrial scenarios by Bellcore. In the following we present an overview of these techniques, making use of work originally presented in [53]. Here the distributed channel selection technique is combined with the centralised carrier to interference ratio threshold checks on the HAP. These are used as an additional form of Connection Admission Control (CAC), in order to improve the QoS and/or capacity, and will allow the exploitation of cell overlap that is readily available with a HAP architecture. These broad techniques could be extended to other forms of admission control, for example with packet oriented communications and a good treatment of this is seen in [54]. Other more detailed work looking at radio resource management for HAPs can be found in [55,56].

First, we will provide a brief overview of the intended HAP architecture and a novel power roll-off approximation that is used to characterise the received power decrease across a cell. This will be followed by a description of the channel assignment strategies that will be used at the HAP and user ends of the links. The performance of the schemes will then be evaluated.

8.5.2   Scenario

It is envisaged that a HAP will serve a coverage area of approximately 30 km radius, corresponding to a minimum elevation angle of 30° as shown in Figure 8.16. The coverage areas will be split into a collection of equal size circular cells on the ground (up to 127) [49], with each cell being centred on a hexagon which forms part of a hexagonal array. Such a configuration will result in a high degree of usable cell overlap and this can be exploited by the algorithms proposed here.

8.5.3   Channel Assignment Strategy

The channel assignment strategies for the HAP and users are best split, as their requirements differ [53]:

Spot Beam (HAP End) – The purpose of assigning channels to the spot beams is to provide spatial separation so that effective channel reuse can be obtained [57]. Fixed channel assignment has been shown to maximise capacity in ideal cases where there are fixed cell sizes, the same number of channels per cell, when the traffic is uniformly distributed, and with users operating in a uniform propagation environment [58]. Such a strategy is entirely appropriate here as the architecture delivers most of these ideal characteristics. The frequency allocations have been split uniformly to provide cluster sizes of 3 or 7, with these further divided into 16 or 7 sub-channels in each cell.

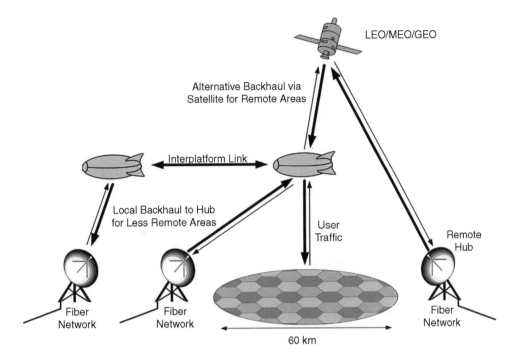

LEO/MEO/GEO

Alternative Backhaul via
Satellite for Remote Areas

Interplatform Link

Local Backhaul to Hub
for Less Remote Areas

User
Traffic

Remote
Hub

Fiber
Network

Fiber
Network

Fiber
Network

60 km

Figure 8.16 A possible HAP architecture (reproduced from [51])

User – The purpose of channel assignment at the user end of the link is primarily
to provide sufficient resources to support the traffic to be carried, rather than to
ensure good spatial channel reuse, since direct user–user interference will be
minimal. The user channel assignment strategy is based around PACS-UB [52]:
the user terminal listens to beacon transmissions from the cells and ranks them in
order of decreasing power. When the user wishes to make a connection it selects
the cell with the highest power (the main spot beam/cell) and if there are sufficient
free sub-channels available (e.g. time slots), owned by cell, and providing
the carrier to interference ratio (CIR) is greater than the specific thresholds
(discussed below), the call is accepted by allocating one of the free sub-channels
with lowest uplink interference measured at the HAP. If either the CIR is too low,
or a cell is full, the cell with the next highest power is used (an overlap spot
beam/cell). The call is blocked if no cell is available to take the connection.

New connections are only admitted to free channels when the CIR exceeds (or
will exceed) the minimum threshold required (17 dB). CIR levels can be measured
at either the HAP end on the uplink, or at the user on the downlink. The HAP and

user can evaluate the CIR of a new connection and a HAP can also determine the impact of a new user on existing connection to see if they fall short of the threshold. Three algorithms are evaluated and compared here as an example of possible advanced radio resource management schemes: No Connection Drop (NCD), No Connection Drop with No Downlink threshold checks (NCD-ND) and No Threshold checks (NT). Each algorithm is discussed in more detail later.

8.5.4 Performance

Cell overlap is an inherent feature of HAP communication systems [49] and the channel assignment schemes can exploit this to improve capacity [51]. The performance of the three channel assignment algorithms, NCD, NCD-ND and NT, with and without the use of cell overlap, for different cluster sizes and a fixed frequency allocation has been assessed by simulation. A list of key simulation parameters is given in Table 8.5.

A virtual base station technique has been used to characterise the propagation environment [51]. This assumes that the received power in dB from the centre of any of the equal-size spot beams (cells) is the same and is given by [51]:

$$P_{rx}(r) = P_{tx} + G_{rx} + G_{rx} - 20\log_{10}(4\pi\sqrt{h^2 + r^2}/\lambda)$$
$$+ \max\{10\,n\log_{10}\{\cos[\arctan(r/h)], S_f\} \tag{8.17}$$

Table 8.5 Simulation parameters for a small-scale scenario (based on [53])

Simulation parameters	Value
Simulation area (circular)	4450 km^2
Cell centre–cell centre	6.3 km
Number of ground-based users	2540
Number of spot beams (cells)	127
Link gains ($P_{tx} + G_{rx} + G_{tx}$)	77 dBm
Platform height (h)	17 km
Antenna roll-off factor (n)	58
Wavelength (λ)	0.01 m
Side lobe floor (S_f)	−40 dBc
Noise floor (N_f)	−130 dBm
Cluster number	3, (7)
Total number of channels in allocation	28
Number of sub-channels per cell	16, (7)
Number of permissible overlap cells	0–6
Mean connection interarrival time (max.)	1000 s
Mean connection duration	50 s
Minimum CIR access threshold	17 dB

where r is the distance away from the centre of the cell with all other parameters specified in Table 8.5. The interarrival times and durations of the connections are exponentially distributed, and users are uniformly distributed over the coverage area.

Performance of the algorithms is compared against theory. The Engset distribution [59] has been used in preference to the Erlang B formula owing to the high node activity factor. Both the nonoverlap and overlap cases have been considered. The nonoverlap case assumes that the there are either 16 or 7 channels (depending on whether the cluster size is 3 or 7) available per cell and that on average there are 20 users in a cell. In practice this will be subject to spatial variations owing to users being distributed randomly across the coverage area. To determine the performance in the overlap case a cluster of cells is considered. It is assumed that all 48 (49) channels are available and there are 60 (140) users per cluster (numbers in brackets relate to cluster 7). These expressions assume that connections are blocked solely based on channel availability rather than inadequate CIR levels. Therefore, they are best seen as upper bounds on performance.

8.5.5 No Connection Drop Algorithm

Previously the CIR performance of the PACS-UB type algorithm has been examined, which controls performance at initial access and on the downlink only [51]. With such an algorithm the CIR on the downlink and uplink during the lifetime of the connection is unbounded, i.e. it can fall much lower than the CIR access threshold. In addition, the uplink CIR at access can also be below the threshold. The NCD algorithm seeks to improve performance by exploiting the unique features of the HAP architecture, particularly the co-location of the base stations on the platform. This allows two additional CIR threshold checks to be performed:

- on the uplink at access;
- on the uplink of all users with connections in progress.

The same threshold value is used (17 dB) and if any of the CIR measurements fall below the threshold the new connection is blocked in the selected slot. The second of these threshold checks is relatively easy to perform because of the co-location of base stations on the HAP. It is much more difficult to perform such checks terrestrially as base stations are normally separated geographically.

8.5.5.1 Cluster Size 7

Figure 8.17 shows the blocking probability as a function of offered traffic for the overlap and no overlap cases. It can be seen how using overlap improves the blocking probability performance. This falls some way short of the theoretical bounds, because these assume no threshold checks (see later).

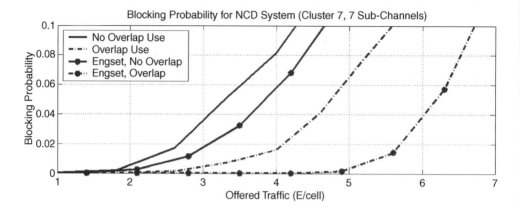

Figure 8.17 Blocking probability for NCD algorithm for a cluster size of 7 with and without cell overlap (reproduced from [53])

To determine the CIR performance CDFs of link quality on both uplink and downlink have been determined at high traffic loads (corresponding to a blocking probability of 5%). These CDFs have been generated from CIR values obtained at initial access and from connections in progress (taken upon a successful arrival in the system). Figure 8.18(a) corresponding to a cluster size of 7 shows that these three threshold checks are sufficient to maintain the CIR above the threshold with no overlap exploitation. This is partly helped by the fact that users are well spaced due to the high reuse number. Figure 8.18(b) shows the corresponding overlap case and now the CIR remains above the threshold on three out of the four cases, only dropping slightly below on downlink continuous measure (this is not subject to a threshold check). The high reuse number again helps maintain the CIR above the threshold, but now users sharing the same resources can be much closer together because they can use the overlapping cells, accounting for the slightly poorer CIR performance. However, this should be offset against the significantly better blocking probability performance shown in Figure 8.17.

8.5.5.2 Cluster Size 3

It is also useful to compare the above with the cluster size 3 performance; here CIR is much more susceptible to the reduced channel reuse distance. The blocking

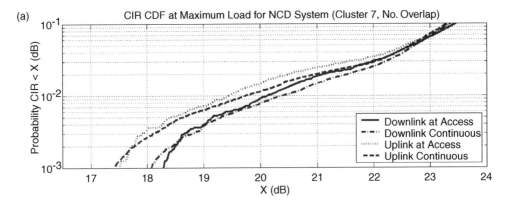

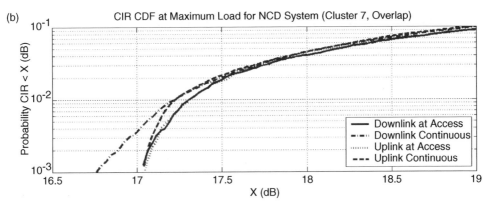

Figure 8.18   Comparison of the CIR performance for the NCD algorithm using a cluster size of 7 (a) without cell overlap exploitation and (b) with cell overlap exploitation (reproduced from [53])

probability performance shown in Figure 8.19, once again shows that overlap exploitation provides a reduction in blocking probability. The actual performance is much worse than the Engset distribution because most connections are blocked due to the CIR threshold, caused by the shorter reuse distance, rather than blocked due to channel unavailability.

Figure 8.20(a) shows CIR performance at access and during the lifetime of the connections when no overlap is used. The CIR is generally lower than in the corresponding cluster size 7 case (see Figure 8.18), primarily as result of the reduced channel reuse distance. Figure 8.20(b) also shows that the three threshold checks, even at the same high traffic loads, are still sufficient to tightly control the CIR.

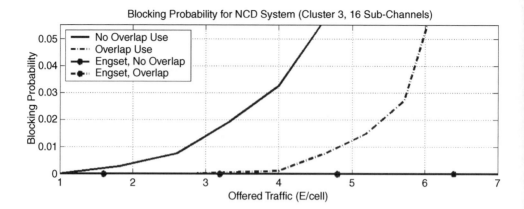

Figure 8.19 Blocking probability for NCD algorithm for a cluster size of 3 with and without cell overlap (reproduced from [53])

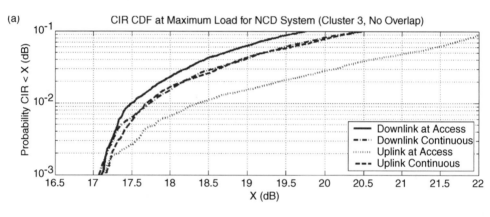

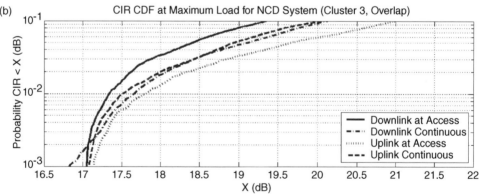

Figure 8.20 Comparison of the CIR performance for the NCD algorithm using a cluster size of 3 (a) without cell overlap exploitation and (b) with cell overlap exploitation (reproduced from [53])

8.5.6 No Connection Drop Algorithm with No Downlink Threshold Detection

The previous section illustrated how three threshold checks were sufficient to effectively control the uplink and downlink CIR even when cell overlap was exploited. A simpler version of the algorithm (NCD-ND), from the user terminal perspective, can be achieved by removing the downlink threshold check at access. Removal of this threshold check also aids the compatibility with the IEEE 802.16 protocol [50], a possible HAP access standard. The performance of this algorithm was assessed for the same user-HAP scenario as the NCD algorithm.

8.5.6.1 Cluster Size 7

Removal of the downlink threshold check has little effect on the blocking probability, with performance being virtually identical to that shown in Figure 8.17. Similarly, there is little change to the CIR performance when cell overlap is not used, with results looking similar to Figure 8.18. It is only when cell overlap is exploited that a slight difference is seen as shown in Figure 8.21. The removal of the downlink threshold check causes the CIR on the downlink to be slightly below threshold in a small number of cases.

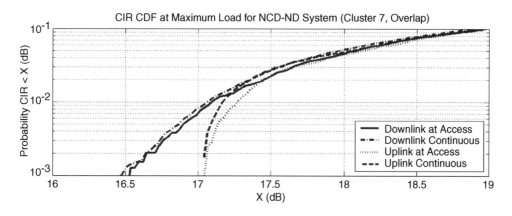

Figure 8.21 Comparison of the CIR performance for the NCD-ND algorithm for a cluster size of 7, with cell overlap exploitation, at access and during a connection (reproduced from [53])

8.5.6.2 Cluster Size 3

In the case of cluster size 3 the blocking probability performance is virtually identical to the corresponding case in the previous algorithm. The CIR performance now is less good than previously, with the downlink CIR being some way below the threshold, even for the no overlap exploitation case as shown in Figure 8.22(a). Such degradations are infrequent and are only likely to happen at high traffic loads, again the CDF relates to the same high traffic loads as previously. Compared with the cluster size 7, this increased degradation is largely a result of the closer reuse distance. This is further exacerbated when cell overlap is used as shown in Figure 8.22(b); now there is even less control on the spacing of users on the same channel, and it is likely that the CIR may be below the threshold on the downlink (very occasionally) even at low traffic loads.

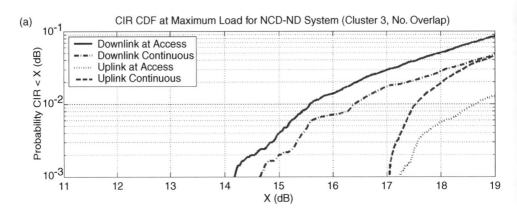

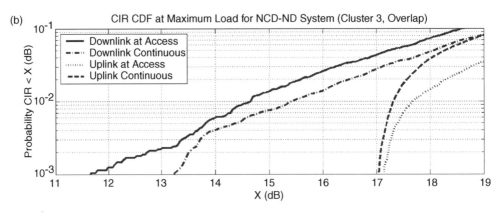

Figure 8.22 CIR of NCD-ND algorithm for a cluster size of 3, (a) without overlap and (b) with overlap (reproduced from [53])

8.5.7 No Threshold Detection

It is useful to ascertain just how much effect the CIR thresholds have on performance. The results shown in Figure 8.23 are for a cluster size of 7. Now the blocking probability is much closer to the theoretical bound given by the Engset distribution.

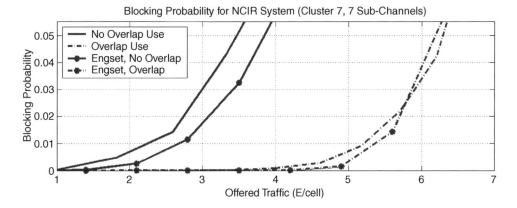

Figure 8.23   Blocking probability for NT detection algorithm for a cluster size of 7 with and without cell overlap (reproduced from [53])

In the no overlap case shown in Figure 8.24(a), the reuse distance is sufficient to protect the CIR at or above 16 dB. In the cases where overlap is used, as shown in Figure 8.24(b), the CIR now varies much more markedly from 0 dB (an artificial lower limit set in the simulation) up to 25 dB. This indicates that without the thresholds, users will accept connections well into overlap regions, in places where they will receive excessive amounts of interference. It is therefore essential that at least one threshold check should be performed if cell overlap is to be exploited effectively.

8.5.8   Discussion

The checks performed by each algorithm are summarised in Table 8.6.

The NCD-ND algorithm despite its poorer performance may still be the most appropriate resource allocation algorithm to select for a HAP system. It has the advantage that most of the intelligence is not at the user end of the link. It is also likely to be more compatible with a possible access standard such as IEEE 802.16. The NCD algorithm should be adopted if tighter control of the CIR is required on the downlink.

The only drawback with both the NCD and NCD-ND algorithms is the degree of processing power required on the HAP to fully implement the scheme. The CIR of all users in all cells must be checked before a new connection is accepted.

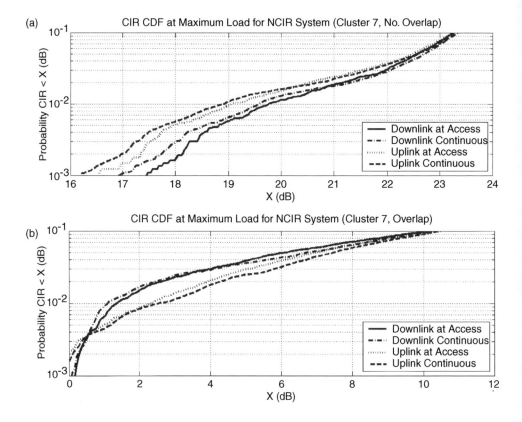

Figure 8.24 CIR of NT detection algorithm for a cluster size of 7, (a) without overlap and (b) with overlap (reproduced from [53])

Table 8.6 Summary of the threshold checks (marked with X) carried out by each of the algorithms

	Check at initial access		Check 'in progress' connections	
	Uplink (HAP)	Downlink (User)	Uplink (HAP)	Downlink (User)
PACS-UB [52]		X		
No Connection Drop (NCD)	X	X	X	
No Connection Drop with No Downlink threshold check (NCD-ND)	X		X	
No Threshold (NT)				

Processing may be carried out in a fully centralised fashion, controlled by one processor, or perhaps more likely, each cell or cluster of cells will have its own processor with a control bus connecting them together. If either of these approaches is considered too complex then it may be possible to restrict the information transfer to a cluster or group of neighbouring cells sharing the same channel.

References

1. J. Mitola, *Cognitive Radio – An Integrated Agent Architecture for Software Defined Radio*, PhD Dissertation, Royal Institute of Technology, Kista, Sweden, 2000.
2. I. F. Akyildiz, W. Y. Lee, M. C. Vuran and S. Mohanty, *NeXT Generation/Dynamic Spectrum Access/Cognitive Radio Wireless Networks: A Survey*, Computer Networks 2006, Vol. 50, No. 12, pp. 2127–2159.
3. P. Likitthanasate, D. Grace and P. D. Mitchell, *Performance of a Spectrum Sharing Game with Path Gain Ratio Based Cost Parameter*, IET Seminar on Cognitive Radio and Software Defined Radios: Technologies and Techniques, London, UK, September, 2008.
4. P. Likitthanasate, D. Grace and P. D. Mitchell, *Spectrum Etiquettes for Terrestrial and High-Altitude Platform-based Cognitive Radio Systems*, IET Commun., July 2008, Vol. 2, No. 6, pp. 846–855.
5. Y. Liu, D. Grace and P. D. Mitchell, *Comparison of CINR-based Cognitive Radio Schemes for Multiple High Altitude Platforms*, Proc. International Workshop on Cognitive Networks and Communications (CogCom 2008), Hangzhou, China, August 2008.
6. S. Bayhan, G. Gur and F. Alagoz, *High Altitude Platform (HAP) Driven Smart Radios: A Novel Concept*, Proc. 3rd International Workshop on Satellite and Space Communications 2007 (IWSSC 2007), Salzburg, Austria, September 2007.
7. G. Giambene and S. Kota, *Cross-Layer Protocol Optimization for Satellite Communications Networks: A Survey*, Int. J. Satellite Commun. Netw., September 2006, Vol. 24, No. 5, pp. 323–341.
8. A. F. Molisch, *Wireless Communications*, John Wiley & Sons, Ltd, Chichester, 2005.
9. U. Drcic, G. Kandus, M. Mohorčič and T. Javornik, *Interplatform Link Requirements in the Network of High Altitude Platforms*, Proc. 19th AIAA International Communications Satellite Systems Conference and Exhibit, Toulouse, France, April 2001.
10. T. Javornik, M. Mohorčič, A. Švigelj, I. Ozimek and G. Kandus, *Adaptive Coding and Modulation for Mobile Wireless Access via High Altitude Platforms*, Wireless Personal Communications, 2005, Vol. 32, No. 3–4, pp. 301–317.
11. CAPANINA Website, http://www.capanina.org/.
12. T. Celcer, T. Javornik, M. Mohorčič and G. Kandus, *Virtual Multiple Input Multiple Output in Multiple High-Altitude Platform Constellations*, IET Commun., 2009, Vol. 3, No. 11, pp. 1704–1715. 10.1049/iet-com.2008.0741.
13. S. Scalise, V. Schena and F. Ceprani, *Multimedia Service Provision On-Board High Speed Train: Demonstration and Validation of the Satellite-Based FIFTH Solution*, 22nd AIAA International Communications Satellite Systems Conference and Exhibit, Monterey, CA, USA, May 2004.
14. D. Gesbert, M. Shafi, Da-shan Shiu, P. J. Smith and A. Naguib, *From Theory to Practice: an Overview of MIMO Space-Time Coded Wireless Systems*, IEEE J. Sel. Areas Commun., April 2003, Vol. 21, No. 3, pp. 281–302.
15. S. Plevel, T. Javornik, M. Mohorčič and G. Kandus, *Empirical Propagation Channel Model for High Altitude Platform Communication Systems*, Proc. 11th Ka and Broadband Communications Conference, Rome, Italy, September 2005, pp. 187–194.
16. D. Gesbert, M. Shafi, Da-shan Shiu, P.J Smith and A. Naguib, *From Theory to Practice: An Overview of MIMO Spacetime Coded Wireless Systems*, IEEE J. Sel. Areas Commun., 2003, Vol. 21, No. 3, pp. 281–302.
17. G. J. Foschini and M. J. Gans, *Layered Space-Time Architecture for Wireless Communication in a Fading Environment when Using Multiple Antennas*, Bell Labs Syst. Tech. J, 1996, Vol. 1, No. 2, pp. 41–59.

18. L. Zheng and D. Tse, *Diversity and Multiplexing: A Fundamental Tradeoff in Multiple Antenna Channels*, IEEE Trans. Inf. Theory, 2003, Vol. 49, No. 5, pp. 1073–1096.

19. E. Telatar, *Capacity of Multi-Antenna Gaussian Channels*, European Trans. Telecommun. ETT, November 1999, vol. 10, pp. 585–596.

20. P. W. Wolaniansky, G. J. Foschini, G. D. Golden and R. A. Valenzula, *V-BLAST: An Architecture for Realizing Very High Data Rates Over The Rich-Scaterring Wireless Channel*, Proc. Int. Symposium on Advanced Technologies, Boulder, CO, September 1998.

21. B. Hassibi and H. Vikalo, *On Sphere Decoding Algorithm. I. Expected Complexity*, IEEE Trans. Sig. Proc., August 2005, Vol. 53, No. 8, pp. 2806–2818.

22. A. Calderbank and S. Diggavi, *Tutorial on Space-Time Coding and Signal Processing for High Data Rate Wireless Communications*, Proc. Int. Zurich Seminar, IZS'02, Zurich, Switzerland, February 2002.

23. S. Alamouti, *A Simple Transmitter Diversity Technique for Wireless Communications*, IEEE J. Selected Areas Commun., Special Issue, 1998, Vol. 16, No. 8, pp. 1451–1458.

24. V. Tarokh, H. Jafarkhani and A.R. Calderbank, *Space-Time Block Codes from Orthogonal Designs*, IEEE Trans. Inf. Theory, July 1999, Vol. 45, No. 5, pp. 1456–1467.

25. M. Rupp and C. F. Mecklenbrauker, *On Extended Alamouti Schemes for Space-Time Coding*, Proc. 5th Int. Symp. on Wireless Personal Multimedia Communications, Hawaii, USA, October 2002.

26. V. Tarokh, N. Seshadri and A. R. Calderbank, *Space-Time Trellis Codes for High Data Rate Wireless Communication: Performance Criterion and Code Construction*, IEEE Trans. Inf. Theory, 1998, Vol. 44, No. 2, pp. 744–765.

27. A.-F. Molisch, *Wireless Communications*, John Wiley & Sons Ltd, Chichester, 2005.

28. P. Horvath, G. K. Karagiannidis, P.-R. King, S. Stavrou and I. Frigyes, *Investigations in Satellite MIMO Channel Modeling: Accent on Polarization*, EURASIP J. Wireless Commun. Netw., 2007, Vol. 2007, Article ID 98942, 10.1155/2007/98942.

29. W. T. Webb and L. Hanzo, *Modern Quadrature Amplitude Modulation*, Pentech Press Ltd, London, 1994.

30. A. J. Goldsmith and S. G. Chua, *Adaptive Coded Modulation for Fading Channels*, IEEE Trans. Commun., 1998, Vol. 46, No. 5, pp. 595–602.

31. K. J. Hole, H. Holm and G. E. Oien, *Adaptive Coded Modulation Performance and Channel Estimation Tools for Flat Fading Channels*, IEEE J. Selected Areas Commun., 2002, Vol. 18, No. 7, pp. 1153–1158.

32. J. Williams, L. Hanzo and R. Steele, *Channel-Adaptive Modulation*, 6th IEE International Conference on Radio Receivers and Associated Systems, London, 1999, pp. 144–147.

33. T. Javornik and G. Kandus, *An Adaptive Rate Communication System Based on the N-MSK Modulation Technique*, IEICE Trans. Commun., 2001, Vol. E84-B, No. 11, pp. 2946–2955.

34. U. Toyoki, S. Sampei, N. Morinaga and K. Hamaguchi, *Symbol Rate and Modulation Level-Controlled Adaptive Modulation/TDMA/TDD System for High-Bit-Rate Wireless Data Transmission*, IEEE Trans. Veh. Technol., 1998, Vol. 47, No. 4, pp. 1134–1147.

35. G. Albertazzi. S. Cioni, G. E. Corazza, M. Neri, R. Pedone, P. Salmi, A. Vanelli-Coralli and M. Villanti, *On the Adaptive DVB-S2 Physical Layer: Design and performance*, IEEE Wireless Commun. December 2005, Vol. 12, No. 6, pp. 62–68.

36. A. Morello and V. Mignone, *DVB-S2: The Second Generation Standard for Satellite Broad-Band Services*, Proc. IEEE, January 2006, Vol. 94, No. 1, pp. 210–227.

37. IEEE 802.16–2004, *IEEE Standard for Local and Metropolitan Area Networks – Part 16: Air Interface for Fixed Wireless Access Systems*, 2004.

38. IEEE 802.16e-2005, *Amendment to IEEE Standard for Local and Metropolitan Area Networks – Part 16: Air Interface for Fixed Broadband Wireless Access Systems – Physical and Medium Access Control Layers for Combined Fixed and Mobile Operation in Licensed Bands*, 2005.

39. ETSI EN 302 307 (V1.1.2), *Digital Video Broadcasting (DVB); Second Generation Framing Structure, Channel Coding and Modulation Systems for Broadcasting, Interactive Services, News Gathering and Other Broadband Satellite Applications*, 2006.

40. ETSI EN 301 790 (V1.4.1), *Digital Video Broadcasting (DVB) Interaction Channel for Satellite Distribution Systems*, European Telecommunications Standard Institute (ETSI), September 2005.

41. 3GPP TS25.211, *Physical Channels and Mapping of Transport Channels onto Physical Channels (FDD)*, version 4.4.0. 2005.
42. 3GPP TS25.855, *High Speed Downlink Packet Access; Overall UTRAN Description*, version 5.0.0. 2005.
43. G. Kandus, M. Mohorčič, M. Smolnikar, E. Leitgeb and T. Javornik, *A Channel Model of Atmospheric Impairment for the Design of Adaptive Coding and Modulation in Stratospheric Communication*, WSEAS Trans. Commun., April 2008, Vol. 7, No. 4, pp. 311–326.
44. M. Smolnikar, T. Javornik and M. Mohorčič, *Channel Decoder Assisted Adaptive Coding and Modulation for HAP Communications*, Proc. IEEE 65th Vehicular Technology Conference (VTC2007-Spring), Dublin, Ireland, April 2007, pp. 1375–1379.
45. M. Smolnikar, T. Javornik and M. Mohorčič, *Target BER Driven Adaptive Coding and Modulation in HAP Based DVB-S2 System*, Proc. 4th Advanced Satellite Mobile Systems (ASMS'08), Bologna, Italy, August 2008.
46. M. Mohorčič, M. Smolnikar and T. Javornik, *Performance Comparison of Adaptive Coding and Modulation in HAP Based IEEE 802.16 and DVB-S2 Systems*, Proc. 26th AIAA International Communications Satellite Systems (ICSSC'08), San Diego, CA, USA, June 2008.
47. M. Smolnikar, T. Javornik, M. Mohorčič and M. Berioli, *DVB-S2 Adaptive Coding and Modulation for HAP Communication System*, Proc. IEEE 67th Vehicular Technology Conference (VTC2008-Spring), Marina Bay, Singapore, May 2008, pp. 2947–2951.
48. ETSI EN 300 421 (V.1.1.2), *Digital Video Broadcasting (DVB); Framing Structure, Channel Coding and Modulation for 11/12 GHz Satellite Services*.
49. J. Thornton, D. Grace, M.H. Capstick and T.C. Tozer, *Optimising an Array of Antennas for Cellular Coverage from a High Altitude Platform*, IEEE Trans. Wireless Commun., May 2003, Vol. 2, No. 3, pp. 484–492.
50. IEEE 802.16e-2005, *Part 16: Air Interface for Fixed and Mobile, Broadband Wireless Access Systems, Amendment 2: Physical and Medium Access, Control Layers for Combined Fixed and Mobile, Operation in Licensed Bands*, 2005.
51. D. Grace, C. Spillard, J. Thornton and T.C. Tozer, *Channel Assignment Strategies for a High Altitude Platform Spot-Beam Architecture*, Proc. 13th IEEE International Symposium on Personal Indoor and Mobile Radio Communications (PIMRC 2002) 2002, Vol. 4, pp. 1586–1590.
52. L.-F. Chang, A. Noerpel and A. Ranade, *Performance of Personal Access Communications System – Unlicensed B*, IEEE J. Selected Areas Commun., May 1996, Vol. 14, No. 4, pp. 718–727.
53. D. Grace, C. Spillard and T.C. Tozer, *High Altitude Platform Resource Management Strategies with Improved Connection Admission Control*, Wireless Personal Multimedia Communications, October 2003.
54. D. Grace, C. Spillard, M. Mohorčič, T.C. Tozer, J. Thornton, K. Katzis, D. Pearce, Yu Ming Lu, J. Bostič, U. Drčič, T. Javornik and G. Kandus, *Resource Allocation Methods and Network Protocols for HeliNet*, HeliNet Doc Ref: HE-064-T4-UNY-RP-01, 31 October 2002.
55. K. Katzis, L. Dong, L. Dinh Dung, D. Grace and P.D. Mitchell, *Resource Allocation and Handoff Techniques for High Altitude Platforms*, CAPANINA Doc Ref: CAP-D23a-WP24-UOY-PUB-01, 18 December 2006, www.capanina.org/documents/CAP-D23a-WP24-UOY-PUB-01.pdf.
56. K. Katzis, *Resource Allocation Techniques for High Altitude Platforms*, PhD Thesis, University of York, York, UK, 2005.
57. W.C.Y. Lee, *Spectrum Efficiency in Cellular*, IEEE Trans. Veh. Technol., May 1989, Vol. 38, No. 2, pp. 69–75.
58. L.J. Cimini, G.J. Foschini C.-H. I and Z. Miljanic, *Call Blocking Performance of Distributed Algorithms for Dynamic Channel Allocation in Microcells*, IEEE Trans. Commun., August 1994, Vol. 42, No. 8, pp. 2600–2607.
59. L. Kleinrock, *Queuing Systems, Vol. 1 Theory*, Wiley-Interscience, 1975, pp. 109.

Part Three

Multiple High Altitude Platforms

9

Multiple HAP Networks

9.1 Why Multiple HAP Constellations?

We have already seen in previous chapters that the ITU-R has allocated a number of frequency bands to HAPS[1] to aid their eventual deployment. There are bands at 48 GHz for worldwide [1] and 31/28 GHz for certain Asian countries [2], with spectrum in the 3G bands also allocated for HAPS use [3].

Efficient spectrum reuse will be required to ensure that such HAP deployments can deliver high spectral efficiencies within these bands. Cellular solutions have been examined in [4,5], specifically addressing the antenna beam characteristics required to produce an efficient cellular structure on the ground, and the effect of antenna sidelobe levels on channel reuse plans [5]. HAPs will have relatively loose station-keeping characteristics compared with satellites, and the effects of platform drift on a cellular structure and the resulting intercell handover requirements have been investigated [6]. Cellular resource management strategies have also been developed for HAP use [7]. Cells can be regularly spaced, as their area and location are substantially unaffected by geography and terrain, and since they all originate from the same HAP this centralisation can be additionally exploited by the resource management strategy.

While it is generally acknowledged that HAPs could offer a higher spectral efficiency than GEO satellites, some scepticism remains over whether HAPs can approach the spectral efficiency of terrestrially based broadband wireless communications. This is based on the assumption that a cellular approach is used with the minimum cell size being limited by the maximum size of the antenna payload that can be accommodated on the HAP. This chapter will illustrate that it is possible to exploit one feature that has been largely overlooked, the fact that user antennas may

---

[1] Here we specifically use the ITU-R abbreviation of HAPS for high altitude platform station.

Broadband Communications via High Altitude Platforms   David Grace and Mihael Mohorčič
© 2011 John Wiley & Sons, Ltd

also be highly directive. This allows spatial discrimination between multiple HAPs located in different parts of the sky, thereby permitting them to share a common spectrum [8,9]. This additional bandwidth reuse, and resulting capacity gain, is dependent on several factors, in particular the number of platforms and the user antenna sidelobe levels. A multiple HAP configuration also provides for incremental roll-out: initially only one HAP needs to be deployed, with all user antennas pointing to the single HAP. As more capacity is required, further HAPs can be brought into service, with new users served by the newly deployed HAPs.

9.1.1 Model of the Multiple HAP System

Figure 9.1 illustrates a multiple HAPs scenario showing a user, main HAP and one interfering HAP both with directional antennas covering a large single cell, with both HAPs being equally spaced around a circle. Here, 'O' is the centre of the coverage area, while the antennas on the main HAP (HAP$_m$) and other interfering HAP (HAP$_{i1}$) point at 'P$_1$' and 'P$_2$', respectively. We give each HAP in the multiple HAP system its own coordinate system for convenience. Each HAP has its 'own x-axis' aligned so as to contain the sub-platform point and the centre of the coverage area. For example, the x coordinate in Figure 9.1 is the 'own x-axis' of the main HAP.

In a multiple HAP system, the system capacity is constrained by interference plus noise. Therefore the carrier to interference plus noise ratio (CINR) is used as a fundamental parameter to assess the system capacity. For a system of N HAPs

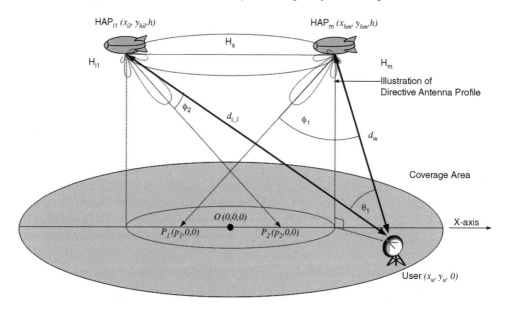

Figure 9.1 Multiple HAPs scenario with a directional antenna (reproduced from [10])

the downlink CINR at an arbitrary point $(x,y,0)$ in the coverage area can be calculated as [11]:

$$\text{CINR} = \frac{C}{N+I} = \frac{P_{Hm}A_{Hm}(\phi_m)A_U(\theta_m)\left(\frac{\lambda}{4\pi d_m}\right)^2}{N_F + \sum_{j \in N \ m} P_{Hi_j}A_{Hi_j}(\phi_j)A_U(\theta_j)\left(\frac{\lambda}{4\pi d_{i_j}}\right)^2} \tag{9.1}$$

where P_{Hm} and P_{Hi_j} are the transmit power from the main HAP and the jth interfering HAP to the user, respectively. The terms $(\lambda/4\pi d_m)2$ and $(\lambda/4\pi d_{i\_j})^2$ are the free space path loss (FSPL) from the main HAP and the jth interfering HAP to the user, respectively. $A_{Hm}(\phi_m)$ and $A_{Hi\_j}(\phi_j)$ are the gains of the main and the jth interfering HAP antennas at angles ϕ_m and ϕ_j away from boresight, respectively. $A_U(\theta_m)$ and $A_U(\theta_j)$ are the receive gains of the user antenna for the main and the jth interfering HAP at angles θ_m and θ_j away from boresight, respectively. N_F is the thermal noise floor, and the transmit power from each HAP is assumed to be identical.

A key aspect of the multiple HAP system is the ability to reuse the radio spectrum by exploiting the directionality of the user antenna. This is relatively straightforward in the case of the mm-wave bands, where such antennas may be highly directional. Thus here we adopt the basic main lobe– sidelobe antenna model discussed in Section 3.2.3, specified as follows which is a slightly modified version of Equation (3.7):

$$A(\theta) = G_{\max}\{\max[\cos^n(\theta), s_f]\} \tag{9.2}$$

where $G_{\max}$ is the maximum gain at boresight, and s_f is a notionally flat sidelobe floor. The variable n controls the rate of roll-off of the antenna profile, thus setting the antenna beamwidth. Thus in Equation (9.1), U and H suffixes have been used to denote the parameters associated with the different antennas. Such an approach has been extensively used in the literature for modelling of multiple HAP systems [8,9,11]. Table 9.1 shows a default set of parameters based on those used

Table 9.1 Default set of parameter values used to assess performance in a multiple HAP system

Parameter	Default value
User antenna roll-off (n_U)	4550 (Beamwidth 2°)
Sidelobe floor (s_f)	−30 dB relative to peak
Coverage area radius	30 km
Platform height (h)	17 km
Transmitter power	0 dBW
HAP and user antenna efficiency	0.8
Noise floor (N_F)	−133 dBW (Equivalent noise temperature is 300 K, bandwidth is 12.5 MHz)
Frequency	30 GHz

elsewhere in the literature [11]. A more comprehensive treatment of multiple HAP performance can be found in [12].

9.2   Multiple HAP Constellation Planning

9.2.1   Multiple HAPs Scenario with Directional HAP Antennas

The main purpose of increasing the number of HAPs serving a common coverage area is to increase the capacity density (i.e. the bandwidth efficiency) [8,9]. Here we restrict ourselves to an analysis with one beam (cell) per HAP, serving the whole coverage area. Previous work [8,9,13] has assumed that an isotropic HAP antenna, and it has been shown that incrementally deploying further HAPs operating in the same allocated spectrum, will provide significant increases on capacity, allowing for the incremental support of future customers. The analysis here looks at the more detailed case using directional HAP antennas.

In order to completely exploit the multiple HAP system, the following aspects are investigated. We study first the effects on CINR across the coverage area for variable antenna beamwidths, spacing radii and pointing offset. We provide a discussion of the spectral efficiency of multiple HAPs using different spacing radii. This is followed by possible ways to control the location of peak power and the CINR behaviour in the multiple HAP system. Finally strategies for the optimum order of population of a single ring multiple HAPs constellation are examined.

9.2.1.1   Effects of Varying the HAP Antenna Beamwidth

Using Equation (9.2), the beamwidth (via n) can be adjusted, in relation to the angle subtended at the edge of the coverage area and the level of power roll-off required from the antenna at the cell edge. This is shown in Figure 9.2.

We examine the impact of the two levels of main lobe antenna roll-off, a roll-off of 3 dB (ψ_{3dB}) and one of 10 dB (ψ_{10dB}) at the subtended angle (ψ_{sub}) [11]. Coverage plots of the CINR received at the user have been generated using Equation (9.1) by assuming that a 'test' user is located at each point in the plot (the antenna of a test user points towards the main HAP), and this is shown in Figure 9.3(a) and (b). Both plots have two peak CINR values outside the coverage area due to the interference from other HAPs counteracting the peak power from the main HAPs. One is near the main HAP (the rightmost HAP) due to the shortest link length; the other is at the other end (on the left) which is outside the 60 km spacing radius where there is less interference from the other HAPs. The CINR contours are affected by the level of roll-off, as expected.

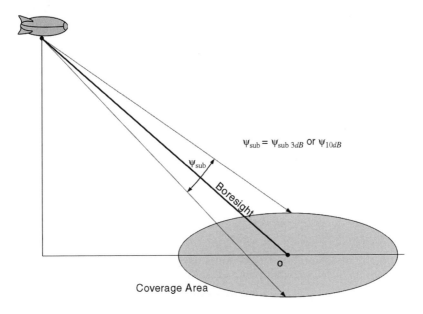

Figure 9.2   Illustration of $\psi_{3dB} = \psi_{\text{sub}}$ or $\psi_{10dB} = \psi_{\text{sub}}$

9.2.1.2 Effects of Variable HAP Spacing Radii

If we assume that the main HAP is located to the right of the coverage area, then Figure 9.4 illustrates the change of the CINR distribution along the x-axis of the coverage area for different spacing radii. The two CINR peaks become more and more separated as the spacing radius changes from 10 to 50 km. The right peak CINR also at the same time increases significantly because the more separate the HAPs, the higher the gain from the main HAP and the lower the level of interference that is received by the users.

Figure 9.5 shows the CDF of the CINR across the coverage area for a range of spacing radii. Performance is similar within the coverage area with most of the change observed at the extremes of the x-axis. One important aspect is seen when the HAPs increase their spacing radii from the inside to the outside of the coverage area. Here, despite the HAPs being located well outside the coverage area, the CINR remains high or even improves over the whole coverage area. This gives the designer significant flexibility as to where to position the HAPs. Here we choose to restrict the spacing radius to less than 50 km, while also making $\psi_{10dB} = \psi_{\text{sub}}$ and operating with a 0 km pointing offset. In such a case the right peak CINR is located within the coverage area as shown in Figure 9.4. This helps to reduce the waste of transmit power on the HAP.

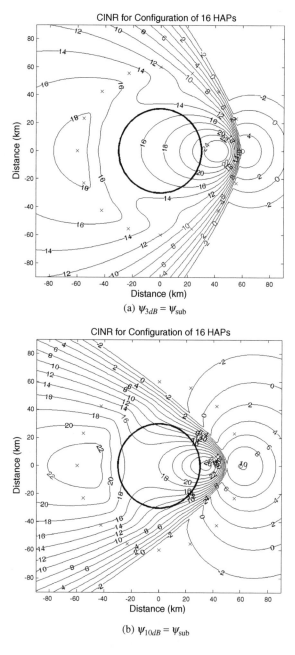

Figure 9.3   Downlink CINR contour plot for a configuration of 16 HAPs using directional antenna (1 main HAP 'o', 15 interfering HAPs 'x', contour labels: CINR (dB), spacing radius 60 km, pointing offset 0 km). Reproduced with kind permission @ Springer Science + Business Media Wireless Personal Communications, Performance of Multiple High Altitude Platforms using Directive HAP and User Antennas, vol. 32, no. 3-4, 2005, G. Chen, D. Grace and T. C. Tozer, Figures 14 and 15

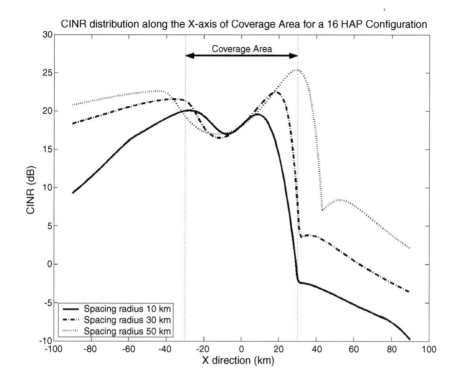

Figure 9.4 CINR distribution along the x-axis of the coverage area for a 16 HAP configuration with different spacing radii (pointing offset 0 km, $\psi_{10dB} = \psi_{sub}$). Reproduced with kind permission @ Springer Science + Business Media Wireless Personal Communications, Performance of Multiple High Altitude Platforms using Directive HAP and User Antennas, vol. 32, no. 3-4, 2005, G. Chen, D. Grace and T. C. Tozer, Figure 21

9.2.1.3 Effects of Variable HAP Pointing Offset

Figure 9.6 illustrates the change of the CINR distribution along the x-axis of the coverage area when different pointing offset values are used (when the main HAP is located to the right of the coverage area). When the pointing offset changes from 0 to −90 km, the left peak CINR increases slightly while the right peak CINR decreases, as expected. The two CINR peaks get closer in terms of location for decreasing values (from 0 to −90 km) of the pointing offset. For the 60 km spacing radii case, the right peak CINR changes by approximately 7 km as the pointing offset changes from 0 to −90 km. This illustrates that the location of peak CINR is relatively insensitive to the HAP antenna pointing offset in a multiple HAPs scenario. However, by varying the pointing offset it is possible to fine-tune the overall CINR and spectral efficiency to improve the system capacity.

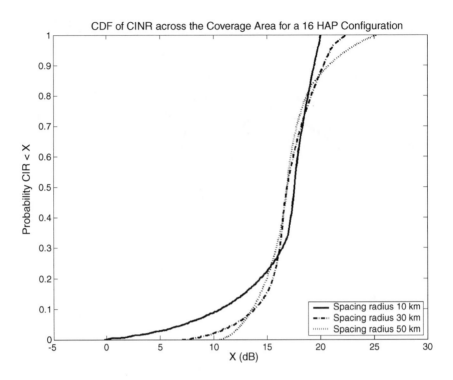

Figure 9.5   CDF of CINR across the coverage area for a 16 HAP configuration with different spacing radii (pointing offset 0 km, $\psi_{10dB} = \psi_{sub}$) (reproduced from [12])

9.2.1.4 Spectral Efficiency in Multiple HAPs Scenario

The CINR can be translated into a theoretical upper bound on spectral efficiency using the Shannon equation[2] [1]:

$$\eta \approx \log_2(1 + \text{CINR}) \qquad (9.3)$$

The aggregate spectral efficiency (η_c) available at points across the coverage area is derived from the summation of that offered by each HAP. That is, for each ground position $(x,y,0)$ there are N 'test' users, whose antenna points directly at a different HAP. The links from each 'test' user to their respective HAP will deliver a specific

---

[2] The Shannon equation assumed that the noise sources are Gaussian – interference will only be approximately Gaussian.

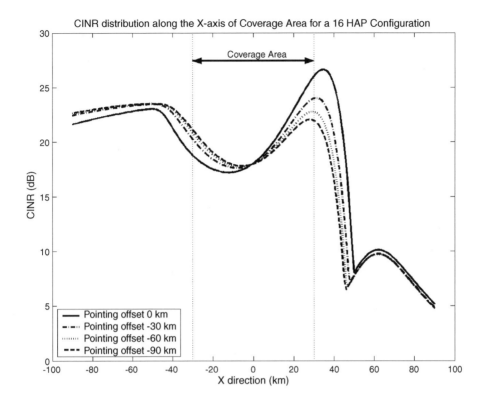

Figure 9.6   CINR distribution along the **x**-axis of coverage area for a 16 HAP configuration with different pointing offset (spacing radii 60 km, $\psi_{10dB} = \psi_{sub}$). Reproduced with kind permission from © Springer Science + Business Media Wireless Personal Communications, Performance of Multiple High Altitude Platforms using Directive HAP and User Antennas, vol. 32, no. 3-4, 2005, G. Chen, D. Grace and T. C. Tozer, Figure 19

spectral efficiency, and these values when summed together yield the aggregate spectral efficiency for each ground position, i.e. [9]:

$$\eta_c \approx \sum_{j \in N} \log_2(1 + \text{CINR}_j) \qquad (9.4)$$

Figure 9.7 shows a CDF of spectral efficiency for three different spacing radii. All three spacing radii yield the same maximum aggregate spectral efficiency of approximately 96 bits/s/Hz, and tend to occur at the centre of coverage area for all 16 HAPs. The CDF curve with spacing radius 50 km is nearly a straight line,

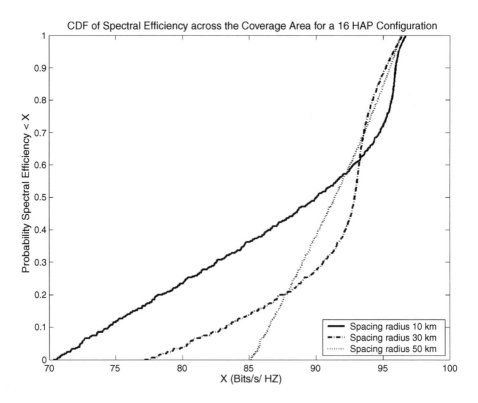

Figure 9.7   CDF of spectral efficiency across the coverage area for a 16 HAP configuration with different spacing radii (pointing offset 0 km, $\psi_{10dB} = \psi_{sub}$). Reproduced with kind permission from © Springer Science + Business Media Wireless Personal Communications, Performance of Multiple High Altitude Platforms using Directive HAP and User Antennas, vol. 32, no. 3-4, 2005, G. Chen, D. Grace and T. C. Tozer, Figure 22

indicating a low spectral efficiency variation. It has a much higher minimum spectral efficiency (85 bits/s/Hz) as a worst case than the other two cases with smaller spacing radii.

Figure 9.8 illustrates the effect on median spectral efficiency across the coverage area for several HAP spacing radii and different numbers of HAPs. This clearly shows that the spectral efficiency increases almost linearly as the number of HAPs increases, up to 16. All three curves are very close to each other, indicating that increasing the HAP spacing radius has little impact on median spectral efficiency. Performance is actually slightly better for larger spacing radii, as a result of the increase in gain for the HAP antenna. Importantly, performance is dominated

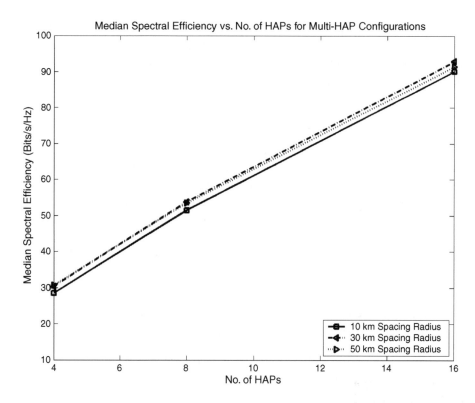

Figure 9.8 Median spectral efficiency **versus number** of HAPs for a set of spacing radii (pointing offset 0 km, $\psi_{10dB} = \psi_{sub}$). Reproduced with kind permission from © Springer Science + Business Media Wireless Personal Communications, Performance of Multiple High Altitude Platforms using Directive HAP and User Antennas, vol. 32, no. 3-4, 2005, G. Chen, D. Grace and T. C. Tozer, Figure 23

by the number of HAPs. Therefore designers should consider deploying HAPs outside the coverage area for the following reasons:

- It ensures that HAPs are unlikely to fall on areas of population in the event of a crash.
- Deploying HAPs outside the coverage area will mean that they are less visible overhead.

However, deploying multiple HAPs outside the coverage area may violate a minimum elevation angle criterion, but such a violation may still be acceptable if line of sight (LOS) coverage from all HAPs is not needed at every location throughout the coverage area in order to yield maximum capacity. Also, it may also be possible to extrapolate performance from these results when frequency bands of sub-10 GHz are used. These do not need the LOS links meaning that the minimum elevation angle is less important. In this situation, the propagation loss due to multipath and shadowing effects needs to be considered as well as the FSPL, considered here.

9.2.1.5 Controlling the Location of Peak Carrier Power

It is possible to determine the relationship between the location of the peak carrier power, HAP's location, and HAP antenna pointing offset. This can be achieved by differentiating the carrier part of Equation (9.1) (numerator) with respect to user location X and set equal to 0. The noise and interference items in the denominator of Equation (9.1) are not related to the carrier power and the user location X:

$$\frac{d(\text{Carrier})}{dX} = \frac{d\left[P_T A_H(\phi) A_U(\theta) \left(\frac{\lambda}{4\pi d_m}\right)^2\right]}{dX} = 0 \tag{9.5}$$

Inserting the antenna profile equation based on Equation (9.2), only $\cos^{n_H}(\phi)$ and $(1/d_m)^2$ in $(\lambda/4\pi d_m)^2$ are related to the user location X, yielding [11,12]:

$$\frac{d(\text{Carrier})}{dX} = \frac{d\left[\cos^{n_H}(\phi) \left(\frac{1}{d_m}\right)^2\right]}{dX} = 0 \tag{9.6}$$

where

$$\cos(\phi) = \frac{[(S-X)^2 + H^2] + [(S-P)^2 + H^2] - (X-P)^2}{2\sqrt{(S-X)^2 + H^2} \cdot \sqrt{(S-P)^2 + H^2}} \tag{9.7}$$

$$d_m = \sqrt{(S-x)^2 + H^2} \tag{9.8}$$

where P is the HAP antenna pointing location along the x-axis, S is the HAP spacing radius and H is the height of the HAP. Finally, the following equation is derived:

$$2(P-S)X^2 + [4S(S-P) + H^2(n_H + 2)]X \\ -2S(S^2 + H^2 - S \cdot P) - n_H \cdot H^2 \cdot P = 0 \tag{9.9}$$

Using the standard quadratic equation formula, Equation (9.9) can be solved as a function of the above parameters, to yield the location of the peak carrier power as follows [11,12]:

$$X = \frac{-B \pm \sqrt{B^2 - 4AC}}{2A} \qquad (9.10)$$

where

$$A = 2(P - S) \qquad (9.11)$$

$$B = 4S(S - P) + H^2(n_H + 2) \qquad (9.12)$$

$$C = -2S(S^2 + H^2 - S \cdot P) - n_H \cdot H^2 \cdot P \qquad (9.13)$$

Equation (9.9) was derived so as to determine the location of peak downlink carrier power of an individual HAP. However, it can also be used to derive further interesting results. For example, the equation can be rearranged in order to fix the location of peak power, enabling us to derive a relationship in terms of the HAP antenna roll-off, pointing offset and spacing radius.

The equation can be derived by fixing the X value in Equation (9.13). For a simple case, we can make $X = 0$ in the equation, changing the equation to:

$$2S(S^2 + H^2 - S \cdot P) + n_H \cdot H^2 \cdot P = 0 \qquad (9.14)$$

This allows the antenna roll-off n_H to be derived:

$$n_H = \frac{-2S(S^2 + H^2 - S \cdot P)}{H^2 \cdot P} \qquad (9.15)$$

where P is the HAP antenna pointing location on the x-axis, S is the HAP spacing radius, and H is the height of the HAP.

Figure 9.9 shows the trend of n_H with respect to the change of the pointing offset such that the peak power can be located at $X = 0$. In practice it is not possible to extend these curves to 0 km pointing offset because that will require an infinite n_H when we use a nonzero spacing radius, while also having the peak power simultaneously located at 0 km. Changing the pointing offset from a large negative value to a small negative value causes n_H to increase dramatically, which means the HAP antenna needs more directivity to compensate for the change in antenna boresight pointing location. Such changes in n_H will yield an antenna of high directivity, meaning that it is unlikely to serve the coverage area well. The spacing radius also has a significant impact. The further away the HAP is from the centre of the coverage area, the greater the n_H required. This is due to the increase in directivity required to compensate for the increase in link length.

Figure 9.10 provides another method to illustrate the relationship between the HAP antenna roll-off, point offset and spacing radius. Rather than use the abstract n_H we use beamwidth and show the subtended angle as well.

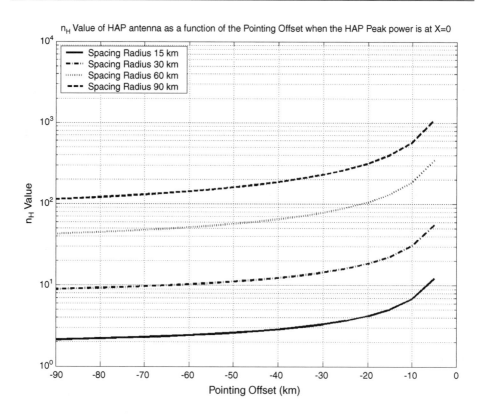

Figure 9.9 Roll-off (n_H) of HAP antenna versus the pointing offset when the HAP peak power is at $X=0$ (reproduced from [12])

For a general case, we can specify any location of peak power along the x-axis and choose other parameters to meet this requirement. We can transform Equation (9.9) in order to get n_H on the left-hand side of an equation, resulting in:

$$n_H = \frac{2(P-S)X^2 + [4S(S-P) + 2H^2]X - 2S(S^2 + H^2 - S \cdot P)}{H^2(P-X)} \qquad (9.16)$$

where X is the location of HAP peak power on the x-axis, P is the HAP antenna pointing offset on the x-axis, S is the HAP spacing radius and H is the height of the HAP.

Choosing the location of the HAP peak power provides the HAP operator with the flexibility to best serve the users on the ground. Here we can give an example. In the previous case, when we used 16 HAPs and chose $\psi_{10dB} = \psi_{\text{sub}}$ ($n_H = 16.5$), spacing radius of 50 km, and a pointing offset of 0 km, we achieved the peak power at 21 km and a peak CINR at 29 km. Now, we want to adjust the HAP antenna roll-off n_H

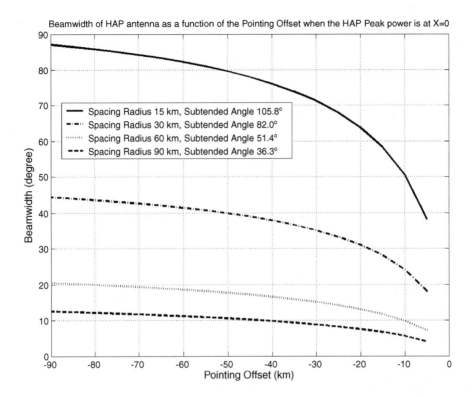

Figure 9.10 Beamwidth of HAP antenna versus the pointing offset when the HAP peak power is at $X = 0$ (reproduced from [14])

whilst keeping the other parameters unchanged in order to position the peak power at 5 km, 15 km, 25 km and 30 km. Following Equation (9.16), n_H can be calculated and this is shown in Table 9.2.

Table 9.2 A range of HAP antenna roll-off n_H according to the change of peak power location

Peak power location (km)	5	15	25	30
HAP antenna roll-off n_H	158	32.9	10.7	5.95
First peak CINR location (km)	16	26	33	37
Value of the first peak CINR (dB)	25.9	25.6	25.2	25.0
Second peak CINR location (km)	−27	−39	−47	−51
Value of the second peak CINR (dB)	29.7	24.9	21.1	19.0

Table 9.2 yields some important results. When multiple HAPs are adopted, there are usually two peak CINRs along the x-axis for each HAP. The first peak CINR refers to the peak CINR that is near to the location of the main HAP. The second peak

CINR refers to the peak CINR that is far from the location of the main HAP. Here we always assume the main HAP is located on the positive x-axis, so the first peak CINR location is always at a more positive position on the x-axis than the second peak CINR location. Normally, the first peak CINR level is greater than the value of the second peak CINR because the location of the first peak CINR is closer to the main HAP, meaning that it has a shorter link length and a greater received power. From Table 9.2, we can see an abnormal case when the peak power location is at 5 km. Here the first peak CINR value is less than the value of the second peak CINR. The reason for this is when every HAP delivers its peak power at 5 km there is significant interference in the centre area meaning that the contribution from the peak power is counteracted by this interference.

Figure 9.11 shows the combined maximum CINRs for 16 HAPs that are calculated by choosing the maximum CINR value among 16 HAPs at every ground location. This is done in order to show the distribution of peak CINR locations of all HAPs related to different peak power positions.

The equation of the combined maximum CINRs is shown as follows:

$$CINRCom\_max (x, y) = max[CINR1(x, y), CINR2(x, y), \ldots, CINR16(x, y)]$$
$$(9.17)$$

Alternatively, we can use the following expression:

$$CINRCom\_max (x, y) = max[CINRj(x, y), j \in N] \qquad (9.18)$$

where N stands for a set of HAPs system, and is equal to 16 here.

Here we only show the one maximum peak CINR. As we mentioned previously, usually the maximum peak CINR is the first peak CINR in Table 9.2, except for the case when the peak power is located at 5 km. Therefore in Figure 9.11(a) this is the second peak CINR. The maximum peak CINR can reach as high as 29.7 dB in Table 9.2 because high n_H value is adopted when peak power is at 5 km. However, this does not mean that the spectral efficiency is better.

The CDFs of the spectral efficiency for different peak power location are shown in Figure 9.12. It shows that the curve with 5 km peak power has the worst spectral efficiency performance, and the curve with the peak power at 25 km has the best performance. The dominant reason for this behaviour is that the closer the HAPs locate their peak powers the more interference the users receive which results in deterioration of spectral efficiency performance. The curve with the peak power at 30 km cannot maintain the highest maximum spectral efficiency compared with the 25 km case because a great deal of power is wasted outside the coverage area. This means that the best spectral efficiency performance tends to occur when the HAPs locate their peak powers at approximately 25 km for a 50 km spacing radius and a 0 km pointing offset.

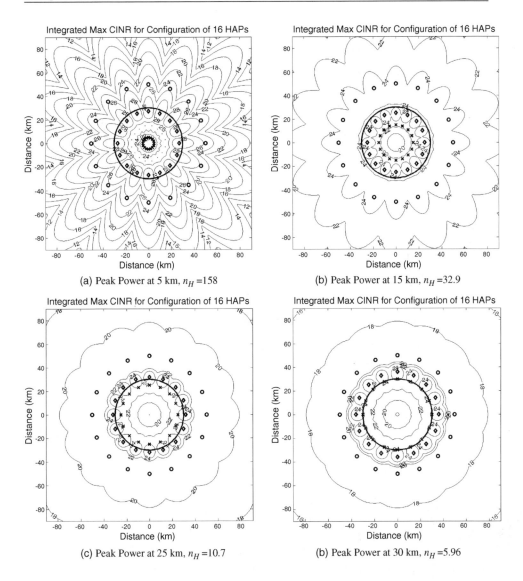

Figure 9.11 Combined maximum CINR for a 16 HAP configuration. Spacing radius 50 km, HAPs marked as 'o', peak power location marked as 'x', maximum peak CINR marked as '◇' (reproduced from [14])

9.2.1.6 Optimum Population of the HAP Constellation

In a general situation, users adopt fixed antennas and always connect to the same HAP, and it is likely the number of users within the coverage area will increase over

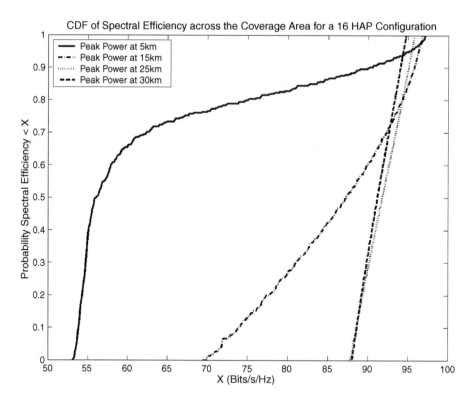

Figure 9.12   CDF of spectral efficiency across the coverage area for a 16 HAP configuration with different appointed peak power. Spacing radius 50 km, pointing offset 0 km (reproduced from [14])

time. This will require the number of HAPs to be gradually increased to meet the system capacity within the coverage area. We have already seen in Chapter 4 how incremental deployment is fundamental to a sensible business model for the rollout of broadband from HAPs, which can be driven by the market need for capacity [15].

From a technical perspective, one possible solution to incremental deployment is to exploit a predetermined HAP constellation, allowing the network operator to increase HAPs one-by-one locating them at a predetermined position in the constellation. Every time a new HAP is introduced into the system, it will change the CINRs of every existing user across the coverage area. The major issue is how to introduce the least increase in interference to existing users when a new HAP is deployed. To keep the analysis simple, a predetermined constellation is used with eight HAPs equally spaced in a circle with a 50 km spacing radius, which is assumed to be an optimal constellation design for eight HAPs in a single ring. Figure 9.13 shows three strategies to deploy these HAPs incrementally in this predetermined circle constellation.

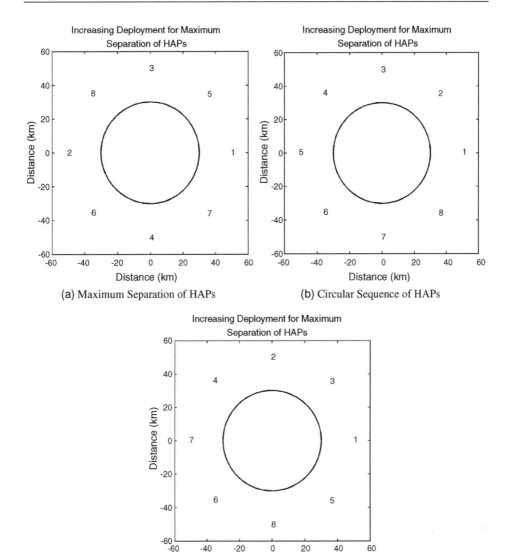

Figure 9.13 Three strategies of increasing deployment of HAPs. The numbers show the order of deployment (reproduced from [14])

Figure 9.13(a) illustrates the techniques of incrementally deploying HAPs with a maximum separation distance from each other, which may mitigate the interference at user points in the coverage area. Figure 9.13(b) shows a simple circular sequence of deployment to deploy HAPs incrementally. Figure 9.13(c) shows the optimum deployment strategy using a minimum spectral efficiency reduction criterion.

The methodology to find the optimum deployment strategy by minimum spectral efficiency reduction is described as follows. Suppose we already have $m - 1$ HAPs deployed in the predetermined final N-HAP constellation. Currently, we need to deploy the mth HAP into the predetermined N-HAP constellation. Instead of using the aggregate spectral efficiency (η_c), we use a vector of total spectral efficiency with m HAPs (η_{total_m}) by combining each η across the coverage area offered by each HAP to each user into the vector as shown in Equation (9.19). The total spectral efficiency with m HAPs (η_{total_m}) is also referred to as 'user spectral efficiency' in this section because it is a vector of the spectral efficiency of all the test users from all the HAPs.

$$\eta_{\text{total}\_m} = [\eta_1, \eta_2, \ldots, \eta_{m-1}, \eta_m] \tag{9.19}$$

Assuming that $m - 1$ HAPs have been deployed, the number of remaining candidate positions (n) to locate the mth HAP is given by:

$$n = N - m \tag{9.20}$$

Therefore depending on the n different positions of the mth HAP, the vector η_{total_m} has n different combinations:

$$\eta_{\text{total}\_m\_j} = [\eta_{1\_j}, \eta_{2\_j}, \ldots, \eta_{m-1\_j}, \eta_{m\_j}] j \in n \tag{9.21}$$

Here we select the median value[3] of the $\eta_{\text{total}\_m\_j}$ to represent the current system performance, the maximum median value among the n different $\eta_{\text{total}\_m\_j}$ shows the minimum system performance reduction (minimum spectral efficiency reduction). Therefore the jth position for the maximum median value of $\eta_{\text{total}\_m\_j}$ is our choice to locate the mth HAP. The process can be described by:

$$[\text{Max. value}_\eta, \text{index}_j] = \text{Max.}[\text{median}(\eta_{\text{total}\_m\_j}, j \in n)] \tag{9.22}$$

where the max. value$_\eta$ is the maximum value for different median $\eta_{total\_m\_j}$, the index$_j$ represents the jth position for the mth HAP where the max. value$_\eta$ is obtained.

We can evaluate the performance of these three strategies for the incremental deployment in terms of the median value of the total spectral efficiency (η_{total}) using:

$$\eta_{\text{Med.}\_m} = \text{Median}(\eta_{\text{total}\_m}, m \in N) \tag{9.23}$$

---

[3] As an alternative to the median value, a high or low percentile can be selected by network operators or regulators to be representative of performance.

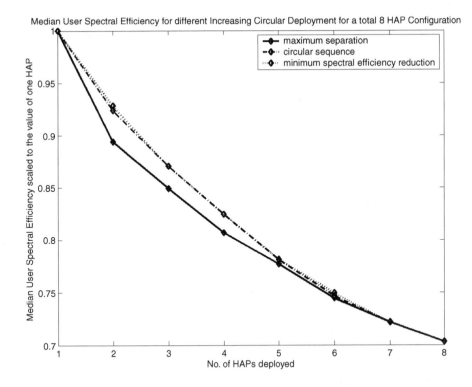

Figure 9.14   Median user spectral efficiency for different incremental deployment for a total 8 HAP configuration. Results have been normalised to the median user spectral efficiency for one HAP, which is 9.4 bits/s/Hz (reproduced from [14])

Figure 9.14 shows the results based on Equation (9.23) normalised to the median spectral efficiency for a single HAP. Compared with the circular sequence deployment, the optimal incremental strategy only shows moderate improvement for the minimum spectral efficiency reduction. The maximum separation deployment strategy has the worst performance using this criterion. From Figure 9.14, there is a large discrepancy when the second HAP is deployed in the constellation. To clearly view the whole CINR behaviour in this case, the CDF plot of user CINR for the first 2 HAP configuration is shown in Figure 9.15.

From the user perspective, the interference received from interfering HAPs is determined by the path loss and the antenna transmission gain of these HAPs. The antenna gain is determined by the angle from the boresight of the interfering HAP. From Figure 9.15, the maximum separation deployment for 2 HAPs has the lowest maximum and median CINR values but the highest minimum CINR value. This phenomenon can be explained as follows. The user positions, which contribute the upper middle parts of the CDF, are located at the area close to the first HAP (main

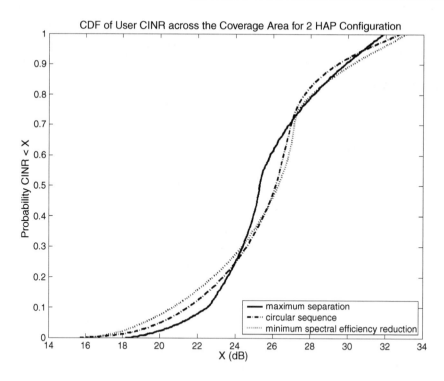

Figure 9.15 CDF of user CINR across the coverage area for the first 2 HAP configuration. The first HAP is the main HAP (reproduced from [14])

HAP). Obviously for these users, the maximum separation deployment can give the longest link length to user from the second HAP. However the angle deviated from the boresight of the interference HAP is very small. Therefore the antenna transmission gain from the second HAP is not significantly reduced compared with the boresight gain. On the other hand, the second HAP in the circular sequence deployment has a much lower link length variation but a higher angle deviation from the boresight of the second HAP. In the calculation of CINR, the interference power that these users suffer from the second HAP under the maximum separation deployment is bigger than under the circular sequence deployment and the minimum spectral efficiency reduction deployment. This results in a lower maximum and median CINR values under the maximum separation deployment.

Figure 9.16 shows the median aggregate spectral efficiency for different deployment strategies for a total 8 HAP configuration. There is an almost pro-rata increase in system capacity (calculated as the median aggregate spectral efficiency) from 1 to 5.68 times the value for a single HAP, for a corresponding increase in the number of HAPs from 1 to 8. There is not much difference among these strategies

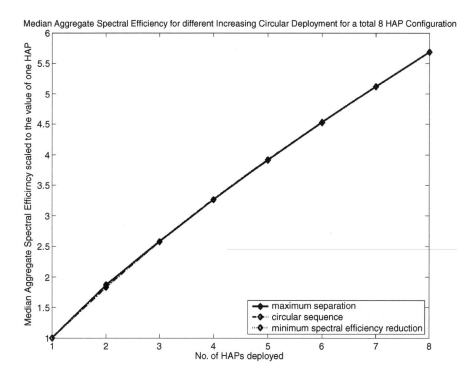

Figure 9.16   Median aggregate spectral efficiency for different increasing circular deployment for a total 8 HAP configuration. Results have been scaled to the median spectral efficiency for one HAP, which is 9.4 bits/s/Hz (reproduced from [14])

compared with Figure 9.14. The reason is that the aggregate spectral efficiency smoothes out discrepancies in spectral efficiency at each test point. It needs to be emphasised that the aggregate spectral efficiency is a method to measure the overall system capacity, the individual user can only benefit from the spectral efficiency from one HAP.

9.3   User Antenna Pointing Error in Multiple HAP Systems

The HAP antenna beamwidth, HAP antenna pointing direction, and the spacing radii of multiple HAPs can be used to control the shape of the footprint within the coverage area of a multiple HAP constellation [4]. However, pointing error associated with the user terminal antenna may degrade system capacity [10]. Here we examine the effect of user antenna pointing error in multiple HAP systems. To do this a truncated Gaussian distribution model is used to describe the pointing error, and different ranges of standard deviations and user antenna beamwidth are used to illustrate the different aspects of the system capacity degradation. To provide an indication of the impact of the pointing error, the concept of virtual points is developed. To aid the system designer, we derive optimum user antenna beam-widths as a function of standard deviation of pointing error, when specific percentiles of the spectral efficiency are required.

9.3.1 Methods for Characterising User Antenna Pointing Error

Imperfect installation and wear and tear after installation will inevitably result in some degree of pointing error for user antennas. This problem may cause a decrease in the received power level from the main HAP and a corresponding increase in interference from other HAPs in the multiple HAP system, which could jeopardise the link budget. User antenna pointing error can be described by the following two models [10]:

- **Error model 1** is a boresight deviation error model that takes into account angular error with respect to boresight from a fixed nominal pointing direction such as ϕ_{dev} in Figure 9.17.
- **Error model 2** is an elevation-azimuth angle error model that uses the deviation from the elevation and azimuth angles from a nominal fixed pointing direction. The user antenna pointing direction can be decomposed into elevation and azimuth angle directions, with the pointing error being represented as the

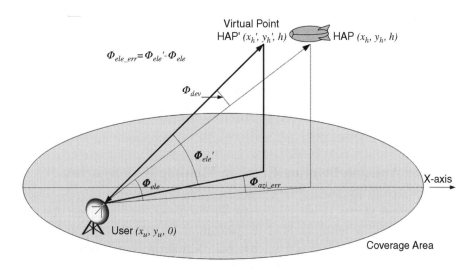

Figure 9.17   User antenna pointing error described as angle deviation from the boresight and as the elevation and the azimuth angle aspects (reproduced from [10])

combination of the elevation and azimuth angle errors as shown in Figure 9.17. Elevation and azimuth angles are often used in user antenna installation, so therefore this model is considered here.

These two models can be considered as representing the fundamental ways in which errors can be specified in a directional antenna system.

It is possible to define a virtual HAP point in order to illustrate the behaviour when subject to pointing error, which in effect describes the assumed position of the HAP from the perspective of the user. In Figure 9.17 the thin solid arrowed line shows the correct pointing direction from user to the HAP, with the bold solid arrowed line showing the incorrect user antenna pointing direction. In error model 1, ϕ_{dev} is the angular difference between these two vectors, whereas in error model 2, these two directions are decomposed into the elevation and azimuth angles, respectively. ϕ_{ele_err} is the elevation angle error for the deviated direction, with ϕ_{azi_err} describing the azimuth angle error. We can imagine that the HAPs are deployed on a virtual plane at a height h, with the intersection of the error line and the virtual plane forming the virtual point (x_h', y_h', h). The virtual point can be treated as being the error position of the real HAP as a result of the user antenna pointing error.

In error model 1, the probability density of the angular deviation can be described by a Gaussian distribution, see Equation (9.24), with the probability density of the angular orientation around the correct direction being described by a uniform distribution as shown in Equation (9.25) [10]:

$$p(\theta) = \frac{1}{\sqrt{2\pi\sigma^2}} \exp\left[-\frac{(\theta-a)^2}{2\sigma^2}\right] \tag{9.24}$$

$$p(\phi) = \frac{1}{2\pi} \qquad 0 \le \phi \le 2\pi \tag{9.25}$$

where σ is the standard deviation, with a corresponding to the mean value. Error model 2 also uses a Gaussian distribution to characterise the elevation and azimuth pointing errors. Assuming angular errors are symmetrical about the correct angle, this mean value a in the Gaussian distribution is always zero. The virtual points generated by the users in the coverage area can be considered as perturbing the locations of HAPs in the constellation as shown in Figure 9.18. (Note, this specular appearance looks similar to variations in a modulation scheme constellation when subject to noise [10].) In Figure 9.18 the 16 HAPs are outside the coverage area, with the circle representing the edge of the coverage area. Assuming that the pointing error is not significant, adjacent virtual points will not overlap, meaning that the interference from interfering HAPs will be limited as the user can still discriminate its main HAP.

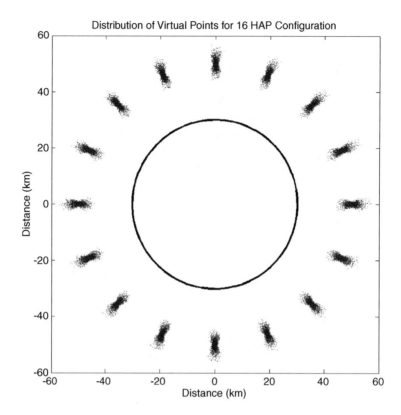

Figure 9.18   Virtual HAP point with 16 HAPs caused by user pointing error – truncated Gaussian distribution, standard deviation 0.5°, HAP height 17 km (Reproduced by permission of © 2007 The IET [10])

The required user antenna pointing accuracy will be a function of the antenna beamwidth. A narrower user antenna beamwidth requires more accurate pointing, otherwise it will cause a reduction in gain meaning that the link budget may not be satisfied, as shown Figure 9.19. For example, the pointing accuracy must be better than 3.2° in order to stay within the main lobe of the 2° beamwidth antenna, assuming a −30 dB sidelobe floor [10].

Therefore it is possible to link the required standard deviation in the pointing error to the user antenna beamwidth, i.e. [10]:

$$\sigma = kU_B \qquad (9.26)$$

where σ is the standard deviation, U_B is the user antenna beamwidth and k represents the fraction of the antenna beamwidth used to characterise the pointing error.

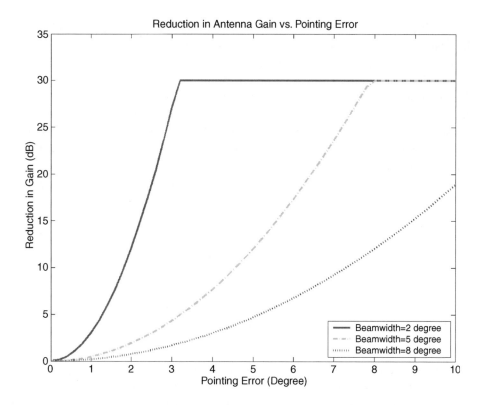

Figure 9.19 Reduction in antenna gain versus pointing error for different antenna beamwidths (Reproduced by permission of © 2007 The IET [10])

If a standard Gaussian distribution is used to model the pointing error there is a certain probability that the range of pointing errors can extend to plus or minus infinity, which means that extreme values would result in the user antenna not pointing towards the HAP at all. Such large errors are unlikely to occur in practice so it makes sense to use a truncated Gaussian distribution to model the pointing errors. Assuming that the random values of x below the value x_L and above the value x_R should be discarded, the truncated Gaussian probability density function can be described as [10]:

$$
p_{TG}(x) = \begin{cases} 0 & -\infty \leq x < x_L \\ \dfrac{p(x)}{\displaystyle\int_{x_L}^{x_R} p(x)dx} & x_L \leq x \leq x_R \\ 0 & x_R < x \leq \infty \end{cases} \tag{9.27}
$$

where $P_{TG}(x)$ is the probability density function of the truncated Gaussian distribution and $p(x)$ is the probability density function of the classic Gaussian distribution. The probability of large pointing errors is assumed to be proportional to the standard deviation of the pointing error, and also that extreme pointing error situations only occupy approximately 5% of the total Gaussian distribution. The erf function or the erfc function can be used to determine whether the truncated boundaries x_L and x_R occupy approximately 5% of the total Gaussian distribution. By assuming the truncated boundaries are symmetric and the mean value is zero, x_R equals $-x_L$, then $-x_L = x_R = x_T$, and thus due to the symmetry of the Gaussian distribution, the portion outside the truncated part in the total Gaussian distribution is [10]:

$$F_T(x_T) = 2\left[1 - \frac{1}{2}\text{erfc}\left(\frac{-x_T}{\sqrt{2}\sigma}\right)\right]$$

$$= 2\left[\frac{1}{2} + \frac{1}{2}\text{erf}\left(\frac{-x_T}{\sqrt{2}\sigma}\right)\right]$$

(9.28)

Here we assume x_T equals 2σ, resulting in the outside portion in the total Gaussian distribution being 4.55% (close to the 5% mentioned above), i.e. we assume that this 2σ maximum can be used to describe the pointing error of installation plus the wear and tear.

Numerical calculation can be used to obtain the CINR performance using Equation (9.1) when subject to random pointing errors [10]. Again in order to determine the CINR over the whole coverage area, 'test' users are deployed on a 1 km grid, with the density of test users large enough to obtain statistically significant results, and that the CINR is not likely to change significantly over the gaps between the 'test' users. The Shannon equation is then used to convert the CINR associated with each point. The default parameter values used in the calculations are listed in Table 9.3 (which is derived from Table 9.1). The HAP antenna roll-off is defined as the specific level of attenuation (in dB) with respect to the angle subtended at the edge of the coverage area. In a multiple HAPs scenario, the subtended angle depends on the position of a HAP. The subtended angle of the coverage area is defined as the angle at the HAP defined by a triangle containing the HAP, with the two other vertices defined by the end points of a diameter across the coverage area, perpendicular to a HAP's 'own x-axis' mentioned previously.

Table 9.3 Default parameters used in user antenna pointing error calculations (based on [10])

Parameter	Default value
HAP antenna pointing offset	0 km (Centre of the coverage area)
HAP antenna roll-off	−10 dB at the angle subtending the edge of the coverage area
Sidelobe floor (s_f) for both HAP and user antenna	−30 dB relative to peak
Noise floor (N_F)	−133 dBW (Equivalent noise temperature is 300 K, bandwidth is 12.5 MHz)
Transmitter power	0 dBW
Antenna efficiency (HAP and user antenna)	0.8
Frequency	30 GHz
Coverage area radius	30 km
Platform height (h)	17 km
Multiple HAPs deployment radius	50 km

9.3.2 Effect of Pointing Error

Figure 9.20 shows the median spectral efficiency per HAP with the elevation, azimuth, elevation-azimuth combined error, and the boresight deviation error, for the two antenna pointing models. In all cases the user antenna beamwidth is 2°, with pointing error truncated to twice the standard deviation from the mean value. As expected all of the median spectral efficiency curves decrease when the standard deviation of pointing error increases. The group of curves for the 16 HAPs is lower than the group of curves for the 2 HAPs, mainly as a result of the 16 HAPs suffering more overall interference.

The boresight deviation error model is the easiest to understand, and we use this as a means of comparison for the more complex models. The elevation angle error can be considered as a special case of the boresight deviation error, except that it happens only in the elevation plane, which means the elevation angle error curves are very close to the baseline curves. By contrast, the curves for azimuth angle error are slightly better than the ones for the elevation angle error as a result of the deviation in the azimuth angle contributing less error in the pointing deviation. The best way to explain this is through suitable special cases. Consider the situation when the user antenna points vertically upwards at a 90° elevation angle, any error in the azimuth direction will not affect the pointing direction. At the other extreme of 0° elevation angle (although impractical), any azimuth error will contribute fully to the pointing error. The elevation angles in between will be

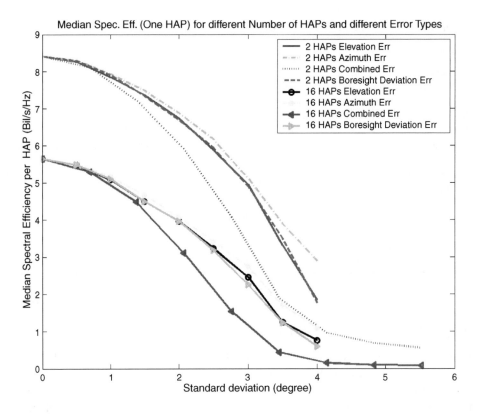

Figure 9.20 The median spectral efficiency for elevation error, azimuth error, combined error and boresight deviation error – user antenna beamwidth 2°, error truncated to 2σ (Reproduced by permission of © 2007 The IET [10])

affected in different proportions. This is because the conventional elevation-azimuth representation is not a truly orthogonal system. In the elevation-azimuth coordinate system, the elevation angle is directly linked to the pointing direction, but the azimuth angle is only indirectly linked. The actual impact of the azimuth angle is determined by a projection of the azimuth angle onto a plane defined by the elevation angle and a line parallel to the X-Y plane as shown in Figure 9.21 [10].

The pointing direction can be projected on to two orthogonal planes, such as plane ABC and the X-Y plane. In such cases, the angles from the pointing direction to these two planes are orthogonal, i.e. the projected angle (ϕ'_{azi}) and the elevation angle (ϕ_{ele}) are orthogonal. However, the pointing direction is specified by the projected angle (ϕ'_{azi}), which is a combination of the azimuth angle (ϕ_{azi}) and the

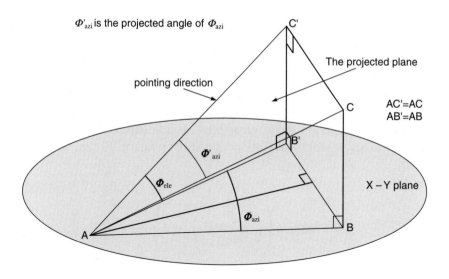

Figure 9.21 Illustration of a projection of the azimuth angle (reproduced from [12])

elevation angle (ϕ_{ele}), as specified by [10]:

$$\cos\phi'_{\text{azi}} = 1-(1-\cos\phi_{\text{azi}})\cos^2(\phi_{\text{ele}}) \tag{9.29}$$

The impact of azimuth and elevation error is also not orthogonal as shown by Equation (9.30), which makes the pointing error much more difficult to analyse because the impact of the azimuth error is elevation angle dependent.

$$\cos\phi'_{\text{azi\_err}} = 1-(1-\cos\phi_{\text{azi\_err}})\cos^2(\phi_{\text{ele}}) \tag{9.30}$$

Equation (9.30) clearly illustrates why the azimuth angle error curves are slightly better than the ones for the elevation angle, as $\phi'_{\text{azi}_{\text{err}}}$ is smaller than $\phi_{\text{azi\_err}}$.

By applying both the azimuth and elevation angle error limit simultaneously to the system, then we have the elevation-azimuth combined angle error. This combined angle error should be larger than the single azimuth and single elevation angle errors. If the standard deviation of the truncated Gaussian distribution is used to quantify the error, the actual standard deviation for the elevation-azimuth combined angle error is also larger than the one for the single azimuth and elevation angle error. For the case where the elevation and azimuth angles are orthogonal to each other, the standard deviation of the combined angle error can be described more simply by [10]:

$$\sigma_{\text{comb\_err}} = \sqrt{\sigma^2_{\text{azi\_err}} + \sigma^2_{\text{ele\_err}}} \tag{9.31}$$

However, the actual standard deviation for the elevation-azimuth combined angle error over the coverage area is hard to determine because of ϕ'_{azi} dependency on ϕ_{ele} for different users over the coverage area. Therefore, an alternative way is to apply an adjustment factor to reflect the standard deviation changes mentioned above. In order to simplify the analysis of the combined results, it is possible to use a simple case of the combined error using the same standard deviation applied to the azimuth and elevation angle error plus an adjustment factor as shown in Figure 9.20. This standard deviation adjustment factor F can be calculated using [10]:

$$
F = \text{median}\left(\sqrt{\left(\frac{\phi'_{azi\_err}}{\phi_{azi\_err}}\right)_i^2 + 1} \right) \quad i \in M \tag{9.32}
$$

where M is the set of users in the coverage area. A typical standard deviation adjustment factor F is 1.38 [10], which is used here, and is a weighted root-mean-square median value that takes into account the varying contributions of the azimuth error across the coverage area.

This combined approach has been chosen to neutralise the impact of the ϕ'_{azi} dependency on ϕ_{ele} over the coverage area, thereby removing the significant complexity associated with deriving a uniform equation for ϕ'_{azi} for the whole system. The curves in Figure 9.20 with the combined angle errors are still worse than either of the single errors even after adjustment.

We focus the rest of the discussion on the boresight error model since it is easier to understand. One main conclusion to draw from the results in Figure 9.20 is that a limit to the standard deviation in pointing error must be applied to the system when a user antenna is installed. This limit should be related to the overall beamwidth of the user antenna. The proportion of the total Gaussian distribution falling between the plus and minus σ is 68.27 % according to Equation (9.28). If we restrict the deviation to σ, which also approximately corresponds with the -3 dB point of the user antenna (half of the beamwidth), then the the power loss for 68.27 % users is limited to 3 dB. It is assumed that any system should be able to deal with this level of capacity degradation. For example, assuming a $2°$ user antenna beamwidth, if we use half of this value, $1°$, for the standard deviation limit in the boresight deviation error model, then this will result in a reduction in median spectral efficiency per HAP of 0.56 bit/s/Hz [10].

It is possible to investigate the effects on spectral efficiency for a certain percentage of users across the coverage area, in order to best illustrate the effect

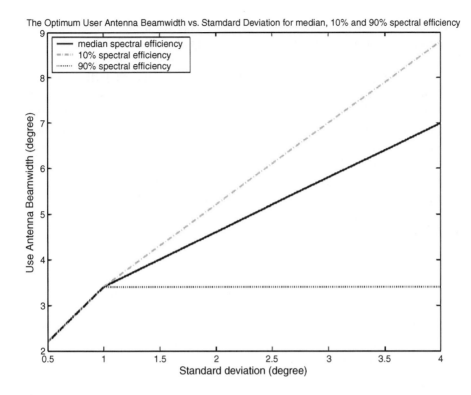

Figure 9.22 The optimum user antenna beamwidth versus standard deviation for median, 10-percentile and 90-percentile spectral efficiency (Reproduced by permission of © 2007 The IET [10])

of pointing accuracy on optimum user antenna beamwidth. Figure 9.22 shows the optimum user antenna beamwidth according to a range of standard deviations of error for median, 10-percentile, median, and 90-percentile user locations. This *x*-percentile spectral efficiency refers to the spectral efficiency achieved by the bottom *x* % of user locations within the coverage area [10]. The results have been obtained using a numerical calculation method. A 1–20° user antenna beamwidth range is examined, with a minimum step size of 0.5° for practical reasons. A more detailed explanation of the calculation method can be found in [12]. Such results can be used as an aid to a system design to choose an appropriate optimum beamwidth, if the degree of likely pointing error is known for an installation. The flat part of the graph for the 90-percentile case is limited by the 0.5° step size of user antenna beamwidth.

These results are based on a LOS propagation model for a mm-wave band multiple HAPs system. If lower frequencies are adopted it would be necessary to incorporate a more sophisticated propagation model for users with low elevation angles, to allow for non-LOS propagation, multipath and shadowing.

It is useful to also consider the dominant factors affecting the system capacity, and this can be done by examining the different constituents of CINR. A reduction in system capacity is caused by a reduction in CINR as a result of pointing error. Figure 9.23 shows the several factors causing the system capacity decrease for the 2 HAPs and 16 HAPs scenarios, respectively. In both cases the reduction in CINR is caused by the reduction of the carrier power of the main HAP. Interference is little affected, mainly due to the fact that interference from the other HAPs mainly falls into the sidelobe of the other antennas [4]. This is an important finding, since it illustrates that a multiple HAPs system is no more affected by antenna pointing error than a single HAP system, providing they are not deployed too close together.

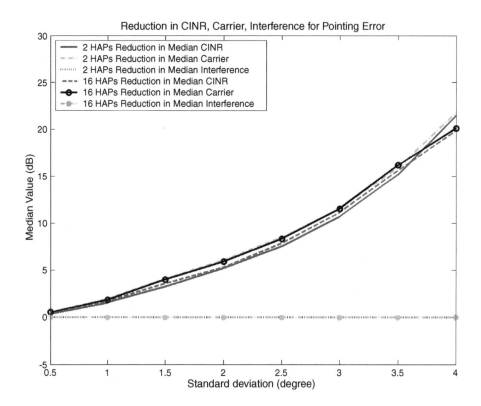

Figure 9.23 Reductions in CINR, carrier and interference for pointing error (Reproduced by permission of © 2007 The IET [10])

9.4   Two-Ring Constellation Design for Multiple HAP Systems

In this chapter we are gradually illustrating the main factors affecting the performance of multiple HAP constellations. Previous constellation designs for multiple HAPs were based on a one-ring constellation. The fixed radius (FR) and fixed arc (FA) of a one-ring constellation has also been investigated in [2]. It shows that the FA has the better performance than the FR in the situation when more than 12 HAPs and wide user antenna beamwidths are used. Therefore there is scope to consider using a more sophisticated constellation rather than the one-ring constellation. Here two-ring constellation design strategies for 16 HAPs are investigated.

9.4.1   Two-Ring Constellations Overview

Previous constellation designs for multiple HAPs have been based on a one-ring constellation where it is relatively easy to analyse the system performance using a number of different parameters. As the number of HAPs is increased in the constellation, a one-ring constellation may no longer be the best arrangement of HAP locations to generate the highest system capacity. In the following analysis, we use 16 HAPs in a constellation, which is big enough to generate quite a sophisticated constellation design. The analysis here is based on two-ring constellations. In the 16 HAPs case, a possible distribution of the HAPs in an outer and inner ring is shown in Figure 9.24.

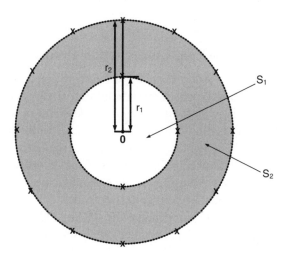

Figure 9.24   Illustration of a two-ring constellation (Reproduced by permission of © 2007 The IET [12])

Here every 'X' position represents a location of a HAP within the constellation, 'o' is the centre of the inner and outer rings as well as the centre of the coverage area, r_1 and r_2 are the radius of the inner ring and outer ring, respectively, S_1 is the surface area within the inner ring and S_2 is the surface area of the shaded area which is between the inner and outer rings.

As shown in Section 9.2.1 [11], multiple HAPs can be deployed beyond the coverage area in order to enhance the system capacity by increasing the HAP antenna directivity and producing less interference from interfering HAPs. However, it may not be possible for every user within the coverage area to see all the HAPs in the sky because of the minimum elevation angle restriction which will be discussed later. This may not cause problems when the number of HAPs is large enough since users can still see several adjacent HAPs to guarantee enough system capacity. Figure 9.24 presents an example of two-ring multiple HAPs deployment, with 12 HAPs in the 30 km radius outer ring and 4 HAPs in the 5 km radius inner ring. The altitude for both of the rings is 17 km. Figure 9.25 shows the number of HAPs that can be seen simultaneously by users for two-ring configuration assuming a 20° minimum elevation angle restriction. Users within a 16 km radius are able to use all 16 HAPs in this situation, with users limited to perhaps just 10 or 11 at the edge of the coverage area.

Considering blockage by tall buildings in urban areas, the 20° minimum elevation angle restriction may need to be applied to every HAP. In this situation, the outer ring radius must shrink to a suitable value to comply with this requirement, which is described later. The default parameters used in this chapter are listed in Table 9.4 (which has been derived from Table 9.1).

9.5 Constraints of Two-Ring Constellation Designs

In order to deliver a better system capacity, a two-ring constellation design must consider four main constraints which are crucial to a HAP system.

The first constraint is the *minimum elevation angle restriction*, which is a major reason to restrict the HAP spacing radius. The frequencies within the mm-waveband require a LOS link, and the lower the elevation angle, especially in urban areas, the more likely a link will be blocked by a building. Thus, in order to make sure that most of the users within the coverage area can communicate without blocking, the minimum elevation angle must be deliberately considered. If the HAP is located at the centre of the coverage area, which means the spacing radius of HAP is 0 km, the maximum radius of the coverage area is calculated using (neglecting the curvature the Earth) [12]:

$$R = \frac{H}{\tan(\gamma)} \tag{9.33}$$

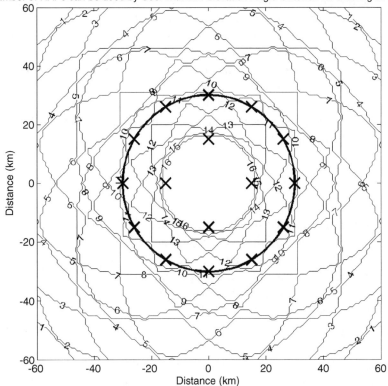

Figure 9.25 Number of HAPs can be used by users with 20° minimum elevation angle criterion for 16 HAPs – marked as 'X'. The inner and outer ring radii are 15 and 30 km, respectively, and the inner and outer ring heights are 17 km (reproduced from [12])

Table 9.4 Default parameters used in calculations for two-ring constellations

Parameter	Default value
Sidelobe floor (s_f)	−30 dB relative to peak
Noise floor (N_F)	−133 dBW (Equivalent noise temperature is 300 K, bandwidth is 12.5 MHz)
Transmitter power	0 dBW
Antenna efficiency (HAP and user antenna)	0.8
Frequency	30 GHz
Coverage area radius	30 km
Platform height (h)	17 km (variable)
Outer ring radius of constellation	16 km (variable)
Inner ring radius of constellation	8 km (variable)

where R is the maximum radius of the coverage area, γ is the minimum elevation angle and H is the height of the HAP. Here a conservative $20°$ minimum elevation angle (γ) is adopted. The maximum radius of coverage area will be 46.7 km when H is 17 km and γ is $20°$, assuming the HAP is at the centre of the coverage area. Since the assumed coverage area has 30 km radius, the elevation angle in this case is greater than the minimum elevation angle.

If the HAP has a nonzero spacing radius, and assuming we still need the HAP to cover the whole coverage area, the maximum spacing radius which the HAP can use can be calculated as follows [12]:

$$S = \frac{H}{\tan(\gamma)} - R'$$

(9.34)

where S is the maximum HAP spacing radius and R' is the radius of the coverage area. For example, the maximum HAP spacing radius is 16.7 km when H is 17 km and γ is $20°$ and R' is 30 km. This means that the HAP can only be deployed within a 16.7 km radius, which significantly limits deployment options. If we need to deploy HAPs further away, constellation design can ensure that the coverage area is covered by one or more (rather than all) HAP, as shown in Figure 9.25.

The second constraint is the *height range* due to the wind speed in the stratosphere. Figure 9.26 repeats the general trend of wind speed profile versus altitude discussed in Chapter 1. In order to obtain a relatively low mean wind speed, a height range roughly from 15 to 30 km is feasible as a general guide. The actual height range depends on the season and location [6,7].

The third constraint is the *minimum angular separation* [9]. Recall that the minimum angular separation is an angle which represents how separate in angle the interfering HAPs are from the main HAP from a user perspective. All separation angles from every interfering HAP to the main HAP are tested and the smallest angle is chosen. Ideally, the power from the interfering HAPs enters into the sidelobe of the user antenna when the minimum angular separation is greater than half of the sidelobe floor beamwidth, defined as the width of the mainlobe at the points at which it terminates and the flat sidelobe floor begins, of the user antenna [9]. The user antenna profile used here is a simplified flat sidelobe model which means that it has no change of impact once the minimum angular separation exceeds half of the sidelobe floor beamwidth. However, in a real antenna model, the amplitude of the sidelobes tapers off in general when the angle deviation from the boresight increases. Therefore, the wider the minimum angular separation, the lower the interference that is produced.

The fourth constraint is what we refer to as the *eclipse effect*. The term eclipse is used in satellite communications. In satellite communications, an eclipse occurs

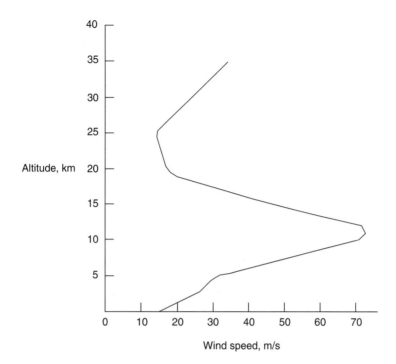

Figure 9.26 Wind speed profile versus altitude. Values vary with season and location but generally follow this rough distribution (reproduced from [16])

when the Earth is between the Sun and the satellite, which shadows the satellite from the Sun, thereby causing insufficient solar power to operate the satellite [17]. An alternative case known as a sun fade is defined as when the Sun is immediately behind a satellite, when the thermal noise created by the sun rivals, and often overpowers, a satellite's output, which results in poor SNR to users [17]. Based on the above definitions, here we still use the term eclipse effect to describe the situation when users cannot discriminate the main HAP from the interfering HAP when one HAP is immediately behind another HAP, or is separated narrowly in angle, as seen by the user. Clearly in such cases, the eclipse region (where users suffer the eclipse effect) should be moved out of the coverage area.

Up to now, all the constellation designs assume that all the HAPs are at the same altitude. However, their heights could be different within the constellation to enhance the system capacity by reducing the interference or expanding the coverage area. A disadvantage of the height variation in two-ring constellation design is that there is an increased probability of the eclipse effect, as shown in Figure 9.27.

The angles θ_{ele_o} and θ_{ele_i} are the elevation angles from A for the outer ring and inner ring HAPs, respectively, and $\theta_{eclipse}$ is the subtended angle from A to B and C.

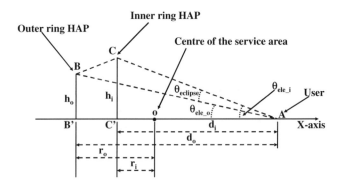

Figure 9.27 Illustration of the eclipse effect when HAPs are at different heights within the constellation (based on [12])

h_i and h_o are the heights of the inner and outer ring, respectively. If θ_{eclipse} is smaller than half of the sidelobe floor beamwidth, at point A, interference enters the main lobe of the user antenna causing the eclipse effect. To solve this problem, for any user point within the coverage area, the following inequalities incorporating minimum elevation angle restriction can be generated [12]:

$$\theta_{\text{ele}\_i} > \theta_{\text{min\_ele}}$$

$$\theta_{\text{ele}\_o} > \theta_{\text{min\_ele}} \qquad (9.35)$$

$$\theta_{\text{eclipse}} = \left|\theta_{\text{ele}\_i} - \theta_{\text{ele}\_o}\right| > 0.5\,\theta_{sf}$$

where $\theta_{\text{min\_ele}}$ is the minimum elevation angle applied for the system and θ_{sf} is the sidelobe floor beamwidth [9]. According to Table 9.4, the sidelobe floor is $-30\,\text{dB}$ relative to the boresight gain. If we do not need this stringent $-30\,\text{dB}$ value applied for the eclipse effect, we can use other values of θ instead of θ_{sf}, such as θ_{10dB} which means there is $10\,\text{dB}$ attenuation relative to the boresight at this angle.

Using the tangent expression, the inequalities in (9.35) can be transformed to the inequalities in (9.36) which allow the height range of each ring to be specified in terms of minimum elevation angle and eclipse effect [12]:

$$h_i > \left[\tan(\theta_{\text{min\_ele}})\right] d_i$$

$$h_o > \left[\tan(\theta_{\text{min\_ele}})\right] d_o$$

$$h_i > \tan\left[\arctan\left(\frac{h_o}{d_o}\right) + 0.5\,\theta_{sf}\right] d_i \qquad (\text{if } \theta_{\text{ele}\_i} > \theta_{\text{ele}\_o}) \qquad (9.36)$$

$$h_o > \tan\left[\arctan\left(\frac{h_i}{d_i}\right) + 0.5\,\theta_{sf}\right] d_o \qquad (\text{if } \theta_{\text{ele}\_o} > \theta_{\text{ele}\_i})$$

If either the height of the inner ring or the outer ring is known, then using inequalities in (9.36) the noneclipse height range of the other ring can be deduced. In other words, the variation of the height of the inner or outer ring affects the movement of the eclipse regions as shown in Figure 9.28 and 9.29.

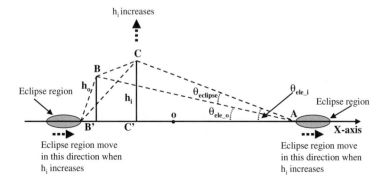

Figure 9.28   Illustration of eclipse region movement when h_i increases (reproduced from [12])

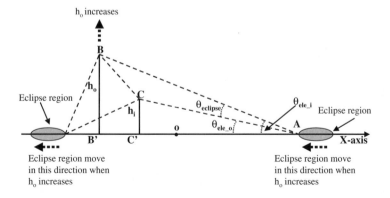

Figure 9.29   Illustration of eclipse region movement when h_o increases (reproduced from [12])

Since we define an eclipse region as the locations where a user will receive interference into the mainlobe, there are two eclipse regions existing to the left and right side of the coverage area, respectively. If we increase the inner ring height h_i as shown in Figure 9.28, the left eclipse region moves towards the centre of the coverage area while the right eclipse region moves away from the centre of the coverage. On the other hand, if we increase the outer ring height h_o as shown in Figure 9.29, the left eclipse region moves away from the centre of the coverage area while the right eclipse region moves towards the centre of the coverage. When the outer ring height is higher than the inner ring height, the right eclipse region may easily appear inside of the coverage area, which makes increasing the outer ring height unworkable.

Figure 9.30 demonstrates in detail the change in location of the eclipse regions by plotting the distance from edges of the regions to the centre of the coverage area as the inner ring height increases from 10 to 30 km. The outer ring and inner ring radii are 16 and 8 km, respectively, in this case. The solid line shows how the right edge of the left eclipse region moves towards the centre of the coverage area when the inner ring height increases. The dash-dot line shows that the left edge of the right eclipse region moves away from the centre of the coverage area. The dotted line is a reference line showing the radius of the coverage area which also represents the edge of the coverage area. The portion of solid line or dash-dot line which is below

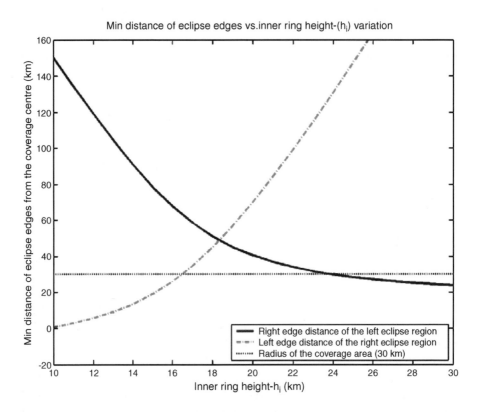

Figure 9.30 Minimum distance of eclipse edges to centre of the coverage area versus inner ring height (h_i) variation $-h_o = 17$ km, $r_o = 16$ km, $r_i = 8$ km, user antenna beamwidth is 2° and half of the sidelobe floor beamwidth is 3.2° (reproduced from [12])

the reference line indicates that the minimum distance of the eclipse region is smaller than the radius of the coverage area. In other words, the eclipse region intrudes inside the coverage area. From Figure 9.30, the inner ring height variation must be within 16.5–23.9 km to keep the eclipse regions out of the coverage area in this particular case.

Figure 9.31 is similar to Figure 9.30 except the outer ring height increases from 10 to 30 km. However, in order to keep the eclipse regions out of the coverage area, the outer ring height variation must be within 12.1–17.6 km, which can only be 0.6 km higher than the 17 km inner ring height. In other words, as mentioned previously, the benefits of this configuration are minimal. The height variation down to 12.1 km may enter into the troposphere, which cannot be used by HAPs. However, some conventional aircraft-based platforms may be able to use the altitude in the troposphere.

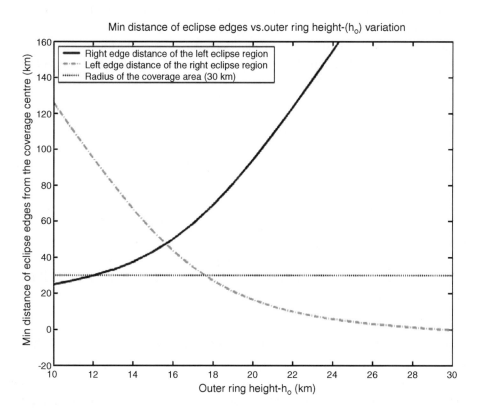

Figure 9.31 Minimum distance of eclipse edges to centre of the coverage area versus outer ring height (h_o) variation – $h_i = 17$ km, $r_o = 16$ km, $r_i = 8$ km, user antenna beamwidth is $2°$ and half of the sidelobe floor beamwidth is $3.2°$ (reproduced from [12])

9.5.1   Constellation Design Strategies

After the discussion regarding the constraints of the two-ring constellations, we now investigate the constellation design strategies. By strategies we mean how to choose values of the three important parameters to implement the two-ring constellation design, the HAP number ratio between the inner and outer rings, optimising the inner and outer radius and height of the two rings, and optimising the pointing offset.

9.5.1.1   HAP Number Ratio Between the Inner and Outer Rings

The first strategy is how to choose the number of HAPs in the inner and outer rings from a predetermined total HAP number. We have two assumptions here. One is that the user density is uniform within the coverage area. The other is that users will try to connect to the nearest HAPs. Based on these assumptions, we can distribute the number of HAPs roughly according to areas of S_1 and S_2 in Figure 9.24. To make it more straightforward, we do not count the area between the outer ring and edge of the coverage area. Therefore the HAP number ratio between the inner and outer ring can be roughly determined by the surface area ratio between the inner ring area and the shaded area specified as [12]:

$$\frac{N_{\text{Inner}}}{N_{\text{Outer}}} = \frac{S_1}{S_2} = \frac{r_1^2}{r_2^2 - r_1^2} \tag{9.37}$$

$$\text{If } r_2 = 2r_1, \text{ then } \frac{N_{\text{Inner}}}{N_{\text{Outer}}} = \frac{1}{3} \tag{9.38}$$

According to Equation (9.38), 4 HAPs should lie in the inner ring, with the remaining 12 HAPs located in the outer ring for a 16 HAP constellation. Such an approach should only be given as a guide to determine the HAP number ratio between the inner and outer ring. For example, it may prove sensible to locate 6 HAPs in the inner ring and 10 HAPs in the outer ring for the 16 HAPs case, if more capacity is need in the inner ring area. This will be investigated later.

9.5.1.2   Optimising the Inner and Outer Ring Radius and Height of Each Ring

The second strategy is how to select the inner and outer ring radius, and the height of each ring. The basic strategy here is to find the optimal minimum angular separation whilst obeying the minimum elevation angle restriction. The minimum angular separation for the user at location (x, y) can be calculated using [12]:

$$A(x, y) = \min[\theta_{m,j}(x, y), j \in N \setminus m, (x, y) \in D] \tag{9.39}$$

where one main HAP is denoted by subscript m in a system of N HAPs. The rest of the N-1 HAPs are the interfering HAPs which are labelled by $j \in N\backslash m$. D represents the coverage area. The $\theta_{m,j}$ is the angular separation as seen by a user at $(x, y, 0)$ between the main HAP and the interfering HAPs [3].

From a user's perspective, 'optimal' means that the minimum angular separation should be as large as possible to reduce the potential interference from unwanted HAPs. Here the overall optimal minimum angular separation is simplified as maximising the minimum value of $A(x, y)$ within the coverage area as follows [12]:

$$A_{\min} = \min[A(x, y), (x, y) \in D] \tag{9.40}$$

where the minimum value of the minimum separation angle of all users over the coverage area is derived. This is then maximised, using [12]:

$$\max[A_{\min}(r_o, r_i, h_o, h_i)] \tag{9.41}$$

subject to $r_i < r_o$, $h_i = h_o \pm \delta$ and inequalities (9.35) and (9.36) (9.42)

In practice, we can find the optimal r_i by fixing (r_o, h_o, h_i) using two-ring constellations to achieve the optimal minimum angular separation. For instance, the parameter (r_o, h_o, h_i) could be chosen as follows. Values r_o and h_o are 16 and 17 km, respectively, in order to obey the minimum elevation angle constraint within the coverage area. Two different values, 17 and 22 km, are considered for h_i. Two HAP number ratios, 4/12 and 6/10, of inner ring/outer ring ratios are investigated.

Figure 9.32 illustrates the CDF of the minimum angular separation for the one-ring case and the four two-ring cases. The four two-ring cases follow the optimal minimum angular separation criterion mentioned previously. The value of the inner ring radius in Figure 9.32 is the optimal r_i when h_i is 17 or 22 km and inner ring/outer ring HAP number ratio is 4/12 or 6/10 for four two-ring cases. The one-ring case acts as a reference using a 16 km ring radius and 17 km ring height which are the same as the outer ring parameters in two-ring cases. The vertical line with the value of 3.2° on the horizontal axis in Figure 9.32 represents the boundary of the half of the sidelobe floor beamwidth. Any minimum angular separation falling into the left side of this vertical line will incur severe interference from interfering HAPs due to their power falling into the main lobe of the user antenna. Fortunately, all of the five CDF curves are on the right side of the vertical line. After the optimisation of the minimum angular separation, three of the four CDF curves for two-ring constellations are always better than the reference one-ring curve. Even in the dash-dot curve which has slightly lower value than the reference one-ring curve in the bottom part, it still shows that over the 82 % of users in the coverage area have better minimum angular separation than for the one-ring case.

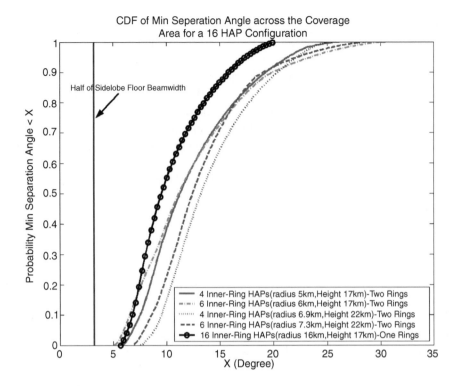

Figure 9.32 CDF comparison of minimum angular separation across the coverage area for two-ring and one-ring constellations in 16 HAP configurations (reproduced from [12])

One notable phenomenon occurs when the inner ring height is the same as the outer ring height (17 km), the smallest value of the minimum angular separation is not much improved compared with the one-ring case at the same height. However, the appropriate change in the inner ring height, for example 22 km in Figure 9.32, does improve the worst case value of the minimum angular separation. Note that despite falling into the sidelobe area, practical systems are likely to have a tapering envelope of the sidelobe floor. So a greater benefit will arise from higher minimum angular separations.

9.5.1.3 Optimal HAP Antenna Pointing Offset

The third strategy is how to choose the HAP antenna pointing offset to achieve better system capacity. Ideally, the CINR should be used as a basic parameter for system capacity. However, the variable interference from the interfering HAPs makes the

CINR difficult to analyse. An alternative way is to investigate the power from the main HAP instead of the CINR. As we discussed above, we can find a combination of (r_o, r_i, h_o, h_i) to make the minimum angular separation bigger than the half of the sidelobe floor beamwidth of the user antenna. Therefore, the interference power always enters into the sidelobe of the user antenna profile, which guarantees that the interference is limited to a very low level compared with the received power from the main HAP. Based on this, the HAP antenna pointing offset analyses can focus on the distribution of the HAP transmit power within the coverage area without worrying about excessive increase in interference delivering a deterioration of the downlink CINR. The trade-off of the power analysis is that the optimal HAP antenna pointing offset is for power not for CINR. However, the discrepancy is reasonably small.

Prior to any detailed analysis, some example CINR contours are given in Figure 9.33 using some typical pointing offsets. Figure 9.33(a) and (b) shows there is a high CINR distributed outside the coverage area when all the HAPs point at the centre of the coverage area. This is a significant waste of power and can bring problems when spectrum sharing with other systems. At the same time, users at the right edge of the coverage in Figure 9.33(b) only get very low CINRs. Figure 9.33(c) and (d) demonstrates noncoverage-area-concentrated CINR contours when the pointing offset is at each HAP's sub-platform point. The pointing offset in Figure 9.33(e) and (f) is a compromise between a pointing offset at the centre of the coverage area and one at the HAP's sub-platform point – in this case the pointing offset is half of the ring radius. This pointing offset shows a good CINR concentration into the coverage area and a relatively good CINR value at the edge of the coverage area. From these examples we realise that the optimal pointing offset exists somewhere between the sub-platform point of a HAP and the centre of the coverage area. Choosing the midpoint is therefore a natural starting point for a potential optimal pointing offset. However, these values are somewhat arbitrary, so deeper analyses follow.

There are two different proposals to find the optimal HAP antenna pointing offset under the two-ring constellation:

- **Idea 1** considers reducing the wasted power outside the coverage area by maximising the received power within the coverage area. This can be achieved by changing the HAP antenna roll-off or changing the HAP antenna pointing offset. Here the latter method is adopted. This is also important from a regulatory perspective as it will reduce interference into neighbouring systems.
- **Idea 2** considers achieving uniformity of the received power distribution by minimising the dynamic received power range within the coverage area in order to achieve an overall CINR improvement.

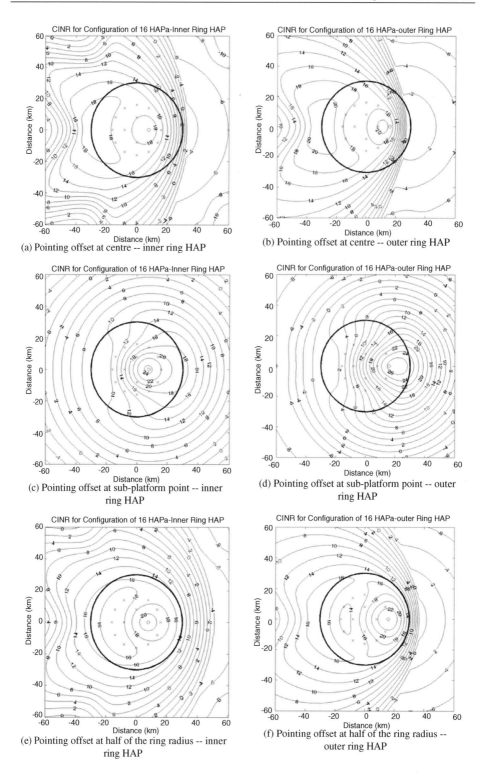

(a) Pointing offset at centre -- inner ring HAP

(b) Pointing offset at centre -- outer ring HAP

(c) Pointing offset at sub-platform point -- inner ring HAP

(d) Pointing offset at sub-platform point -- outer ring HAP

(e) Pointing offset at half of the ring radius -- inner ring HAP

(f) Pointing offset at half of the ring radius -- outer ring HAP

The HAP antenna profile is assumed to be sufficiently wide to allow the main beam to cover the coverage area in this chapter. Therefore the following differential equation shows a theoretical way to find a suitable pointing offset for idea 1. If the antenna profile has many local peaks, another approach must be used [12]:

$$\frac{d\left[\iint_D \text{Power}(x, y, p) \, dxdy\right]}{dp} = 0 \tag{9.43}$$

Here Power refers to the power received by a user from a HAP as a function of the user position (x, y) and the HAP antenna pointing offset p. The integration of power is carried out over the coverage area D. The maximum total power can be achieved by setting the differential result to zero.

Equation (9.43) cannot be solved analytically due to excessive complexity. Therefore an alternative approximate computational method is adopted, based on the following equations:

$$\text{Power}_{\text{sum}}(p) = \sum_{x,y \in D} \text{Power}(x, y, p) \tag{9.44}$$

The optimisation then yields the optimum pointing offset p_{max} for which:

$$\text{Power}_{\text{sum\_max}}(p_{\text{max}}) = \max[\text{Power}_{sum}(p)] \quad p \in [-60 : 1 : 30] \tag{9.45}$$

where p_{max} is an optimal pointing offset for the maximum sum of the received power.

Figure 9.34 shows the optimal pointing offset versus the HAP location according to the criterion of the maximum sum of the received power (in dB). Three different HAP antenna roll-offs, 10 dB, 6.5 dB and 3 dB, at the edge of the coverage area are evaluated in Figure 9.34. These three antenna roll-offs give identical curves under the 1 km pointing offset steps, which means the trend of curves is independent of antenna roll-off. The optimal pointing offset increases at first and then decreases according to the increment of the HAP location from 0 to 60 km with the turning point of 25 km. This phenomenon can be explained as follows. When the HAP

Figure 9.33 Example contours of CINR for different pointing offsets – outer ring radius 16 km, inner ring radius 8 km, main HAPs marked as 'o' and other HAPs marked as 'x' (reproduced from [12])

location is moved away from the centre of the coverage area, the HAP antenna pointing offset has to move away from the centre to compensate for the received power loss at an area somewhere between the sub-platform point and the boundary of the coverage area. The shape and the location of this area keep changing according to the change of the HAP location. However, this area disappears when HAP location is greater than 25 km which is close to the boundary of the coverage area. That means that the pointing offset of the HAP antenna needs to move towards the centre of the coverage area to compensate for the received power loss within the coverage area. The two vertical dashed lines in Figure 9.34 indicate optimal pointing offsets 3 and 6 km for the default values of HAP locations 8 and 16 km (inner ring and outer ring radii), respectively. The curves in Figure 9.34 are stepped. The reason for this phenomenon is the 1 km discrete step change in the HAP pointing offset (a smaller step-change would result in smoother curves).

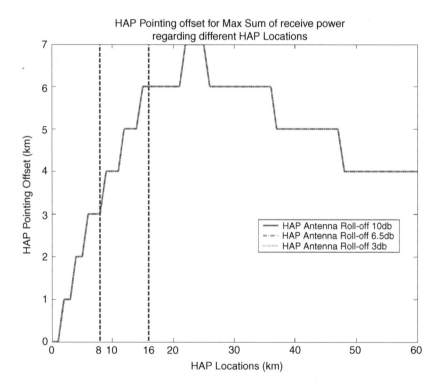

Figure 9.34 Optimal HAP pointing offset for the maximum sum of received power regarding different HAP locations (reproduced from [12])

Equations (9.46)–(9.48) illustrate the methodology to find the optimal HAP antenna pointing offset for minimum dynamic range of received power (idea 2) [12]:

$$\text{Power}_{\max}(p) = \max[\text{Power}(x, y, p)] \qquad (x, y \in D) \qquad (9.46)$$

$$\text{Power}_{\min}(p) = \min[\text{Power}(x, y, p)] \qquad (x, y \in D) \qquad (9.47)$$

$$\text{Power}_{\min\_dyn\_range}(p_{\min\_dyn\_range}) = \min[\text{Power}_{\max}(p) - \text{Power}_{\min}(p)] \qquad (9.48)$$
$$p \in [-60, 30]$$

where $p_{\min\_dyn\_range}$ is the sought optimal pointing offset for a minimum dynamic range of received power.

Figure 9.35 shows the optimal pointing offset versus the HAP location according to the criterion of minimum dynamic range of received power. Three different HAP antenna beam roll-offs are also considered, and within these three roll-offs, the common trend of the optimal pointing offset is to increase slowly at first and then

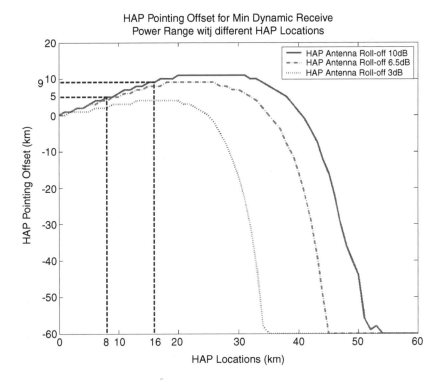

Figure 9.35 Optimal HAP pointing offset for minimum dynamic received power range regarding different HAP locations (reproduced from [12])

decrease rapidly. In the rapid decrease part, the pointing offset values that are smaller than -30 km (which is outside of the coverage area) should be abandoned because the waste in transmit power is significant. Two vertical and two horizontal dashed lines in Figure 9.35 indicate optimal pointing offsets 5 and 9 km for HAP locations 8 and 16 km (inner ring and outer ring radii), respectively.

In order to compare the system performance comprehensively, we choose a two-ring constellation design with different pointing offset criteria and a one-ring constellation with pointing offset designed for minimum dynamic power range. The two-ring constellation parameters (r_o, r_i, h_o, h_i) used here are set to (16 km, 8 km, 17 km, 17 km) with 4/12 inner ring/outer ring HAP number ratio in the two-ring case. Four different sets of pointing offset parameters are used as shown in Table 9.5 including the optimal pointing offsets for idea 1 and 2. We use the parameter sets 1 to 4 to refer to different parameter sets of pointing offset in this section.

Table 9.5 Pointing offset parameters in two-ring constellations

	Pointing offset (km) – inner ring HAP	Pointing offset (km) – outer ring HAP
1. Pointing offset for maximum received power in the coverage area	3	6
2. Pointing offset for minimum dynamic power range	5	9
3. Pointing offset equals half of the inner/outer ring radius	4	8
4. Pointing offset at the centre of the coverage area	0	0

Although parameter set 3 represents arbitrary pointing offset values, it is still worth including to investigate its effects on performance. The pointing offset used in the one-ring constellation is the same as the pointing offset of outer ring HAPs in parameter set 2, which is 9 km. Figure 9.36 and 9.37 illustrate four CDF curves of downlink CINR for a chosen HAP at the inner and outer ring, respectively, in two-ring constellations plus one CDF curve for a HAP in a one-ring constellation. Comparing the four CDF curves in two-ring constellations in Figures 9.36 and 9.37, the worst case CINR value for parameter set 4 is much lower than for the other three parameter sets. This means that when all the inner and outer ring HAPs point at the centre of the coverage area it does not distribute the HAP transmit power well and hence some users have low CINR. The curves for the other three parameter sets in Figure 9.36 and 9.37 are considerably closer together. However, the CDF curves for parameter sets 2 and 3 are slightly better than the curve for parameter set 1 which has

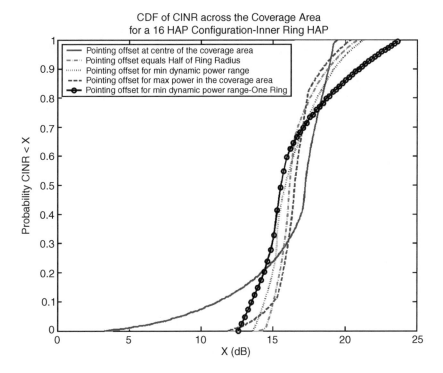

Figure 9.36 CDF of CINR across the coverage area for a 16 HAP configuration-inner ring HAP (reproduced from [12])

a slightly lower worst case CINR in Figures 9.36 and 9.37. There are no major advantages and disadvantages shown between parameter sets 2 and 3. From Table 9.5, discrepancies between these two parameter sets are only 1 km. This can explain why the CINR plots are close together for these two. However, parameter set 2 is designed for the minimum dynamic power range criterion and parameter set 3 is just arbitrarily chosen for half of the inner/outer ring radius. This means that on some occasions, parameter set 3 can be used as a quick and simple approximation (i.e. as rule of thumb) to parameter set 2 in a constellation design.

Figure 9.36 and 9.37 also show the CDF curves of CINR in the one-ring constellation with pointing offset designed for minimum dynamic power range. These one-ring CDF curves are close to curves with parameter sets 1, 2 and 3 in two-ring constellations. In Figure 9.36, the best case of the CDF curves for two-ring constellations are slightly lower than the CDF curve for one-ring constellations due to the larger interference caused by the adjacent inner ring HAPs. However, it does

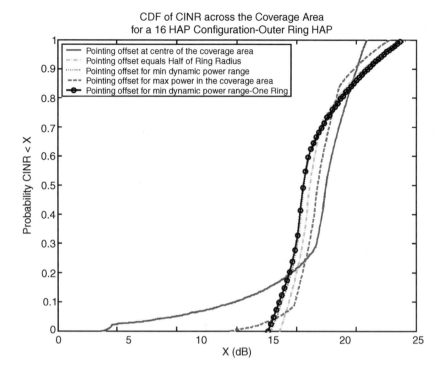

Figure 9.37   CDF of CINR across the coverage area for a 16 HAP configuration-outer ring HAP (reproduced from [12])

not mean that the two-ring constellations are detrimental to system capacity measured in terms of the spectral efficiency, when compared with the one-ring constellation as shown in Figure 9.38. These five CDF curves have the same maximum value of spectral efficiency but different minimum values. If we take the one-ring CDF curve as the reference line, the CDF curves for parameter sets 2 and 3 in two-ring constellations have a very close minimum value when compared with the reference line. Furthermore, the shape of the CDF curves shows the spectral efficiency of parameter set 3 is better than that of parameter set 2; and both of them are always better than the reference line. For parameter sets 1 and 4, about 20 % and 42 % of the bottom part of the CDF curves are worse than the reference line. After these comparisons, parameter set 2 could be used as a design guide of the pointing offset of HAP antenna. In this particular two-ring constellation, parameter set 3 (pointing offset equals half of the inner/outer ring radius) has the best performance. However, it lacks a theoretical justification which only makes it a simple empirical method which may provide good system performance in other situations. Here the pointing offsets optimised for power analysis (parameter sets 1 and 2) do not have the best system performance. The reason of this is the discrepancy between the power analysis and the CINR analysis mentioned before.

Figure 9.38 also demonstrates that a proper pointing offset selection in a two-ring constellation makes capacity evenly spread across the coverage area. From the curves in Figure 9.38, parameter set 4 has the largest variation of the spectral efficiency across the coverage area which means that the capacity in some areas is considerably less than the capacity in other areas. The other three parameter sets give a relatively even spread of the capacity over the coverage area. The variation also can be quantified using the standard deviation. The mean spectral efficiency of the four parameter sets ranges from 87 to 90 bits/s/Hz which shows no major difference. However, the standard deviation ranges from 2.6 bits/s/Hz (parameter set 2) to 8.8 bits/s/Hz (parameter set 4) which is a significant variation. It should be noticed that parameter set 2 which is designed for minimum dynamic power range has the lowest standard deviation of spectral efficiency.

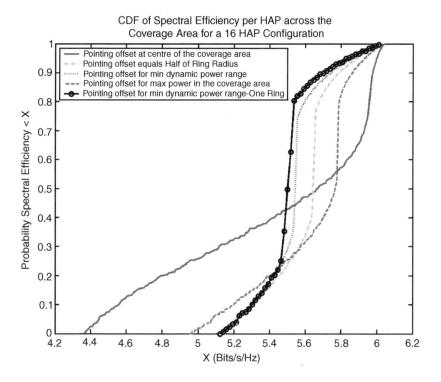

Figure 9.38 CDF of spectral efficiency across the coverage area for a 16 HAP configuration (reproduced from [12])

References

1. Recommendation ITU-R F. 1500, *Preferred Characteristics of Systems in the Fixed Service Using High-Altitude Platform Stations operating in the Bands 47.2–47.5 GHz and 47.9–48.2 GHz*, International Telecommunications Union, Geneva, Switzerland, 2000.
2. M. Oodo, R. Miura, T. Hori, T. Morisaki, K. Kashiki and M. Suzuki, *Sharing and Compatibility Study between Fixed Service Using High Altitude Platform Stations (HAPs) and Other Services in 31/28 GHz Bands*, Wireless Personal Communications, 2002, Vol. 23, pp. 3–14.
3. Recommendation, ITU-R M. 1456, *Minimum Performance Characteristics and Operational Conditions for High Altitude Platform Stations providing IMT-2000 in the Bands 1885–1980 MHz, 2010–2025 MHz and 2110-2170 MHz in the Regions 1 and 3 and 1885–1980 MHz and 2110–2160 MHz in Region 2*, International Telecommunications Union, Geneva, Switzerland, 2000.
4. B. El-Jabu and R. Steele, *Cellular Communications Using Aerial Platforms*, IEEE Trans. Veh. Technol., May 2001, Vol. 50, pp. 686–700.
5. J. Thornton, D. Grace, M. H. Capstick and T. C. Tozer, *Optimising an Array of Antennas for Cellular Coverage from a High Altitude Platform*, IEEE Trans. Wireless Commun., May 2003, Vol. 2, No. 3, pp. 484–492.
6. J. Thornton and D. Grace, *Effect of Lateral Displacement of a High Altitude Platform on Cellular Interference and Handover*, IEEE Trans. Wireless Commun., August 2002, Vol. 4, No. 4, pp. 1483–1490.
7. D. Grace, C. Spillard, J. Thornton and T. C. Tozer, *Channel Assignment Strategies for a High Altitude Platform Spot-Beam Architecture*, Proc. 13th IEEE International Symposium on Personal Indoor and Mobile Radio Communications (PIMRC 2002), 2002, Vol. 4, pp. 1586–1590.
8. D. Grace, J. Thornton, G. P. White and T. C. Tozer, *Improving Spectrum Utilisation for Broadband Services in the mm-Wave Bands through the Use of Multiple High Altitude Platforms*, presented at IEE Seminar - Getting the Most Out of the Radio Spectrum, London, UK, 2002.
9. D. Grace, J. Thornton, G. Chen, G. P. White and T. C. Tozer, *Improving the System Capacity of Broadband Services Using Multiple High Altitude Platforms*, IEEE Trans. Wireless Commun., 2005, Vol. 4, pp. 700–709.
10. G. Chen, D. Grace and T. C. Tozer, *Evaluation of the Effects of User Antenna Pointing Error in Multiple High-Altitude Platform Systems*, IET Commun., June 2007, Vol. 3, No. 3, pp. 424–429.
11. G. Chen, D. Grace and T. C. Tozer, *Performance of Multiple High Altitude Platforms using Directive HAP and User Antennas*, Int. J. Wireless Personal Communications - Special Issue on High Altitude Platforms, 2005, Vol. 32, No. 3–4, pp. 275–299.
12. G. Chen, *Capacity Enhancement Using Multiple Platforms*, PhD Thesis, University of York, York, UK, 2006.
13. D. Grace, G. Chen, G. P. White, J. Thornton and T. C. Tozer, *Improving the System Capacity of mm-Waveb and Broadband Services Using Multiple High Altitude Platforms*, presented at IEEE GLOBECOM 2003, San Francisco, USA, 2003.
14. G. Chen, D. Grace and T. C. Tozer, *Controlling the Received Power, Interference and Capacity in Multiple HAP Configurations*, presented at WPMC 2004, Abano Terme, Padova, Italy, 2004.
15. D. Grace and P. Likitthanasate, *A Business Modelling Approach for Broadband Services from High Altitude Platforms*, presented at ICT 2006, Funchal, Portugal, 2006.
16. L. J. Ehernberger, *Stratospheric Turbulence Measurements and Models for Aerospace Plane Design*, NASA Technical Memorandum 104262, 1992.
17. B. G. Evans, *Satellite Communication Systems*, 2nd Edn, Peter Peregrinus Ltd, 1991.

10

Networking Implications of Using Multiple HAP Constellations

10.1 Network Protocols

Utilisation of multiple HAPs, discussed in Chapter 9, has important implications for the overall network architecture as well as for the networking protocols. The network architecture of a HAP system generally resembles the network architecture of a satellite system, but in terms of network protocols, HAP networks tend to be more similar to terrestrial wireless networks, especially in terms of interworking requirements, mobility management and load balancing.

The most important requirement for a HAP network architecture is efficient interworking with existing ground networks. A decision regarding this requirement has significant implications on the network protocols, especially on those supporting mobility of nodes in the case of mobile operating scenario. At the present and in the foreseen future, Internet Protocol (IP) technology seems to be playing the key role in the global convergence process of different sectors, e.g. telecom, data communication, radio and television. The doctrine 'keep it simple' for the design and implementation of IP-based networks has significantly contributed to their great popularity and success. The overwhelming popularity of IP technology naturally leads to IP-based HAP networks. Even more, an all-IP approach is preferred, as explained in the following.

10.1.1 IP Foundations

The essential attributes of an all-IP network are end-to-end IP connectivity all the way to the user nodes and IP-based control functions, including the mobility

Broadband Communications via High Altitude Platforms   David Grace and Mihael Mohorčič
© 2011 John Wiley & Sons, Ltd

handover procedures and routing. All-IP networks rely entirely on IP, from the user node to the border routers towards external networks. Note that the all-IP term is used for many emerging wireless networks. While most of them use IP for transport purposes only, the all-IP term, as used here, should be understood as using IP in the native mode. Duality – transport only or native – can be hidden in upper layers, however choosing the native IP mode can have a significant impact on HAP network efficiency and performance.

For instance, ATM was chosen as the transport technology in UMTS, but higher-layer protocols demonstrate the UMTS openness towards a pure IP solution. IP is actually a Layer-3 protocol which proves that these networks are also moving towards all-IP networks. However, a pure IP solution is a major shift from current third generation mobile networks where the last hop IP router is GGSN and native transport technology is not IP based.

Numerous reasons justify all-IP architecture. Using IP technology as the foundation of HAP networks makes engineering and economic sense. Today, IP is the prevalent fixed networking standard. The philosophy embodied in the IP has made it ubiquitous. The main aspects of this philosophy are simple and stateless networks, with complexity pushed to the edge. The IP network is modular, with open interfaces placed along functional boundaries. A single routing network that would support all mobile and fixed users, public and private, would be the preferred choice for network operators from an economic as well as engineering perspective. Finally, at the user level, it is expected that all end user applications will be IP based. With the adoption of all-IP architecture applications available on fixed networks will inherently be available on mobile networks, without their characteristics being impaired by specific mobile protocols. The same will apply to terminals where only an IP protocol stack will be necessary. It is the native use of IP that more readily allows for building an efficient network regardless of access technologies. As the coverage of native IP increases, the wireless-specific protocols are pushed farther towards the access segment.

10.1.2   Mobile IP Protocol

In order to discuss various aspects of mobility management in HAP networks, we have to be familiar with the basics of Mobile IP (MIP). A short introduction is given in the following.

The MIP protocol defines a set of entities and procedures that enable a mobile node to retain its home address on the move, without requiring changes in the intermediate routing nodes. Routing to the mobile node (MN) is done partially to the home address (HoA) and partially to the care-of address (CoA), that is a unicast routable address associated with a mobile node visiting a foreign micro-mobility area. While the mobile node is away from home, it registers its current CoA with the home agent (HA). The agent intercepts packets from a correspondent node (CN) destined for the mobile node's HoA, encapsulates them, and tunnels them to the mobile node's registered CoA. The CN is the node with which the MN is communicating. MN has to report any change in current CoA by sending a binding update (BU) to its HA.

Registration and packet forwarding in MIPv4 is managed through a foreign agent (FA), whereas in MIPv6, the FA is not needed. A FA is actually a router for mobile nodes on a visited network that operates as the end-point for tunnels from the HA and provides CoA on behalf of the MNs. In the radio resource perspective, the use of a FA is more effective because in MIPv6 tunnels need to be established over the radio interface, while in MIPv4 this is not the case.

MIPv6 [1] builds on IPv6 transport technology, which gives it an advantage over MIPv4 [2]. Integrated route optimisation functionality is one of the benefits of MIPv6. MIPv4 in its basic form suffers from the triangular routing problem, where all packets sent to a MN while away from home are intercepted by its HA and tunnelled to the MN; thus taking a long detour. In MIPv6 the CN can send IP packets directly to the MN after the optimisation step takes place. The optimisation is triggered by the MN. A binding update, similar to the update that is sent to the HA, is used to inform the CN about the MN's current location. Note that several messages must be exchanged in this process because it involves a return routability test and binding acknowledgement.

The basic MIP includes the movement detection procedures and registration with the HA that can trigger handover. However, this cannot be achieved in a fast way. Every time the mobile node gets a new local IP address, movement detection and registration steps must be completed. It is the MN that initiates these processes. Movement detection latency and registration latency are serious limitations for real-time communications. The overall latency may be large, since the movement detection mechanisms in MIP are based on either the expiration of the lifetime of foreign agent advertisements, or on the comparison of the address prefix of two different agent advertisements. The registration latency may be even larger as the HA may be located anywhere in the Internet.

10.1.3 Hierarchical MIP

In order to improve MIP scalability, hierarchical mobility architecture was introduced [3,4] on a micro-mobility level. Without any hierarchical structure the number of binding updates (BUs) increases proportionally as the network grows and the number of MNs increases. Efficient mobility management should keep network load low and provide optimum routing of packets. Hierarchical MIP

(HMIP) technology reduces frequent location registrations and the time needed for handovers.

The protocol differentiates the intra-site mobility from the inter-site mobility. A site can be an ISP network, a company network, a set of LANs or even a single LAN. A new node, called the Mobility Anchor Point (MAP), can be located at any level in a hierarchical network of routers. It can be viewed as a local HA for the site. The MN obtains the on-link care-of address (LCoA) and regional care-of address (RCoA). The RCoA is an address of the mobile node in the MAP domain. Note that in MIP only CoA, an equivalent to LCoA, is allocated and used. Before registering the RCoA with the HA and correspondent nodes, the MN registers with the MAP in order to establish a binding between the RCoA and LCoA.

When a MN moves within the site, a new LCoA is allocated on its new access point. Because the RCoA remains constant, only local binding updates are required within the MAP area. As a result, a foreign correspondent node, i.e. a CN that does not reside within the same MAP domain, is aware only of MN's inter-site mobility. As a consequence, all inter-site traffic is routed through MAP, which is not always optimal. In order to cope with this problem, the specification allows several MAPs to be deployed for the same area of nodes. Furthermore, MAP can be placed within the border router to improve path efficiency.

Intra-domain route optimisation is proposed to eliminate path deviations from its optimum, which would occur if all intra-site traffic were routed through the MAP [3]. Correspondent nodes that reside within the same site as the MN get special treatment. They are referred to as local CNs. The BUs to these nodes carry LCoA instead of RCoA. Registration updates are sent more frequently, i.e. on every point-of-attachment change. In this way an optimal route can be established between the CN and the MN within the site. The RFC 4140 [4] is more restrictive as to which CN the MN sends a BU with LCoA as the source address. The CN must be attached to the same link as the MN in order to receive the BU with LCoA. This is not route efficient if the MAP is placed away from the network edge.

A mechanism is proposed to utilise network resources more efficiently by the choice of MAP. In a way, it can be considered as a substitute for intra-domain optimisation. The alternative use of more than one MAP is permitted for redundancy and as an optimisation for the different mobility scenarios experienced by MNs. MAPs are used independently of each other. In order to achieve as optimal route as possible, the most suitable MAP should be selected for local CNs. This will avoid sending all packets via the distant MAP, hence resulting in more efficient routing. The MN may need some sophisticated algorithms to be able to select the most appropriate MAP. The IETF specifications only provide a default algorithm for selecting the most distant and furthest available MAP, which does not offer any intra-domain route optimisation capability.

Distinguishing a local CN from one further away has not been addressed yet. The problem is not easy to solve because very little information is available to MNs about the network topology in which they roam. In fact, the MAP option in router advertisement messages includes only the distance vector from the MN, the preference for the particular MAP, the MAP's global IP address and the MAP's subnet prefix. Currently only CNs on the same link are recognised as local CNs. On the basis of the MAP's subnet prefix it would be possible to identify CNs within the same MAP domain from CN's RCoA, providing that CNs are also MNs.

The multi-level hierarchy of MAPs is not supported by the RFC 4140 [4], in contrast to the proposal in [3], because it is not required for a better handover performance, which is one of the reasons for introducing HMIP in the first place. Moreover, it is prohibited to select more than one MAP and to force packets to be sent from the higher MAP down through a hierarchy of MAPs, because this may add forwarding delays and affect the robustness of IP routing.

10.2 Mobility Management in HAP-Based Communication Systems

Mobility management in HAP-based communication systems is as complex as in any existing wireless networks. In general, complex problems should be partitioned into a set of smaller problems that can in turn be solved more efficiently. A well known approach is to split the problem of routing to a MN, which is in the core of any mobility management, into micro- and macro-mobility parts. Apart from that division, access mobility is frequently considered as the lowest level of mobility. While micro- and macro-mobility divide the network layer mobility events into those that can be handled locally and those with global impact on the route, the access mobility is a link layer task within the scope of a single HAP. In third generation networks, access mobility refers to the methods and protocols that ensure mobility within the scope of a single Radio Network Controller (RNC), which corresponds to a single HAP in multiple HAP constellations.

Another important issue that influences the mobility management procedures is how the HAPs are distributed between network operators. The simplest scenario is that one network operator operates all the HAPs in a particular region. In this case the user roams to that network and is connected to the same operator while being within the coverage of the HAP system. However, it can be expected that there will be more than one HAP network operator with different numbers of HAPs. Some of them will be inter-connected via terrestrial links but also via interplatform links.

10.2.1   Access-Level Mobility

MIP as the highly developed macro-mobility management solution and its various micro-mobility extensions do not attempt to solve access control to a link. Note that MIP was originally developed in wired networks to support relatively infrequent node migrations and therefore lacks some functionality to handle issues inherent to the wireless networks. Authentication, authorisation and accounting are such functionalities that need to be addressed on the access level. The mobile network itself has to allow access to the resources for the MN before higher layer protocols actually consume radio resources.

Mobility within a single HAP should be implemented on a link layer, i.e. as access mobility. Loose coupling with network layer mobility enables exploitation of the specific HAP access technology. This is not possible if IP mobility is employed at cell level. Access level mobility is expected to take care of movements across the cell boundaries for the cells that are beamed from a single platform. In this way the entire HAP coverage area is seen as a single link by the network layer entities.

10.2.2   Micro-Mobility

In order to minimise the interplatform handoff latency problem, IP-based micro-mobility protocols can be employed. Several alternatives are possible, mainly in the form of an extension of the MIP. The mesh interconnection of HAPs and the presence of bandwidth limited backhaul links require careful selection and configuration of the micro-mobility protocol.

Cellular IP [5], Hawaii [6] or TeleMIP [7], propose a form of a tree-like hierarchy of local routers. The infrastructure nodes in these schemes are arranged in a strict inverse tree structure beneath the root router, i.e. every node is a child of one and only one other node, and may be a parent of zero, one or more other nodes. As the platforms are connected in a mesh topology, routing possibilities would be significantly limited if a tree-topology is superimposed. Furthermore, a tree-like topology of nodes consisting of several levels mapped on a HAP constellation would most probably require involvement of backhaul links in the tree of routers. In this case, a route optimisation issue arises for HAP-to-HAP communication. Namely, backhaul forwarding for this particular type of traffic is difficult to be avoided without a customised solution.

The second group, i.e. the tunnel-based micro-mobility schemes, performs registration and a different kind of encapsulation in hierarchical fashion. A concatenation of tunnels is created over the local access network. A typical representative of this group is the HMIP [8,9]. In older IPv4 networks, regional registration performs a similar function.

Mechanisms in both groups of micro-mobility schemes define a kind of Anchor Node (AN) that is responsible for micro-mobility within the area. For instance, in HMIP the AN is called the MAP. There are two alternatives where to locate the AN within the HAP network. If the AN is placed on the ground, backhaul links are included in the domain tree. The AN is extensively involved into message exchange within the micro-mobility domain, which puts additional load on scarce radio resources on the backhaul link. However, change of IP address when the user moves into the coverage area of a next platform is not needed, meaning that global path updates are less frequent. The degree of location anonymity is higher. One must be aware that route optimisation and processing of ICMP messages reveals the MN's point of attachment to the rest of the world with the precision of a single AN location.

On the other hand, putting the AN on the platform increases the load on the selected platform and reduces the interplatform communication efficiency. Route optimality is improved because the AN, which is a fixed point on any route to/from the micro-mobility area, is placed on the network edge. Furthermore, the network scalability is higher, because new HAPs may be added to the network without any impact on micro-mobility management in the rest of the network.

10.2.3 Macro-Mobility

While micro-mobility enables transparent mobility within a limited area, macro-mobility procedures make possible transparent movements of a MN between the micro-mobility areas on the global scale. Due to address change, all transport layer connections will break down unless appropriate mechanisms are introduced.

Several macro-mobility protocols have been proposed in recent years. Although all of them have the common goal of location transparency, they differ from each other as they operate in different protocol layers. For instance, Session Initiation Protocol (SIP) operates at the application layer [10], while MIP is a network layer protocol [1,2]. SIP is limited by the performance of TCP or UDP over wireless links. MIP is considered as the most promising IP macro-mobility mechanism. It defines a set of entities and procedures that enable a MN to retain its HoA on the move, without requiring changes in the intermediate routing nodes.

In case of multiple platform networks, time-critical operations, such as interplatform handover, cannot be handled just by macro-mobility protocol. Namely, MIP suffers from several weaknesses that require introduction of micro-mobility support in networks.

10.2.4   Types of Mobile Users

Mobile users are considered an important end-user segment of the HAP-based communication system. For the purpose of this chapter, mobile users can be individual subscribers or collective nodes. A typical representative of the latter type of node is a vehicle, i.e. bus, train, ship or automobile, where the passengers access the network through a local wireless or wired network.

The main difference between the end-user mobile terminal and the terminal on the vehicle is that the latter is expected to function as a mobile router. A mobile router allows roaming of an entire network. Often, the passengers will require global mobility, in which case they will have to use their own MIP, in addition to the MIP performed by the mobile router. So-called two-level mobility management will take place.

Efficient mobility support in the HAP network should handle mobility of the vehicles as well as that of the passengers. The limited radio spectrum should be used as efficiently as possible as the wireless link will often be the bottleneck. Unnecessary use of backhaul links must be avoided. Note that user data do not necessarily pass through the ground core network – only user-to-HAP and HAP-to-HAP links may be involved.

10.2.5   Network Mobility

The concept of mobile networks, where a group of nodes move together and access a global network through a mobile router, has been widely recognised as a viable future mobility scenario. The Network Mobility Working Group (renamed the Mobility EXTensions for IPv6 Working Group in 2007) has proposed a network mobility (NEMO) technology that allows mobile routers to roam within foreign networks in a manner similar to end-nodes [11,12].

The key component of a mobile network is a mobile router (MR), which provides a communication link to the rest of the world. Mobility management for end-nodes is well defined in current MIP specifications but the support for mobile networks has not been addressed. This kind of support will be needed for various types of communication devices moving on vehicles. The support is particularly important in HAP-based communication networks.

MIPv4 and MIPv6 have been designed for end-node mobility. Network mobility was not completely taken out of the specifications. The designers of MIPv4 claim that it could support mobile networks equally as MNs. This may be true for MIPv4; however, in [3] the authors show that MIPv6 cannot support network mobility without some changes. Recently, MIP extensions for mobile routers have been studied within the IETF Network Mobility Working Group [11,12].

Mobility principles, similar to those for end-nodes, apply to mobile routers. There is an HA that intercepts all packets for an entire network domain and forwards them to the MR. Note that the MR's sub-network address prefix is bound to the MR's CoA, which enables tunnelling sub-network traffic to a remote location. This differs from MIP for end-nodes, where a single address is bound to the CoA. The MR implements a movement detection mechanism and performs registration with its HA.

10.2.5.1 Intra-domain path optimisation

When multiple MRs are present within the site, the intra-site route optimisation issue arises. In contrast to MIP for end-nodes, path optimisation for MRs remains a largely open challenge, although several proposals exist that deal with the problem [13–20]. In [21] a proxy MAP solution is proposed for the scenario with the HMIP as a micro-mobility protocol, which is the only solution that also targets the intra-site route optimisation problem.

Path optimisation for MRs is a challenging task due to multiple-level or nested mobility. The problem can be illustrated by an example. Let us assume that an MN, served by the MR, registers with its HA because it requires global mobility. The first data packet from the CN to the MN is routed to the MN's home network. The MN's HA tunnels the packet to the MN's CoA, which is in the name space of the MR's home domain. Therefore, the tunnel from the MN's HA to the node itself passes through the MR's HA, where it is further encapsulated, due to a tunnel that connects the MR's HA with the MR. The resulting two-level MIP architecture is illustrated in Figure 10.1. The inner architecture is required because of the mobility

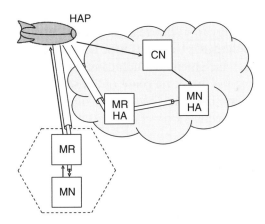

Figure 10.1   Two-level MIP routing with CN in a global network

of the router mounted on the vehicle, while the outer architecture guarantees global mobility of the passenger's mobile terminal.

Triangular routing, as seen in Figure 10.1, is inefficient as regards backhaul link utilisation. Without route optimisation, HAP-to-HAP and even intra-HAP communications unnecessarily use the scarce radio resources on the backhaul link. This is illustrated in Figure 10.2.

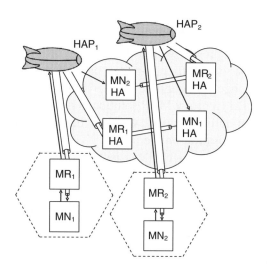

Figure 10.2   Two-level MIP routing for HAP-to-HAP traffic

Top-level MIPv6 optimisation removes MN's HA from the communication path; however it cannot skip the MR's HA. The traffic between the MN and the CN passes through the MR's HA, which is clearly inefficient. In Figures 10.3 and 10.4 packet flow is illustrated after the first-level route optimisation takes place. All packets between MNs still traverse backhaul links in both directions, even when both nodes are within the same HAP network coverage. This holds even for nodes that do not require global mobility.

Several proposals for addressing the problem of route optimisation for MRs have been made by the Internet community that are also applicable to a HAP network. An extensive survey of route optimisation solutions can be found in [17]. Prefix Scope Binding Updates (PSBUs) [13] establish a many-to-one relationship between the set of nodes that are serviced by the MR and its CoA. The MR sends PSBUs to all CNs that communicate with the MR or with any node on the mobile network that the MR

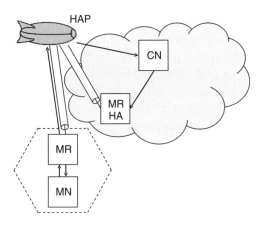

Figure 10.3   Two-level MIP routing after first-level route optimisation (CN in a global network)

is serving. The binding message tells its recipients to use CoA of the MR for all packets with the address within the mobile network prefix. The MR deduces that PSBU should be sent to the originator of the packet if the packet is received from its HA via the tunnel. Home agents and CNs should therefore be adapted to handle the PSBU-based route optimisation.

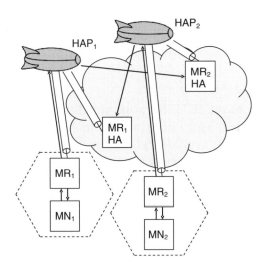

Figure 10.4   Two-level MIP routing after first-level route optimisation (CN in a HAP network)

In [14], the reverse routing header method was proposed. The method is based on so-called heart-beat messages, which are sent periodically by MNs and MRs to HAs. On its path the information is added to a message that enables the packets to be sent directly to the root-MR. The method produces a large number of control messages in the network and requires changing the existing MIP mechanisms. It was further extended and used as a basis for a solution proposed in [20].

Optimised Route Cache Management Protocol (ORC) [18] introduces new functionality in some Interior Gateway Protocol (IGP) routers. Binding information is scattered over the network. Path optimisation is possible only if the ORC router is available in the same network as CNs.

In [22–24] route optimisation burden is transferred to the nodes within the mobile network. On every MR reattachment a new CoA is allocated to each node and then communicated to CNs, while [19] defines Distributed Home Agent Protocol, which authorises the MR to act as the HA for the local MSs.

In [25], route optimisation in nested mobile networks is achieved by exchanging address information between root-MR and MNs or sub-MRs. The mechanism is very similar to the mechanism in HMIP, where an MN obtains location information from the MAP through router advertisements and forms RCoA that actually contains the information about the MAP location. In this proposal, MN registers once to a root-MR and once to the HA. The root-MR memorises MN's location within the nested mobile network while the MN's HA binds MN's home address with the location of the MR. The MN's HA can bypass the MR's HA by sending packets directly to the MR's current location. The root-MR has enough information to relay the packets to the MN. Because the latter approach maps well on the IETF hierarchical mobility management, it was further extended [15] and published as an Internet draft [16].

10.2.5.2 Proxy Mobility Anchor Points

In [21] an extension of the IETF Network Mobility Group's route optimisation solution [16] is proposed, which maps well on multiple HAP constellations. The Proxy Mobility Anchor Point (PMAP) functionality is defined in order to enable intra-domain route optimisation, even when MRs are present within the network. The PMAP mechanism is completely transparent to MIP entities, i.e. MNs, MRs, HAs and CNs, which communicate with PMAPs as with a regular MAP node; therefore network providers may choose to support PMAP functionality or not. The functionality must be implemented in the access network nodes within the HMIP domain. The root-PMAP allocates MAP information, which is communicated to the MNs via router advertisements, in the same way as a conventional MAP would do. In addition, a multicast address is allocated and added to the RA messages as a new PMAP option.

10.2.5.3 Proxy Home Agents

The inter-domain route optimisation issue may be solved even if HMIP-based route optimisation for MRs is not implemented in the HAP network. Similar proxy functionality may be applied in the form of Proxy Home Agents (PHAs). However, this solution requires extensions in MRs that are expected to roam the HAP network. Optimal interplatform routes may be achieved even when mobility management for MRs does not include support for route optimisation. PHAs adopt an idea of distributing routing information in selected parts of the network in order to speed up route convergence to its optimal path. In the following we provide a short description of PHA functionality.

PHA functionality should be implemented in HAP nodes. The PHAs perform similar tasks as HAs in home networks. The PHAs are used only by MRs that roam in the HAP network.

A mechanism allows a MR to discover the nearest PHA on a visited link. This may be achieved in a similar way as MNs discover MAP nodes in HMIP, where a new option is introduced in router advertisements. In addition to sending a registration request to its HA, the MR would register, i.e. send a binding update, with the nearest PHA, which would in turn multicast the registration request to the PHA group.

Each PHA maintains a binding cache. Upon receiving the binding update, the PHA creates a new entry for the MR or updates its existing entry, if such an entry already exists. The entry binds the MR's home network prefix with its CoA. The entry is not removed from the binding cache until the expiration of the lifetime period or until a cancellation is received.

HAP intercepts all packets that are addressed to the MR's network while it is serving as the PHA for the MR. Like the HA, the PHA establishes a tunnel to the MR and forwards intercepted packets.

The MR continues to notify, i.e. sends binding updates to the advertised PHA, until it stays within the HAP coverage. As soon as the MR leaves the HAP network it de-registers with the last known PHA and withdraws any further notifications. The cancellation is, like other messages to the PHA, multicast to the PHA group.

Suppose that both a CN and a MN access the network through the MRs that are roaming in a HAP network. The first few packets from the CN to the MN are routed through the MN's HA. After the first-level route optimisation, the CN is notified about the MN's CoA, which is within the MR's home domain. Further packets from the CN to the MN are destined to the MR's HoA and, therefore, intercepted by the PHA on the access HAP and tunnelled to the MR's CoA. Backhaul links and the MR's HA are excluded from the packet flow between these two nodes. The exchanged packets take an optimal path because they are intercepted at the very border of the network. Existing IP routing mechanism forwards the packets to the MR's current location based on the knowledge of the network topology, which includes the knowledge of the alternative routes.

The proposed mechanism was evaluated in terms of backhaul link utilisation and flow performance [26]. Simulations were performed for the environment with no second-level optimisation, for the HMIP-based optimisation with MAP placed on the ground, and for the PHA-based optimisation. Results resemble that of the PMAP solution, with slight differences in actual timing.

10.3 Mobility and Backhaul Load Reduction Techniques

10.3.1 Placement of Home Agents

In this section an alternative placement of the MR's HA is investigated as an approach to alleviate the triangular routing problem and reduce backhaul load as much as possible. This could be rewarding for communications that, due to some reason, cannot use route optimisation mechanisms. We must be aware that route optimisation support for mobile networks is not a current priority of the Network Mobility Group. Route optimisation support is planned as an extension to the basic mobility support for MRs. Even after the standardisation, the extensions would not be required to be implemented. Moreover, some proposed route optimisation schemes require involvement of the end-nodes within the mobile network, which cannot solve the problem for mobility unaware of local fixed nodes. The described HA placement optimisation is possible because of the localised mobility of MRs.

Normally, the MR's home network is located in the company headquarters. When a vehicle, for example a train, is at home station, its MR is attached to a fixed network, where the HA is located. Once in operation the train registers from time to time over its radio link with a series of wireless access points.

Due to relatively large coverage area, which can be expanded even further by a network of interconnected platforms, it is expected that many public transport vehicles will access the HAP network most of the time. If the MR's HA is located in the company headquarters, traffic is routed through the home network. This is clearly inefficient due to deviation from the optimal best-effort type of path, eventually resulting in a triangular routing problem for communications that cannot use route optimisation mechanisms. HAP backhaul links are additionally penalised because even HAP-to-HAP and inter-HAP communications are tunnelled to ground-based HA. In this case, it would be better to place the HA on the platform and make the access network to be the home network for such MRs.

Unfortunately, the availability of the user-to-HAP link depends on direct visibility from the vehicle to the platform. Tunnels, bridges, hilly terrain, canyons and other obstacles block the line of sight (LOS). In order to avoid communication blackouts, the MR can have multiple roaming interfaces. This would allow for connection to a variety of wireless links, operated by different network providers. The dual interface architecture for MIP handover during non line of sight (NLOS) conditions is illustrated in Figure 10.5.

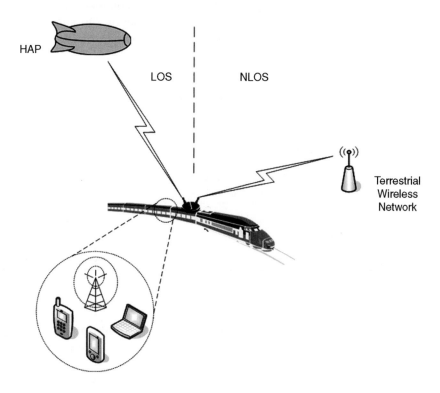

Figure 10.5   Dual interface architecture for MIP handover during NLOS conditions

However, the advantage of a HAP-based home network is not so straightforward in this scenario. During blackouts the packets are rerouted to a new location by the HA, which means that packets from a distant CN are unnecessarily forwarded up and down along the backhaul link. There is a third alternative as regards HA placement, which would avoid the above effect. By placing HA in the ground station (GS), i.e. the ground end of the backhaul link, one can avoid the above situation. Unfortunately, in this case HAP-to-HAP and inter-HAP communications are routed through the GS. Obviously, the proper choice of the HA location depends on the HAP visibility conditions.

The decision on the MR's HA placement in the HAP network is not straightforward. It depends on many factors. The first factor is backhaul link availability. If its capacity is high, it would not get congested easily. Therefore, placing the HA on the platform is preferred. On the other hand, if capacity is low, the choice of solution with the lowest backhaul link load is essential. Otherwise, the communication delays would increase. The location of CNs has to be taken into account. In order to make a proper decision an analysis should be done on expected communication patterns.

The optimal placement for different objective functions is summarised in Table 10.1. For every combination of scenario and criteria, HA locations are given in the order of preference.

Table 10.1 Optimal HA placement in terms of throughput and backhaul link loads

	CN in global network	CN in HAP network
Throughput	1. HA in the GS	1. HA on HAP
	2. HA on HAP (if backhaul link load is low)	2. HA in the GS
	3. HA in the global network (if backhaul link load is low)	3. HA in the global network
Backhaul link load	1. HA either in the GS or in the global network	1. HA on HAP
	2. HA on HAP	2. HA either in the GS or in the global network

10.3.2 Multihoming Support

If only one HA is supported per MR, choosing its optimal location depends on many factors, as seen in the previous section. However, there is a possibility to have more than one HA per MR. That can be achieved by having a multihomed mobile network. It is expected that providing a permanent access to mobile networks, such as those on trains, would require the use of several interfaces and technologies, which supports the idea of multihomed mobile networks.

NEMO basic support does not prevent MRs from being multihomed. There are many definitions of multihoming. A multihomed router is a router with multiple HAs, generally one per egress interface. According to [27], the MR is multihomed if either multiple prefixes are advertised on a visited link or the MR is equipped with multiple interfaces, while in [28] a host is said to be multihomed if it has several IP addresses to choose between. In general, a multihomed router may have many HAs, more than one HoA as well as many CoAs. It can have one or more interfaces. A mobile network is multihomed in the case of hosting a multihomed MR or multiple classical MRs.

The consequence of having multiple HAs is the existence of multiple bi-directional tunnels between the mobile network and its HAs. This may eventually result into multiple prefixes being advertised to the MNs. Currently, there are neither requirements nor a standardized protocol defining how to use several tunnels inside a MR.

The main advantages of having multihomed routers in HAP networks are the following:

1. Providing connection redundancy and fault recovery. This can be used as a way to deal with NLOS intervals, and would result in permanent and ubiquitous access.
2. Smoother handoff between HAP and terrestrial wireless systems.
3. Gains in terms of optimal routes.
4. Load sharing and load balancing between multiple HAPs or between HAP and terrestrial access network.

In the following we identify the multihoming configuration that best suits the HAP network, and point out some related issues critical to deployment of multi-homing in HAP networks.

10.3.2.1 Configuration Taxonomy

Taxonomy (iface, HoA, CoA) is suggested in [29] to identify different possible multihoming configurations for end-nodes, where iface stands for the number of interfaces, HoA represents the number of home addresses and CoA the number of care-of-addresses. In the case of mobile networks the use of taxonomy (MR, HA, MNP) is suggested in [27], where MR indicates the number of MRs, HA indicates the number of HAs associated with the entire mobile network and MNP the number of prefixes advertised in the mobile network. Here we suggest the use of (MR, HA, MNP, iface, CoA) taxonomy to fully describe a certain NEMO configuration among all possibilities. The HAP important 5-tuples are:

(1,1,1,N,1):   This is a classical set-up with a single MR and one HA, advertising a single prefix to the end-nodes. Multiple (N) interfaces are available, but only one CoA may be active at any given moment. In the HAP network, this configuration can be used as a baseline solution to the NLOS problem. The configuration does not solve excessive use of backhaul link due to inefficient routing through MR's HA, and the dilemma, where to put the HA still remains. Some kind of priority scheme should be followed by the MR in tunnel selection for the outbound traffic, based on triggers from the link layer. Inbound traffic is less protected from loss due to HA's inability to establish the current state of wireless segment in each tunnel. Having at least one bi-directional tunnel available at any point in time between the mobile network and the fixed network is required but not a sufficient condition for transparent communication. Fast acquisition of tunnel state information is equally important to prevent loss of data. Both the inbound and outbound traffic must be diverted over another bi-directional tunnel once NLOS conditions occur.

(1,1,N,N,N):　Multiple advertised prefixes allow end-nodes being multihomed; however, that does not mitigate problems due to HAP unavailability. Connections must be torn down and established from scratch because the transport layer in the end-nodes does not allow dynamic address change. Stream Control Transmission Protocol (SCTP) is an exception. Due to multihoming capabilities at transport layer SCTP may be able to continue without disruption. Overall, advertising multiple prefixes in a mobile network, e.g. one per MR's egress interface, brings no particular advantage to the train-to-HAP scenario.

(1,N,1,N,N):　In this configuration multiple HAs exist to support a single MR. Nodes served by the MR are not aware of multihoming as they see only one prefix. The HAs may be located in different parts of the network; however, they should advertise a route to the same prefix. Due to lack of route aggregation in neighbouring routers this may lead to routing table growth in part of the global routing infrastructure. The configuration is advantageous in the HAP network because it solves the backhaul link issue. If we put one HA on the platform and the other on the ground, inter-HAP traffic is automatically routed through the platform-based HA, while the ground-based HA intercepts global traffic. As a consequence, near optimal routes are used while the MR roams close to one of the HAs. In order to support smooth handover between the agents due to NLOS periods, coordination between the two HAs is required. There are several inter HA protocols proposed for this task, the HAHA protocol being one of them [30].

(N,1,1,N,N):　A train, as an example of a moving network, is composed of multiple cars, which are dynamically added or removed according to the needs. Therefore, it is possible that multiple MRs will be attached together in a single mobile network. Configuration (N,1,1, N,N) assumes MRs that serve a single prefix. A common HA is provided for all MRs. Each MR may be itself multihomed using different egress interfaces. A set-up with some MRs connected to HAP and some to a terrestrial wireless system is less likely because it would be difficult to guarantee train modularity unless each train car would contain both types of routers.

(N,1,N,N,N): This configuration has an advantage over (N,1,1,N,N) due to multiple prefixes being advertised within the mobile network. Multiple prefixes may be advertised by all MRs or, more likely, different prefixes are advertised by different MRs. MRs can still be multihomed, which may provide route optimality.

The remaining more general multihomed configurations do not bring additional benefits to the mobile networks accessing HAPs.

10.3.2.2 Handling NLOS Conditions

With the increasing demands for wireless resources, the use of frequency spectrum for wireless networks is steadily moving from the congested low GHz bands to higher frequency ranges. Differences in the radio propagation properties in the higher frequency bands reflect in the link layer protocols. Furthermore, an influence on the higher layer protocols, such as MIP, is also expected. Due to the high penetration loss caused by the obstacles, communication during NLOS conditions becomes impossible. The HAP network is an example of such a communication system because of the allocated frequency bands at 48/47 GHz and at 31/28 GHz. In these bands a channel is available only if there are LOS conditions. NLOS conditions cause complete loss of connectivity. This could pose serious problems for mobile users blocked by various obstacles, such as buildings, hilly terrain, tunnels, etc.

Extended MIP functionality includes fast movement detection in order to inform the HA and correspondent nodes about the reattachment to a new link as soon as possible. Several actions must be completed by the MN in this process. The node must detect the movement, form a new CoA and inform the HA by sending a binding update.

Current MIP lacks specific support for sudden loss of connectivity. Loss of connectivity is usually referred to as a link failure. Unless NLOS–LOS transition triggers are provided to the network layer, a link failure followed by the link reestablishment is treated by the network layer as a regular break-before-make type of handover. This would unnecessarily trigger the actions described above instead of just resuming the network traffic.

The NLOS–LOS transition should not be treated as a topological movement and, therefore, should not trigger IP layer movement detection. Link layer should differentiate cell handover from the LOS/NLOS conditions and provide specific triggers to the IP layer. However, this is not a property of all communications and has not been included in the MIP specifications, which applies to all link layers. Multihoming extensions offer an opportunity to standardise adequate triggers.

10.3.2.3 Route Optimality

Excessive use of backhaul link is a major drawback in the HAP network if MRs lack route optimisation functionality. The problem emerges for the HAP-to-HAP communications, in which data packets unnecessarily traverse backhaul link twice in order to follow the path through the MR's HA. If localised quasi-deterministic movement of the entire network is expected, a careful selection of the MR's HA can improve communication performance to some extent. However, if multiple HAs are introduced, the need for route optimisation support may be completely eliminated. Route optimisation is achieved if both HAs advertise the same prefix, i.e. the mobile network prefix. Note that the route optimisation for end-nodes must still be operational.

There is a shortcoming of advertising the same prefix in different parts of the network. Advertising multiple routes to the MR adds a burden to routing tables as multiple routes to the same prefix are managed by the routing nodes. A similar issue arises in some route optimisation proposals for MRs. In [18] multiple Optimised Route Cache (ORC) routers advertise a proxy route of the mobile prefix. Uncontrolled growth of the routing tables is prevented by not allowing route inter-exchange by any exterior gateway protocol such as BGP. However, in order to provide the route optimisation feature, the IGP domain must host its own ORC router. Solutions to the routing table growth problem are currently being investigated in the research community. For example, the extent of the multihoming contribution on routing table size is, among other factors, studied in [31].

10.3.2.4 Open Issues

One of the open issues in routing protocol implementation supporting high speed moving vehicles is a path selection algorithm in the case of multiple bi-directional tunnels. The mode of operation may be primary-secondary or peer-to-peer. In the first case one tunnel has precedence over the other at all times, except when it is unavailable, e.g. NLOS conditions. The other mode can be beneficial in the case of comparable costs of using a particular access technology. Efficient fault detection mechanisms are necessary to recover in a timely fashion. While link level trigger may be used in a MR, the way the MR's HA is informed remains an open issue. Lack of a standardised tunnel failure detection mechanism requires further research.

Continuous transmission of heartbeat messages, explicit notifications, frequent binding updates and other mechanisms need to be reconsidered.

A multi-router site, e.g. a train with one MR per car, needs some form of router coordination. The coordination should cover advertising of the same prefix and relaying between MRs everything that needs to be relayed in case of a router failure. Basically, there are two kinds of problems associated with the prefix delegation in the case of multiple HAs or MRs. First, there is a question how multiple HAs would delegate the same prefix to the mobile network. The multiple routers case poses a second question regarding the mechanism for MRs synchronisation in order to advertise the same prefix. Prefix delegation is under investigation by the MEXT working group.

As a consequence of multiple interfaces, multiple CoAs are assigned for the same prefix. This opens a multiple binding problem.

If a mobile network splits because two MRs go apart, the only available prefix will then be registered by two different MRs on different links. The problem emerges for the HA, which has no way to establish which node with an address configured from the prefix is attached to which MR. Forcible removal of the prefix from one or all MRs is a possible solution to the problem. Other solutions should be investigated in the future.

10.3.3   MN Movement Predictability

The majority of micro-mobility protocols assume complete randomness of MN movement. However, MNs can move according to certain repeating patterns, which is especially true in HAP-to-train operating scenario. The frequency of repetitions may vary in the order of hours or days, up to weeks, months or even years. Illustrative examples of repeating patterns of an average person are travel paths to work, sport activities, weekend trips, vacations, etc. In some cases, the movement patterns may be well determined. Public transport vehicles that would carry MRs such as buses and trains move along very deterministic travel paths. If movement patterns repeat, the prediction of future moves on a particular journey will, to some extent, be possible.

The most notable micro-mobility approach, HMIPv6, besides HA, contains an additional entity called the MAP. In principle, HA manages the location of the MN on the macro-level, while MAP keeps track of the precise location of the MN on the micro-level. As MAP is always present in the routing path and MN may choose among different MAPs, the MAP selection technique affects the efficiency of the protocol significantly.

The most basic MAP selection algorithm, proposed by the authors of the HMIPv6 protocol, always selects the MAP that is furthest in terms of routing hops. The intention of such an approach is to minimise the number of required MAP changes, thus reducing signalling overhead and handover delays. The two most commonly recognised drawbacks of furthest MAP selection are high load burden on the most distant MAPs and unnecessary signalling delays for the MNs which move in the scope of nearer MAPs. To overcome these drawbacks, more efficient MAP selection algorithms have been proposed [32–37].

For the HAP networks a novel approach, based on the future movement prediction capabilities of MNs was proposed [38]. Using information about future MAP availability, MNs can choose MAPs which better fit their moving patterns and thus assure better service. The simulation results show lower average distance from chosen MAPs and lower average number of changed MAPs, which implicitly leads to better conditions for QoS provision and resource usage efficiency. Furthermore, MAP selection can be improved even if the prediction of future movement is not absolutely accurate.

References

1. D. Johnson, C. Perkins and J. Arkko, *Mobility Support in IPv6*, RFC 3775, June 2004.
2. C. Perkins, *IP Mobility Support for IPv4*, RFC 3344, August 2002.
3. C. Castellucia, *HMIPv6: A Hierarchical Mobile IPv6 Proposal*, ACM Mobile Comput. Commun. Rev., 2000, Vol. 4, No. 1, pp. 48–59.
4. H. Soliman, F. Castellucia, K. E. Malki and L. Bellier, *Hierarchical Mobile IPv6 Mobility Management (HMIPv6)*, RFC 4140, August 2005.
5. A. Campbell, J. Gomez, S. Kim, A. Valko, C. Wan and Z. Turanyi, *Design Implementation and Evaluation of Cellular IP*, IEEE Pers. Commun., 2000, Vol. 7, No. 4, pp. 42–49.
6. R. Ramjee, K. Varadhan, L. Salgarelli, S. Thuel, S. Wang and T. L. Porta, *HAWAII: A Domain-Based Approach for Supporting Mobility in Wide-Area Wireless Networks*, IEEE/ACM Trans. Netw., June 2002, Vol. 10, No. 3, pp. 396–410.
7. S. Das, A. Misra, P. Agrawal and S. Das, *TeleMIP: Telecommunication Enhanced Mobile IP Architecture for Fast Intra-Domain Mobility*, IEEE Pers. Commun., 2000, Vol. 7, No. 4, pp. 50–58.
8. C. Castellucia, *HMIPv6: A Hierarchical Mobile IPv6 Proposal*, ACM Mobile Comput. Commun. Rev., 2000, Vol. 4, No. 1, pp. 48–59.
9. H. Soliman, F. Castellucia, K.E. Malki and L. Bellier, *Hierarchical Mobile IPv6 Mobility Management (HMIPv6)*, RFC 4140, August 2005.
10. N. Banerjee, W. Wu, K. Basu and S. Das, *Analysis of SIP-Based Mobility Management in 4G Wireless Networks*, Comput. Commun., 2004, Vol. 27, No. 8, pp. 697–707.
11. V. Devarapalli, R. Wakikawa, A. Petrescu and P. Thubert, *Nemo Basic Support Protocol*, Internet Draft, draft-ietf-nemo-basic-support-01.txt, NEMO Working Group, 2003.
12. T. Ernst, *Network Mobility Support Goals and Requirements*, Internet Draft, draft-ietf-nemo-requirements-02.txt, IETF, 2004.
13. T. Ernst, A. Olivereau, L. Bellier, C. Castelluccia and H. Lach, *Mobile Networks Support in Mobile IPv6 Internet Draft*, draft-ernst-mobileip-v6-network-03.txt, NEMO Working Group, 2002.
14. P. Thubert and M. Molteni, *IPv6 Reverse Routing Header and its Application to Mobile Networks*, Internet Draft, draft-thubert-nemo-reverse-routing-header-04.txt, NEMO Working Group, 2004.
15. Y. Takagi, H. Ohnishi, K. Sakitani, K. Baba and S. Shimojo, *Route Optimization Methods for Network Mobility with Mobile IPv6*, IEICE Trans. Commun., 2004, Vol. E87-B, No. 3, pp. 480–489.

16. H. Ohnishi, K. Sakitani and Y. Takagi, *HMIP Based Route Optimization Method in a Mobile Network*, Internet Draft, draft-ohnishi-nemo-ro-hmip-00.txt, NEMO Working Group, 2003.
17. E. Perera, V. Sivaraman and A. Senevirtane, *Survey on Network Mobility Support*, ACM Mobile Comput. Commun. Rev., 2004, Vol. 8, No. 2, pp. 7–19.
18. R. Wakikawa, S. Koshiba, K. Uehara and J. Murai, *ORC: Optimized Route Cache Management Protocol for Network Mobility*, Proc. 10th *International Conference on Telecommunications ICT, Tahiti Papeete, French Polynesia*, February 2003, Vol. 2, pp. 1194–1200.
19. E. Perera, A. Seneviratne and V. Sivaraman, *OptiNets: An Architecture to Enable Optimal Routing for Network Mobility*, Proc. *International Workshop on Wireless Ad-hoc Networks, Oulu, Finland*, May 2004.
20. J. Na, J. Choi, S. Cho and C. Kim, *A Unified Route Optimization Scheme for Network Mobility*, Proc. *International Conference on Personal Wireless Communications, Delft, the Netherlands*, September 2004, Vol. 3260, pp. 29–38.
21. R. Novak, *Proxy MAP for Intra-Domain Route Optimization in Hierarchical Mobile IP*, IEICE Trans. Commun., February 2006, Vol. E89-B, No. 2, pp. 472–481.
22. J. Jeong, K. Lee, J. Park and H. Kim, *ND-Proxy Based Route and DNS Optimizations for Mobile Nodes in Mobile Network*, Internet Draft, draft-jeong-nemo-ro-ndproxy-02.txt, IETF, 2004.
23. K. Lee, J. Jeong, J. Park and H. Kim, *Route Optimization for Mobile Nodes in Mobile Network Based On Prefix Delegation*, Internet Draft, draft-leekj-nemo-ro-pd-02.txt, IETF, 2004.
24. E. Perera, R. Hsieh and A. Seneviratne, *Extended Network Mobility Support*, Internet Draft, draft-perera-nemo-extended-00.txt, IETF, 2003.
25. T. Kniveton, J. Malinen, V. Devarapalli and C. Perkins, *Mobile Router Tunneling Protocol*, Internet Draft, draft-kniveton-mobrtr.txt, NEMO Working Group, 2002.
26. D. Luong, M. Mohorčič, T. Van Do, R. Novak, A. Švigelj, M. Berioli, A. Vilhar, C. Fortuna, A. Glicser, N. Do Hoai and G. Buchholcz, *Network Architecture and Protocols,* Deliverable D27, CAPANINA (FP6-IST-2003-506745), November 2006.
27. C. Ng, E. Paik and T. Ernst, *Analysis of Multihoming in Network Mobility Support*, Internet Draft, draft-ietf-nemo-multihoming-issues-02.txt, NEMO, 2005.
28. N. Montavont, R. Wakikawa, T. Noel and T. Ernst, *Problem Statement of Multihomed Mobile Node*, Internet Draft, draft-montavont-mobileip-multihoming-pb-statement-00.txt, IETF, 2003.
29. N. Montavont, R. Wakikawa, T. Ernst, C. Ng and K. Kuladinithi, *Analysis of Multihoming in Mobile IPv6, Internet Draft*, draft-montavont-mobileip-multihoming-pb-statement-04.txt, IETF, 2005.
30. R. Wakikawa, V. Devarapalli and P. Thubert, *Inter Home Agents Protocol (HAHA)*, Internet Draft, draft-wakikawa-mip6-nemo-haha-01.txt, NEMO, 2004.
31. T. Bu, L. Gao and D. Towsley, *On Routing Table Growth*, ACM SIGCOMM Comput. Commun. Rev., 2002, Vol. 32, No. 1, p. 11.
32. K. Kawano, K. Kinoshita and K. Murakami, *A Mobility-Based Terminal Management in IPv6 Networks*, IEICE Trans. Commun., October 2002, Vol. E85-B, No. 10, pp. 2090–2099.
33. K. Kawano, K. Kinoshita and K. Murakami, *Multilevel Hierarchical Mobility Management in Densely Meshed Networks*, IEICE Trans. Commun., July 2006, Vol. E89-B, No. 7, pp. 2002–2011.
34. Y. Xu, H.C.J. Lee and V.L.L. Thing, *A Local Mobility Agent Selection Algorithm for Mobile Networks, Proc.* IEEE International Conference on Communications ICC, Anchorage, AK, USA, May 2003, Vol. 2, pp. 1074–1079.
35. T. Kumagai, T. Asaka and T. Takahashi, *Location Management Using Mobile History for Hierarchical Mobile IPv6 Networks*, IEICE Trans. Commun., September 2004, Vol. 8 7-B, No. 9, pp. 2567–2575.
36. X. Hu, J. Song and M. Song, *An Adaptive Mobility Anchor Point Selection Algorithm for Hierarchical Mobile IPv6 Proc.* IEEE International Symposium on Communications and Information Technology ISCIT, Beijing, China, October 2005, Vol. 2, pp. 1148–1151.
37. S. Pack, M. Nam, T. Kwon and Y. Choi, *An Adaptive Mobility Anchor Point Selection Scheme in Hierarchical Mobile IPv6 Networks*, Elsevier Computer Commun., October 2006, Vol. 29, No. 16, pp. 3066–3078.
38. A. Vilhar, R. Novak and G. Kandus, *MAP Selection Algorithms Based on Future Movement Prediction Capability in Synthetic and Realistic Environment*, J. Commun. Softw. Syst., 2008, Vol. 4, No. 2, pp. 122–130.

Index

Broadband Communications via High Altitude Platforms   David Grace and Mihael Mohorčič
© 2011 John Wiley & Sons, Ltd